AF691969

international® AIR POWER REVIEW

AIRtime Publishing
United States of America • United Kingdom

international® AIR POWER REVIEW

Published quarterly by AIRtime Publishing Inc.
US office: 120 East Avenue, Norwalk, CT 06851
UK office: CAB International Centre, Nosworthy Way, Wallingford, Oxfordshire, OX10 8DE

© 2002 AIRtime Publishing Inc.
Tu-142MZ cutaway © Aleksey Mikheyev
F-105 cutaway © Mike Badrocke/Aviagraphica
Photos and other illustrations are the copyright of their respective owners

Softbound Edition ISSN 1473-9917 / ISBN 1-880588-46-3
Hardcover Deluxe Casebound Edition ISBN 1-880588-47-1

Publisher
Mel Williams

Editor
David Donald e-mail: airpower@btinternet.com

Assistant Editors
John Heathcott, Daniel J. March

Sub Editor
Karen Leverington

US Desk
Tom Kaminski

Russia/CIS Desk
Piotr Butowski, Zaur Eylanbekov e-mail: zaur@airtimepublishing.com

Europe and Rest of World Desk
John Fricker, Jon Lake

Correspondents
Argentina: Jorge Felix Nuñez Padin
Australia: Nigel Pittaway
Belgium: Dirk Lamarque
Brazil: Claudio Lucchesi
Bulgaria: Alexander Mladenov
Canada: Jeff Rankin-Lowe
France: Henri-Pierre Grolleau
Greece: Konstantinos Dimitropoulos
India: Pushpindar Singh
Israel: Shlomo Aloni
Italy: Luigino Caliaro
Japan: Yoshitomo Aoki
Netherlands: Tieme Festner
Romania: Danut Vlad
Spain: Salvador Mafé Huertas
USA: Rick Burgess, Brad Elward, Mark Farmer (North Pacific region), Peter Mersky, Bill Sweetman

Artists
Mike Badrocke, Piotr Butowski, Zaur Eylanbekov, Keith Fretwell, Aleksey Mikheyev, Mark Rolfe, Andrey Yurgenson, Andrey Zhirnov

Designer
Zaur Eylanbekov

Controller
Linda DeAngelis

Origination by Universal Graphics, Singapore
Printed in Singapore by KHL Printing

All rights reserved. No part of this publication may be copied, reproduced, stored electronically or transmitted in any manner or in any form whatsoever without the written permission of the publishers and copyright holders.

International Air Power Review is published quarterly in two editions (Softbound and Deluxe Casebound) and is available by subscription or as single volumes. Please see details opposite.

Acknowledgments

We wish to thank the following for their kind help with the preparation of this issue:

Mike Gorn/NASA Dryden
Mike Hardaway, Rapid Terrain Visualization Program Manager
Eric Hehs
Dennis R. Jenkins
Tony Landis
Dr Craig Luther/AFFTC
Mike Machat
Jay Miller
Christian Moscoso, Rapid Terrain Visualization Technical Manager
Ken Neubeck
Terry Panopalis
Mick Roth
Gordy Williams Collection
Museum of Flight (Seattle)
San Diego Aerospace Museum

The editors welcome photographs for possible publication but can accept no responsibility for loss or damage to unsolicited material.

Subscriptions & Back Volumes

Readers in the USA, Canada, Central/South America and the rest of the world (except UK and Europe) please write to:
AIRtime Publishing, P.O. Box 5074, Westport, CT 06881, USA
Tel (203) 838-7979 • Fax (203) 838-7344
Toll free 1 800 359-3003
e-mail: airpower@airtimepublishing.com

Readers in the UK & Europe please write to:
AIRtime Publishing, RAFBFE, P.O. Box 1940,
RAF Fairford, Gloucestershire GL7 4NA, England
Tel +44 (0)1285 713456 • Fax +44 (0)1285 713999

One-year subscription rates (4 quarterly volumes), inclusive of shipping & handling/postage and packing:

Softbound Edition
USA $59.95, UK £48, Europe EUR 89, Canada Cdn $99,
Rest of World US $79 (surface) or US $99 (air)

Deluxe Casebound Edition
USA $79.95, UK £68, Europe EUR 123, Canada Cdn $132,
Rest of World US $99 (surface) or US $119 (air)

Two-year subscription rates (8 quarterly volumes), inclusive of shipping & handling/postage and packing:

Softbound Edition
USA $112, UK £92, Europe EUR 169, Canada Cdn $187,
Rest of World US $148 (surface) or US $188 (air)

Deluxe Casebound Edition
USA $149, UK £130, Europe EUR 236, Canada Cdn $246,
Rest of World US $187 (surface) or US $227 (air)

Single-volume/Back Volume Rates by Mail:

Softbound Edition
US $16, UK £11.95, Europe EUR 19.95, Cdn $25.50 (plus s&h/p&p)

Deluxe Casebound Edition
US $20, UK £13.95, Europe EUR 23, Cdn $31 (plus s&h/p&p)

All prices are subject to change without notice.
Canadian residents please add GST. Connecticut residents please add sales tax.

Shipping and handling (postage and packing) rates for back volume/non-subscription orders are as follows:

	USA	*UK*	*Europe*	*Canada*	*ROW (surface)*	*ROW (air)*
1 item	$4.50	£4	EUR 8	Cdn $7.50	US $8	US $16
2 items	$6.50	£6	EUR 11.50	Cdn $11	US $12	US $27
3 items	$8.50	£8	EUR 14.50	Cdn $14	US $16	US $36
4 items	$10	£10	EUR 17.50	Cdn $16.50	US $19	US $46
5 items	$11.50	£12	EUR 20.50	Cdn $19	US $23	US $52
6 or more	$13	£13	EUR 23.50	Cdn $21.50	US $25	US $59

Volume Six
Autumn/Fall 2002

CONTENTS

Major Features planned for Volume Seven

Focus Aircraft: Boeing F-15 Eagle, **Warplane Classic:** Republic P-47 Thunderbolt Part 2, **Variant File:** Grumman OV-1 Mohawk, **Technical Briefing:** Sukhoi Su-25 'Frogfoot' upgrades, **Photo-feature:** Sri Lankan Air Force, **Type Analysis:** Fiat (Aeritalia) G91, **Special Feature:** Zhukovskiy test centre

PROGRAMME UPDATE

Lockheed Martin F-22

Lockheed Martin has completed all planned static testing of the F-22A and demonstrated that the airframe can withstand forces that are 50 per cent higher than those expected throughout its service life. The full-scale static testing, which was conducted on Raptor 3999, one of two non-flying airframes, began at the Marietta facility during 1999. Since then, the airframe has been subjected to loads applied by numerous hydraulic rams designed to simulate the forces the aircraft encounters during flight. This testing is done primarily to demonstrate the structural integrity of the airframe. The contractor has also completed an airframe lifetime fatigue test designed to replicate a 20-year airframe service life, equivalent to 8,000 simulated flight hours, using a second airframe, known as Raptor 4000. Each lifetime equivalent of testing involves more than 1.2 million stress events simulating aircraft manoeuvres of up to 9 *g*. Fatigue testing of Raptor 4000 will continue and is scheduled to simulate a total of two lifetimes of fatigue evaluation (equivalent to 16,000 flight hours). The data accumulated from these tests will support long-term efforts to extend the F-22A's service life beyond 20-years.

Raptor 04 underwent climatic trials in early 2002, seen above in the McKinley test chamber with bay doors open and fuel tanks fitted to the optional underwing hardpoints. Meanwhile, Raptor 05 was involved in tests in an ECM environment, provided by F-16s carrying ALQ-167 jamming pods (right).

Having been retained at Marietta for maintenance testing, Raptor 08 joined the Edwards fleet, following Raptor 09 to complete the EMD test force.

PROJECT DEVELOPMENT

International

JSF partnerships escalate

Considerable impetus has recently been gained by Lockheed Martin Aeronautics' Joint Strike Fighter programme, from expansion of its original US/UK Level I team membership by several other countries. Canada led the way on 7 February with $150 million 10-year allocations to become a Level III partner, as a 1-2 per cent contribution to the F-35's decade-long $24 billion System Development and Demonstration (SDD) phase. R&D recoupment costs of some $5-8 million per aircraft could result in subsequent procurement contracts.

On 28 May, Denmark became the third international participant, also at Level III, for $125 million, with similar rights to influence elements of JSF design and capabilities, and supply parts and systems. Level III partners can also bid for significant production work-shares in up to 3,000 planned JSF aircraft procurement.

Similar moves followed from the Netherlands, which on 10 June joined the JSF SDD phase, as the first Level II member, committing $800 million towards replacing 85 F-16s from around 2012, costing some 5.5 billion euros ($5.5 billion). Norway signed a NKr1,016 billion ($137 million) MoU on 20 June for Level III SDD participation, towards its requirement for 48 new combat aircraft.

Italy followed suit on 24 June, from a Level II JSF MoU, with a 4 per cent SDD share, and full rights of a 5 per cent partnership. Italy will invest $1.028 billion in JSF development between 2002-12, for procurement of 100-180 Block 3 F-35s. Selected from the three F-35 variants, these would supplement the AMI's planned Eurofighter purchases.

Australia announced its intention in late June of acquiring up to 100 F-35s, as a proposed $A300 million ($172 million) 10-year Level III SDD partner. JSFs could meet its Air 6000 requirement for new combat aircraft from around 2012 onwards. Turkey joined the JSF programme on 11 July, with a $175 million Level III SDD commitment, to which Israel also urgently aspired, given the required funding.

Lockheed Martin at Fort Worth, Texas, is leading JSF development with its principal partners, Northrop Grumman and BAE Systems. Two separate but interchangeable propulsion systems are being developed by Pratt & Whitney and General Electric/Rolls-Royce.

Between 8 and 12 April 2002 the German Eurofighter prototype, DA1, was at Jever for supersonic handling and inflight-refuelling trials from a buddy pod-equipped Tornado of JBG 38. At the end of the trials DA1 had accumulated 283 flight hours in 340 sorties. In May it performed daily at the ILA 2002 airshow at Berlin-Schönefeld.

Eurocopter Tiger rolled out

Eurocopter recently rolled out the first example of the Franco-German Tiger attack helicopter at its facility near Munich, Germany. The helicopter is the first of 80 UHT (Unterstützungs-hubschrauber Tiger) multi-role platforms on order for the German Army. Delivery of the first aircraft to the Heeresflieger will take place in December 2002. The first of France's 10 HAP (Hélicoptère d'Appui et de Protection) escort and fire-support variants will be delivered to the Armée de Terre in June 2003. Subsequent aircraft for France comprise 70 HAC (Hélicoptère Anti-Char) anti-tank platforms. Although the two partner countries have a combined requirement for up to 427 of the attack helicopters, Eurocopter recently announced that they expect production will be reduced to 120 aircraft for each of the customers.

More doubt shed on A400M

While further progress with the A400M military transport aircraft programme has been delayed by German parliamentary complications, its administration was rationalised by transfer earlier this year of Airbus Military Company (AMC) management from EADS Spain to Airbus HQ in Toulouse. A new Euro Prop International (EPI) consortium was also formed by Rolls-Royce, SNECMA, MTU and ITP, to develop a 7457-9694-kW (10,000-13,000-shp) three-shaft turboprop for the A400.

The new TP400D6 engine will be jointly designed and built by the four-nation EPI team to meet uprated A400M power and electrical output requirements. Its development schedule is tight, despite a prototype A400M first flight date rescheduled to mid-2007, for initial deliveries from late 2008. In contrast, a demonstrator version of Pratt & Whitney Canada's competing PW800-based turboprop, also competing for A400M installation, has been undergoing ground testing for over a year.

Russia

Sukhoi to develop Russia's JSF

An industrial group headed by Sukhoi was selected on 26 April by the Military Science Committee of the Russian Federation Air Forces (VVS), to develop a fifth-generation combat aircraft. Sukhoi's proposals for the Perspektivnyi Aviatsionnyi Kompleks Frontovoi Aviatsyi (PAK FA), or Prospective Aviation Complex for Frontal Aviation requirement, were chosen in preference to those from MiG Russian Aircraft Corporation (RSK MiG). MiG will partner Sukhoi in PAK FA development, however, together with the Yakovlev Design Bureau (OKB). PAK FA replaced the earlier Mnogofunktionalniy Frontovoi Istrebitel (MFI) multi-role tactical fighter requirement, for which the MiG OKB developed the massive but little-flown Article 1.44 prototype.

Sukhoi's PAK FA design will be derived from its extensive Su-27 series development, plus technology from recent flight research with its forward-swept heavy-weight Su-47 Berkut demonstrator prototype. Preliminary indications, however, are that the PAK FA will probably be slightly smaller

Bell-Boeing V-22 Osprey

The Bell-Boeing MV-22B Osprey returned to the air at NAS Patuxent River, Maryland, on 29 May 2002 for the first time since being grounded 18 months prior. During the short flight test pilots conducted a vertical take-off and, while hovering the tilt-rotor aircraft at an altitude of 30 ft (9.1 m), conducted a series of basic manoeuvres. The MV-22B test aircraft (BuNo. 164942) to resume flying has received improvements to its hydraulic and flight control software systems. During a second flight, conducted later on the same day, the pilots performed a full conversion to airplane mode at level flight speeds of 250 kt (463 km/h). Over the upcoming 12 months seven modified MV-22s will participate in the flight test programme and a total of 1,800 flight test hours is scheduled. Four additional aircraft will undergo testing with the USAF at Edwards AFB, California. The entire fleet was grounded in December 2000 following two crashes that killed 23 Marines. The Marine Corps hopes to place the Osprey in service by December 2003.

and of conventional tapered-wing configuration, with no canards, and of similar layout to the Lockheed Martin F-22. Its features will include low radar observability, dry thrust supersonic cruise, high agility from 2-D thrust vectoring, up to 12 internal weapons bays, and short-field operating capability. Inclusion of Yakovlev in the PAK FA team may indicate that a STOVL version of the new Russian combat aircraft is also contemplated.

An initial PAK FA development contract of only $1.5 billion for Sukhoi is programmed for a prototype first flight in 2006, and production from 2009-10. International development is mooted by the Russian government for this ambitious project, probably with current major foreign customers, notably China and India.

United States

Military S-92 prototype flying

The first military variant of Sikorsky's S-92 helicopter rejoined the flight test programme on 22 March 2002 after receiving major structural enhancements and systems upgrades. They include additional gross weight capability, an extended main rotor shaft and new main rotor servos, and an upgraded electrical system, redesigned cockpit layout, improved landing gear, upgraded drive train, optimised vibration system and airframe modifications. The aircraft had previously compiled 221 flight hours in its original configuration. As part of the modifications, the cabin was extended 16 in (40.6 cm) and a redesigned production configuration tail pylon was installed. The new pylon is shorter and the horizontal stabiliser is relocated.

UCAR teams selected

The Defense Advanced Research Projects Agency (DARPA) and US Army have selected four teams to participate in the first phase of the Unmanned Combat Armed Rotorcraft (UCAR) programme. Each team will receive approximately $3 million for the 12-month concept development and system trades phase. The teams comprise: Lockheed Martin Systems Integration/Bell Helicopter; Boeing's McDonnell Douglas subsidiary; Northrop Grumman/MD Helicopters/BAE Systems/L3 Communications; Sikorsky Aircraft/Raytheon. As part of phase I each of the teams will produce a system capabilities document and a detailed development plan for follow-on UCAR programme phases. DARPA will subsequently select two contractors for a nine-month preliminary design phase, followed by a system development phase that will yield two prototype vehicles. The winning system will be delivered to the US Army in 2009. A production UCAR system will be fielded in 2010. Ultimately, the programme will demonstrate the feasibility of using the UCAR to conduct armed reconnaissance and attack missions.

On 1 July 2002 Pilatus test pilot Bill Tyndall took the prototype PC-21 next-generation basic trainer aloft on its first flight from the company airfield at Stans. The first flight lasted 1 hour 13 minutes, although the gear remained down throughout. Later in the month the aircraft appeared at the Fairford and Farnborough air shows, following a fault-free initial flight test phase.

Third VTUAV prototype flies

The third prototype of the US Navy/Northrop Grumman RQ-8A Fire Scout vertical take-off and landing tactical unmanned air vehicle system (VTUAV) successfully completed its first two flights at Naval Air Warfare Station China Lake, California on 19 May 2002. These flights tested the system's ability to take off, fly and land autonomously. The initial flight successfully verified the ability to modify the system's fully autonomous, pre-programmed mission by executing override commands that caused the UAV to enter a hover and land. The second flight demonstrated a completely autonomous mission that included vehicle start, take-off, flight, landing and shutdown. The RQ-8A's flight control system is derived from the contractor's RQ-4A Global Hawk

The future of airpower? On 22 May 2002 Boeing's X-45A UCAV demonstrator took to the air for the first time at Edwards AFB, completing a 14-minute hop during which a speed of 195 kt (361 km/h; 224 mph) and altitude of 7,500 ft (2286 m) were achieved.

long-endurance, high-altitude reconnaissance system, which has repeatedly demonstrated its ability to conduct autonomous missions of more than 30 hours' duration over thousands of miles. The Fire Scout system is currently in low-rate initial production (LRIP) for the Navy.

Comanche returns to flight

The No. 2 prototype of the RAH-66 Comanche made its first flight with a newly installed mission equipment package (MEP) software and new LHTEC T-800-LHT-801 engines on 23 May 2002. Each of the improved engines has an output of 1,563 shp (1166 kW), and provides a 17 per cent increase in power over the earlier -800 model. Prototype No. 2 was taken off flight status in May 2001 so that the engines could be installed and for integration testing of the MEP. It will continue its flight tests throughout the summer in support of continued development of the core MEP functions, flight controls development work and night vision pilotage system (NVPS) development testing.

Boeing's YAL-1A Airborne Laser flew for the first time with nose turret and roof fairing at McConnell AFB, Kansas, on 18 July. The laser and tracking equipment had not yet been installed.

X-47A UCAV ready to fly

Northrop Grumman's recently completed several tests in preparation for the first flight of its X-47A Pegasus experimental unmanned combat air vehicle scheduled for mid-2002. Included was the third engine run test which was accomplished in preparation for the X-47A's first autonomous engine run. Once the final integration tests have been completed the UCAV will be transported to the Naval Air Weapons Station China Lake, California, where taxi tests and first flight will be conducted. The company-funded prototype is designed to demonstrate aerodynamic qualities suitable for autonomous aircraft-carrier flight operations. The company will use the results of the Pegasus demonstration effort in its work on a naval unmanned combat air vehicle (UCAV-N) for the Defense Advanced Research Projects Agency (DARPA) and the US Navy. This programme will demonstrate the feasibility of using an unmanned system to conduct sea-based surveillance, strike and suppression of enemy air defences missions.

Upgrades and Modifications

Angola

ELINT 707 nears service

Conversion is nearing completion at Ben Gurion airport in Israel of a Boeing 707 equipped for electronic intelligence and surveillance operations from a $20 million Angolan contract, which also included aircraft acquisition and refurbishment costs. Most of the ELINT and electronic warfare equipment is being supplied and installed by Israel's Elisra and LR Avionics companies. Also involved is IAI's Taman division, which is supplying stabilised long-range day/night optical and IR sensors.

Cyprus

'Hind' upgrade plans

News emerged at the Farnborough show of a Rostvertol PLC helicopter group contract to upgrade a dozen or so recently-delivered Mil Mi-35P twin 30-mm (1.18-in) cannon-armed attack helicopters in the Greek-Cypriot National Guard. Some of the enhancements incorporated in Rostvertol's upgraded Mi-35M project are being installed in the Cypriot Mi-35Ps, from the local Defence Ministry contract.

Weight-saving measures include replacement of the 'Hind's' original retractable undercarriage by a simpler fixed installation, plus shortened stub-wings with built-in stores-loading systems. Limited avionics improvements include provision of night-vision goggles, and associated cockpit and exterior lighting.

Egypt

Chinook upgrades extended

Following earlier $400 million US authorisations for Boeing/MDH to upgrade 35 Egyptian army AH-64As to AH-64D Longbow standards in March 2001, $155 million in Foreign Military Sales funding was granted in May to modify six EAF Agusta/Meridionali-built CH-47C medium-lift helicopters to enhanced CH-47D International Chinook configuration. Deliveries will start by the year-end, with five more in 2003.

Egypt originally purchased 15 CH-47C Chinooks in 1980. Three will not be modernised, and may eventually revert to maintenance training. Contracts to modernise Egypt's remaining six CH-47Cs are expected by the year-end, allowing completion of two more upgrades in 2003 and four in 2004, to supplement four new-build EAF CH-47Ds delivered from 1999.

Russia

Upgraded Frogfoot

Russia's Sukhoi design bureau has begun preliminary flight tests of its upgraded Su-25SM attack aircraft. The aircraft, which was modified at Kubinka, has already completed two flights and is receiving test instrumentation in preparation for flight tests. The modification of a second aircraft will be completed by July 2002. As part of the modification the aircraft, received a 'glass' cockpit and will be capable of delivering precision guided munitions.

Seen during handling trials with dummy 1,000-lb (454-kg) bombs, XX108 is the first Jaguar to be fitted with the Rolls-Royce Adour Mk 106 engine, with which the RAF's Jaguar fleet is being re-engined.

Slovakia

MiG-29 upgrades

While Requests for Information for 18 new multi-role combat aircraft and up to a dozen advanced trainers were expected from the Slovakian government late last summer, RSK MiG received a $56 million contract in July to overhaul and upgrade 18 of Slovakia's MiG-29 fighters. Costs were expected to be offset mostly from Russia's long-standing $1+ billion trade debts owed to Slovakia.

Local industry, led by Letetske Opravovne in Trenchin, will undertake much of the contract under RSK MiG supervision, including installation of NATO-interoperable communications and operational equipment. New aircraft equipment plans are expected to give priority to advanced trainers, for which compatibility will be required with the selected combat aircraft type.

Spain

F-5B upgrade progress

Following flight development started last October at EADS Military Aircraft Spain's (MAS) Getafe facility in Madrid with two prototype Spanish air force (EdA) Northrop SF-5B two-seat combat trainers incorporating advanced avionics packages, formal EdA acceptance of these aircraft took place at Getafe on 17 July. After more than 50 test-flights, programme chief Coronel Jesús Zuazu Nagore handed over the two aircraft to the commander of Ala 23, at Talavera la Real (Badajoz), Coronel Luis Miguel Llanos Morán.

These were the first of four EdA prototype SF-5Bs being upgraded by EADS-CASA Military Aircraft and Indra, in conjunction with Israel Aircraft Industries' (IAI) Lahav division as prime contractor, since July 2000. New digital avionics and structural enhancements are involved in this $19 million prototype life-extension programme, now stretched to include an additional $30 million contract to upgrade another 18 CASA-built SF-5Bs.

These are being operated as AE.9s, alongside three single-seat Northrop SF-5As (A.9s) and four SRF-5A (AR.9) tactical-recce aircraft in 23 Wing (Ala 23) from Talavera, for advanced and lead-in fighter training. Completion of all four upgraded EdA SF-5B prototypes is expected this year.

United States

First equipped MH-60R flies

The first Sikorsky MH-60R Seahawk multi-mission helicopter equipped with a fully operational weapon system flew for the first time at Lockheed Martin's facility in Owego, New York, on 4 April 2002. Currently three MH-60R aircraft are undergoing

Still in use at the CEV's Cazaux facility is this Mirage IIIR. The three-aircraft Mirage fleet, which includes another IIIR and a IIIE fitted with a IIIR nose, is used for chase and photo duties during weapon tests, especially of the Apache stand-off missile.

In late 2002 the Brimstone anti-armour weapon is to be released for service with the RAF's Tornado fleet. This GR.Mk 1 is seen in May during the final handling tests with 12 missiles carried on the underfuselage hardpoints. Four ALARMs are also carried.

development and operational testing at the Naval Air Warfare Center, Aircraft Division at NAS Patuxent River, Maryland, and three additional aircraft are undergoing integration and test at the Owego facility. The MH-60R mission and weapon systems comprise the AN/AQS-22 airborne low frequency dipping sonar, new electronic support measures (ESM), forward looking infra-red, datalink and acoustic equipment. The service plans to buy about 270 MH-60Rs for initial operational capability in 2004.

UH-1Y and AH-1Z progress

First flown in December 2001, the first upgraded UH-1Y Huey helicopter recently achieved 190 kt (352 km/h) in a shallow dive and 166 kt (307 km/h) in straight and level flight. In addition, the helicopter has also been flown 25 kt (46 km/h) sideways, 20 kt (37 km/h) backwards and manoeuvring up to 30° bank. Known as Y-1, the aircraft has made more than 50 flights and accumulated over 44 hours of flight time. To date, all of the flight tests have been conducted at Bell's Flight Research Center in Arlington, Texas, however, Y-1 and its follow-on ship, Y-2, currently in the final stages of pre-flight testing, will join the AH-1Z prototype at NAS Patuxent River, Maryland, for further testing by the H-1 integrated test team.

The first AH-1Z, Z-1, completed its initial flight in December 2000 and since that time has accumulated more than 250 hours of flight time. The AH-1Z has already demonstrated a maximum airspeed of 222 kt (411 km/h) and 160 kt (296 km/h) cruise. In addition, the AH-1Z has been flown to an altitude of 10,000 ft (3048 m) and been flown at 45 kt (83 km/h) in backward and sideways flight. Bell Helicopter delivered the second of three AH-1Z prototypes to the Naval Air Warfare Center Aircraft Division at NAS Patuxent River, Maryland on 3 May 2002. The aircraft, known as Zulu Three, is actually the third prototype built. Zulu Two is currently undergoing flight test at Bell's facility in Arlington, Texas. After reassembly, Zulu Three will join the first prototype with the Integrated Test Team. The aircraft will eventually support weapons testing at Naval Air Weapons Station China Lake, California.

Army accepts first CH-47F

Boeing recently delivered the first remanufactured CH-47F Chinook helicopter to the US Army. The engineering and manufacturing development (EMD) prototype was initially flown from the Boeing flight test centre in Wilmington, Delaware, to Fort Campbell, Kentucky, and later ferried to Fort Rucker, Alabama. The two EMD CH-47Fs have been undergoing flight and system validation testing with the contractor since 2001. Additional flight testing, along with maintainability validation and aircrew training, will be carried out at Fort Rucker. Boeing will begin modifying the first CH-47D to the latest configuration at its facility in Philadelphia, Pennsylvania, in January 2003 and the initial production deliveries will follow in mid-2004.

Greyhound life extension

Naval Air Depot North Island has begun a structural life enhancement programme (SLEP) that will extend the service life of the US Navy's fleet of 36 C-2A Greyhound carrier on-board delivery (COD) aircraft. The current life span of the C-2A is 10,000 flight hours and/or 15,000 landings and the SLEP will extend these totals to 15,000 hours and/or 36,000 landings. The latter number is significant because the aircraft will typically reach its limit on landings before flight hours. The programme includes structural enhancements, rewiring and standard depot level maintenance (SDLM). Although the first of two validation and verification aircraft has already been inducted into the SDLM phase, the entire modification sequence will likely take a full year to complete. The depot plans to deliver six aircraft per year once it reaches full production.

A US Army Special Operations MH-47E undergoes electronic systems compatibility tests, principally of its defensive suite.

CIP Viking fielded

The 'Checkmates' of VS-22 at NAS Jacksonville recently became the first squadron to operate S-3Bs modified under the communications improvement programme (CIP). The CIP provides the Viking with improved UHF, VHF and satellite communications and close air support capability.

Zimbabwe

'Hinds' adapted for night attack

Four twin 30-mm (1.18-in) cannon-armed Mil Mi-24P 'Hind' attack helicopters operated by the Air Force of Zimbabwe (AFZ) are being upgraded by Rostvertol for night/all-weather operations. Apart from night-vision systems and cockpit lighting, US-supplied imaging infra-red/TV 445MGII sighting sensors and displays, with Garmin 115L GPS inputs, are being installed.

In Russia, Rostvertol is undertaking further Mi-24P development from incorporation of some Mi-35M airframe and avionics installations. Apart from weight-saving measures, NVG systems and simplified hydraulics, the resulting Mi-24PN features Russian BREO-24 radio equipment, and a Zarevo night thermal-imaging sight, plus standard day weapons-guidance. State development of the Mi-24PN has now reached an advanced stage.

Aussie Hornets – HUG Phase 2

In a ceremony at RAAF Williamtown on 5 March 2002, the first F/A-18 Hornet was inducted into the second phase of the Hornet Upgrade Programme (HUG). Phase One conferred upgraded communications and navigation systems (ASN-172 INS, ARC-210 jam-resistant radios, a sixth multiplex bus), an interim RWR software upgrade (APX-11), a mission computer upgrade to XN8+, and wiring to enable the use of the AIM-132 ASRAAM WVR missile. This phase is almost complete, with 58 of the 71 aircraft completed by the time that the first aircraft for Phase Two was inducted. Phase Two is to be completed in three stages:

2.1 – Integration of Raytheon AN/APG-73 Radar, Enhanced Interference Blanker Unit (EIBU) and software upgrade to 15C (Operational Flight Program). Due for completion in mid-2003.

2.2 – Integration of a Joint Helmet-Mounted Cueing System (JHMCS), ALE-47 CMDS, Upgraded Tactical Data Link, upgrade to 19C OFP, Cockpit full-colour displays and tactical digital moving map system. Due for completion by late 2006

2.3 – EWSP upgrade (ALQ-165 ASPJ, ALQ-214[V]4 or equivalent) and RWR upgrade (ALR-67[V]3 or equivalent. This phase was originally a component of 2.2 but several rethinks within RAAF circles have caused it to be pushed out to 2.3. At the time of writing, the requirements are yet to be defined, and the above equipment is what was originally specified. Completion of this phase is scheduled for 'later in the decade'.

Phase Three will be a structural modification programme, in two parts, that will ensure that the Hornet will remain operational until the planned retirement date of 2012:

3.1 – A minor structural refurbishment programme, to take place between 2003 and 2005

3.2 – A major structural refurbishment and modification, including major component replacement. This will be accomplished between 2007 and 2010.

The Hornet (as well as the F-111) is due to be replaced from 2012 by the Lockheed Martin F-35 under Project AIR 6000, and the HUG is designed to enable the Hornet to remain operationally viable until that time.

Nigel Pittaway

ARDU F/A-18B A21-101 was the main test platform for the ASRAAM missile, and is seen here carrying six missiles. Note the photo-calibration markings and test camera installation under the rear fuselage.

Despite the poor quality, this photo shows an Indian Navy Il-38 'May' carrying a Sea Eagle missile on a pylon aft of the wing trailing edge. The ability to carry the weapon was added in a recent upgrade.

The RAAF has reaffirmed plans to keep its fleet of elderly de Havilland DHC-4A Caribous in service with No. 38 Squadron at Amberley until at least 2010, but has dropped plans to re-engine them with turboprop engines. A project to replace the Caribou (Light Tactical Airlift Capability, or AIR.5190) has long been underway, and was narrowed to a choice between the Alenia-Lockheed Martin C-27J and CASA C.295 before the project was once again placed under review following Australia's involvement in East Timor in 1999.

Procurement and Deliveries

Austria

Eurofighter selected

Accelerated Austrian air force (OeLk) requirements to replace 23 ageing SAAB 35Oe Draken air defence fighters resulted in selection on 2 June of the Eurofighter Typhoon, over the JAS 39C Gripen and new or surplus F-16s. Contract negotiation then started with the European Aeronautics Defence & Space Company (EADS) to acquire up to 24 air defence Typhoons.

Agreed industrial offsets, reportedly totalling 200 per cent of the estimated 1.79 billion euro ($1.79 billion) total purchase price, were key features of both Eurofighter and Gripen bids. Austria is further expected to receive Luftwaffe help with Typhoon service entry, including transfer of four single- and two two-seat Eurofighters, from mid-2003. In an accompanying substantial weapons package, Austria will choose between the European MBDA Meteor and Raytheon's AIM-120C AMRAAM for its beyond-visual-range AAM requirements.

OeLk buys RAF Hercules

Long-standing OeLk requirements for tactical transport aircraft are being met from government-level acquisition of three RAF surplus Lockheed C-130K Hercules. Austrian government agreement for C-130K procurement was signed in Vienna on 22 May with the UK MoD's Disposal Services Agency.

About half the RAF's current C-130Ks, delivered from 1966, are being replaced by 25 C-130Js and stretched C-130J-30s. Lockheed Martin has repurchased some C-130Ks for resale, within the UK C-130J procurement contract, and has already disposed of several to overseas customers, after refurbishment. A Lockheed Martin C-130K bid was among other commercial approaches to Austria, but the US company is not directly involved in OeLk Hercules procurement.

Although a 40 million euro ($40 million) Austrian C-130K contract then remained to be signed, the UK MoD said that Marshall Aerospace had already started major maintenance and upgrades on the aircraft concerned (XV181, 291 and 292), in Cambridge. Austria will receive its first C-130K on 17 December, and the others next year.

Angola

Tucanos acquired

Six EMBRAER EMB-312 Tucano turboprop trainers withdrawn from Peruvian service around 1997 are being refurbished, following their purchase by the Angolan People's air force (APAF) from a $4.8 million contract. Deliveries should be completed this year.

Bahrain

BAE Hawk selected

In a further extension of previous military training agreements with BAE Systems, the Bahrain Amiri air force announced in June selection of the Hawk as its future advanced trainer. BAE reached agreement with the BAAF in July 2000 to study prospects for establishing a similar military basic and advanced flying training academy to the NFTC in Canada, for which a contract was finalised this January. Six Hawks are initially expected to be ordered, to supplement an initial three Slingsby T.67 Firefly primary trainers, for Bahrain's new air academy, which will presumably be open to foreign military students.

Canada

Challengers ordered

The Canadian government announced plans to acquire two Bombardier Challenger 604 aircraft for use by the Administrative Flight Services Fleet at a cost of $100.9 million. This cost will be partially offset by the sale of two older CC-144 Challenger models that are currently in service. Operated by 412 Transport Squadron at MacDonald Cartier IAP in Ottawa, Ontario, the Administrative Flight Services Fleet is owned by the Department of National Defence (DND) and supports the air travel requirements of the military and government officials. The Challenger 604s, which will enter service in late-2002, are capable of using shorter runways than current models and flying non-stop to Europe.

Cormorants replace CH-113As

Retirement of the four 38 year-old tandem-rotor Boeing Vertol CH-113A Labradors operated by No. 442 Transport & Rescue Squadron in 19 Wing at Comox, British Columbia, on SAR roles for many years, began this spring, following their replacement by new EH101 AW 520 (CH-149) Cormorant helicopters. Five Cormorants had each flown the 11,000+ km (nearly 6,000 nm) on delivery from AgustaWestland in Italy to Comox, to re-equip 442 Squadron as an OTU, following initial CAF air and ground crew conversion in Italy.

Delivery of the 10 remaining Cormorants is due next summer from the $C790 million ($501 million) contract, to replace CH-113s in three other CAF units, comprising 413 Squadron at 14 Wing in Greenwood, Nova Scotia; 424 Squadron in 8 Wing at Trenton, Ontario; and 103 Rescue Unit in 9 Wing at Gander, Newfoundland.

An ASW/ASuW version of the Cormorant, similar to RN Merlin HM.Mk 1s, is also a strong contender for the CAF's new maritime helicopter programme (MHP). CAF procurement is planned of 28 Sikorsky Sea King replacements, costing up to $C2.9 billion ($1.84 billion), complete with mission systems, over about eight years. The EH101 and Sikorsky S-92 have now been short-listed, although delays in the issue of Requests for Quotations (RFQs) for platform selection were expected.

China

Sukhoi orders extended

More than 30 Su-30MKK two-seat multi-role fighters have recently been delivered to the AF/PLA from KnAAPO, and at least 10 Su-27UBK combat trainers of 28 ordered from IAPO. These are from AF/PLA contracts worth about $2 billion signed in late 1999, which also included further KnAAPO-built Su-27SK kits of parts for Shenyang Aircraft Plant assembly.

Another 20 Su-30MKKs, again with new, mainly Russian Ramenskoye avionics, but lacking canards or thrust-vectoring, are expected to follow, from a $1.8 billion contract signed in July 2001. The first of these was flown from the KnAAPO factory on 9 March, and more Su-27UBKs have been delivered from IAPO, which has regained their production from KnAAPO by Russian government decree.

Israel has settled its dispute with China over the US-imposed cancellation of a $1 billion AEW contract, by the return in June of the Ilyushin Il-76 scheduled for IAI conversion, and $350 million in compensation.

Recent South American deliveries include a new Gulfstream IV (above) for Chile's Grupo de Aviación 10, which replaces a Gulfstream III traded in as part payment. The Ecuadorian navy has bought four ENAER ECH-51B (T-35B-6) Pillán trainers (left), delivered at the end of May 2002.

VC10 at 40

On 29 June 2002 the RAF's VC10 tanker/transport fleet commemorated the 40th anniversary of the type's first flight. The service's association with the majestic airliner began in July 1966, with the delivery of the first VC10 C.Mk 1s (Type 1106) to No. 10 Squadron. From May 1984 ex-airliners converted into tankers were allocated to No. 101 Squadron, comprising five K.Mk 2s (Type 1101, ex-Gulf Air) and four K.Mk 3s (Type 1154, ex-East African Airways). Subsequently, five K.Mk 4s (Type 1151, ex-British Airways) were added to the fleet, while the surviving C.Mk 1s were given two-point tanker equipment to become C.Mk 1Ks.

Today both squadrons continue to operate a dwindling number of aircraft from Brize Norton. No. 10 Sqn has 11 C.Mk 1Ks, while No. 101 has four K.Mk 3s and four K.Mk 4s. The fleet has recently been working hard over Afghanistan, where it drew praise from US Navy receivers. Replacement is due around 2008 by a Private Finance Initiative (PFI) scheme. Two consortia are bidding for this FSTA contract, offering the Boeing 767 and Airbus A330.

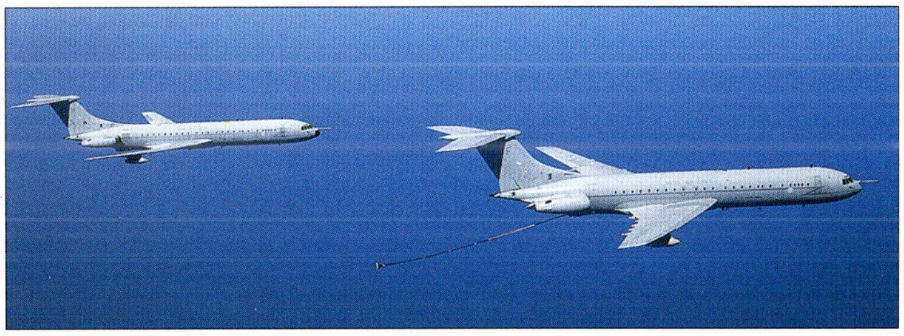

As well as routine UK military training sorties, the VC10 fleet occasionally supports Eurofighter test flights from Warton. Above a K.Mk 3 refuels the black-painted DA2, which carries an anti-spin chute apparatus for envelope expansion work. The VC10 fleet, including No. 10 Sqn's C.Mk 1Ks, has all been painted in the low-visibility grey scheme displayed by these two No. 101 Sqn K.Mk 3s seen in June 2002 (right).

Czech Republic

Gripen contract hiatus

Czech air force (CLPO) plans for the proposed CK60.2 billion ($1.8 billion) acquisition of 24 JAS 39 Gripen multi-role combat aircraft were abruptly halted before the June elections, when approval was vetoed by a senate vote on 31 May. Two days earlier, the retiring Czech government had agreed to defer contract signature until late September, for which Gripen International extended its conditions.

Planning for Gripen procurement was continued by the Czech Defence Ministry, however, which claimed that the alternative was to be left without air defences in 2004, when first deliveries were due. For the new Czech government, a key feature of the Gripen contract is its CK80 billion worth of industrial offsets, representing about 150 per cent of total procurement costs. These are being covered by CK58.5 billion in 15-year credits offered by BAE Systems for the 24 aircraft, plus CK18.4 billion for spares and equipment.

Retirement in March of the 10 remaining CLPO Sukhoi Su-22M-4 'Fitter-K' ground-attack fighters reduced its reported combat strength to only eight MiG-21MFs from 20 in service in 2000, of which just four are now serviceable. But 42 of 72 single-seat Aero Vodochody L-159A ALCA advanced light combat aircraft ordered by the CLPO had been delivered by 26 July, for initial training at Namest na Oslavou air base in south Bohemia. Their weapons integration, however, then still remained uncompleted.

The first two-seat L-159B light combat aircraft/advanced trainer began flight development on 1 July, and after 23 sorties appeared statically at the Farnborough air show. Two have been ordered by the Czech government, and the L-159B is being offered for export, to countries including India, for an $8.5 million fly-away unit cost.

On 10 June 2002 Sikorsky rolled out the first of nine S-70A-42 Black Hawks for the Austrian air force at its Stratford, Connecticut, plant. The S-70s were ordered in the wake of the Galtür avalanche, and greatly enhance the Austrian capability to respond to similar disasters. After delivery from August to December 2002, they will be operated by 1. Staffel/Fliegerregiment 1 at Tulln-Langenlebarn and, as well as domestic requirements, will support international peacekeeping operations.

New transports

The Czech Republic has agreed to take delivery of three Antonov An-70s in a deal to settle $1.1 billion in Soviet-era debt. The airlifters will be delivered in late-2005 or early-2006. Capable of carrying up to 77,000 lb (35000 kg) of cargo, flying approximately 2,700 nm (5000 km), the An-70 is being jointly developed by Russia and the Ukraine. The first of the medium-haul aircraft will be delivered in 2004 and the Russian and Ukrainian Air Forces are expected to acquire 164 and 65 aircraft, respectively.

France

More transport equipment

Following initial AdA procurement of 15 EADS-CASA CN-235-100 twin-turboprop tactical transports, delivery is scheduled later this year of two more included as options in the original 1991 contract. These were ordered earlier this year, and options for three more CN-235s were also taken up in a new French contract, increasing total AdA CN-235 procurement to 20, for delivery from mid-2003.

Eurocopter also started deliveries this spring to French Army Aviation (ALAT), from a 152 million euro contract, of 32 10-passenger EC 145 utility helicopters, powered by twin Turboméca Arriel 1E2 turboshaft engines. Derived from the MBB/Kawasaki BK 117C-1, the EC-145 is equipped with Thales 'glass' cockpit instrumentation, for ALAT operation in SAR, medevac and support roles. The French gendarmerie has also ordered eight EC 145s.

VFA-115's F/A-18Es share the bow of USS Abraham Lincoln with F/A-18Cs from VFA-25 and VFA-113 during the Super Hornet's first cruise in June 2002.

Georgia

Bell UH-1s supplied

Ten ex-US Army Bell UH-1 utility helicopters, plus two from Turkey, have been transferred to Georgia, where a small US advisory group is supervising a $64 million military aid programme. US special forces have also joined with Georgian troops in searching for Al Qa'eda terrorists and rebels from adjacent Chechnya, in the isolated Panski Gorge area.

India

ALH deliveries begin

After extensive development delays, the first two HAL Dhruv Advanced Light Helicopters (ALHs) were accepted on 20 March by Indian Army Aviation from initially limited production. Deliveries of six more ALHs were scheduled to the army (one), IAF (two), Indian navy (two), and the Coastguard (one) by the year-end. Deputy Chief of Army Staff, Lt Gen R.S. Kadyan said procurement was planned of 120 ALHs, including 60 weapon system-integrated versions from 2004 onwards. More power was being sought for the ALH, from joint development by HAL and Turboméca of the 900-kW (1,207-shp) Ardiden or Shakti turboshaft engine.

Apache news: the first AH-64D for the Republic of Singapore was handed over on 17 May 2002 (above), while on 10 June 2002 in another ceremony at Mesa, Arizona, the Koninkiljke Luchtmacht (Dutch air force) accepted its 30th and last AH-64D (below) from a 1995 order. Deliveries began in April 1998, following two years of operating leased US Army AH-64As. Meanwhile, the IDF/AF announced that its AH-64Ds, of which nine (eight new and one rebuild) are on order, is to be named Sharaf (serpent). The AH-64As currently in service are known as the Peten (python).

Indonesia

SF-260s received from Singapore

The first six of 19 SIAI-Marchetti SF-260 Warrior piston-engined primary trainers formerly operated by the Singapore air force were handed over to the TNI-AU at Halim air base in July. Originally delivered from 1971, the SF-260s had been used mainly for SAF recreational flying following the 1999 transfer of its flight training organisation and SIAI-Marchetti S-211 jet-trainers to Australia.

Indonesian naval aviation (TNI-AL) plans to acquire Mi-17 transport helicopters and eight smaller Mi-2s from a Rosoboronexport contract have reportedly been deferred because of Russian credit problems.

Ireland

New turboprop trainers sought

Current year defence budget economies of some 40 million euros ($40 million) were partly accommodated by cancellation of Irish Air Corps plans to acquire three Sikorsky S-92 medium-lift helicopters costing $62 million, for SAR and military transport missions. Some 27 million euros, however, were then allocated for replacement of the IAC's six Fouga Magisters, retired in 2000, by eight new tandem-seat types for advanced training and light ground-attack.

These will also take over the roles of seven IAC SIAI-Marchetti SF-260WE and six Cessna FR172 light trainers, for which evaluations are expected of the EMBRAER EMB-314M ALX Super Tucano, Pilatus PC-9; and Raytheon T-6A. Selection is possible by the year-end, for initial 2004 deliveries.

Israel

New SIGINT system

Israel Aircraft Industries' ELTA subsidiary is to develop a new airborne signals intelligence (SIGINT) system that will be installed on three Gulfstream V business jets that will be delivered to the Heyl Ha'Avir or Israel Air Force/Defence Force.

Italy

AMI expands tactical transport

Having withdrawn from the Airbus A400M project in late 2001, Italy is continuing its tactical transport force expansion through alternative and cheaper programmes. On 2 July, the Italian government accepted the first two of 10 stretched Lockheed Martin C-130J-30s (USAF CC-130Js) ordered for the Aeronautica Militare Italiana.

With 12 more Hercules 2 delivered or on order, the Italian C-130J configurations are among the most advanced yet developed. They are also the first receiver-tanker versions built, for greater strategic or tactical transport mission flexibility, in replacing Aeritalia G222s in the AMI's 46th Air Brigade at Pisa. After the UK, with 25 C-130Js, Italy is Europe's second largest Hercules 2 operator, with 22 aircraft on order from 118 C/CC-130Js then sold, and 86 delivered.

On 27 June, the AMI signed an Alenia Aeronautica contract for the first five of 12 C-27J Spartan twin-turboprop tactical transports it requires, plus seven options. With two Rolls-Royce Allison AE2100D2 engines driving six-bladed Dowty composite propellers, and new digital 'glass' cockpit avionics derived from the C-130J, the C-27J was developed jointly with Lockheed Martin from the well-proven G222/C-27A. It retains the same dimensions, but with substantially improved performance.

Delivery of the AMI's first C-27J is expected by mid-2005, with the final aircraft due in late 2006. Actual contract value and production schedules await finalisation, but will be worth over $200 million to Lockheed Martin. As the first air force to operate both the C-130J and the C-27J, the AMI will benefit from their avionics and propulsion commonality, to replace its 30 or so G222s.

Japan

Navy orders EH-101

AgustaWestland is expected to be awarded a $1.5 contract to supply the Japanese Maritime Self Defence Force with up to 13 EH-101 helicopters but could order as many as 80 aircraft. Japanese company Kawasaki Heavy Industries will assemble the helicopters at its factory near Tokyo.

Kuwait

AH-64D procurement progresses

Long-planned acquisition by Kuwait of 16 Boeing AH-64D Longbow Apache day/night attack helicopters, following a lapsed 1998 FMS LOA, was revived in April, with US DoD notification of a proposed $2.1 billion KAF procurement package. Apart from the AH-64Ds, with eight Lockheed Martin/Northrop Grumman AN/APG-78 Longbow millimetric-wave fire-control radars, the revised proposal included four spare T700-GE-701C turboshafts, 96 Boeing/Lockheed Martin AGM-114L3 and 288 AGM-114K3 Hellfire fire-and-forget ATMs. Contract signature was expected by the year-end.

Malaysia

Super Hornet selected

Official US sources were quoted in May as confirming selection of the two-seat Boeing/MDC F/A-18F Super Hornet to equip an RMAF multi-role combat aircraft squadron. Eight two-seat F/A-18Ds have been operated by No. 22 Skuadron of the RMAF at Butterworth since 1997, and a contract for 12-16 F/A-18Fs, retaining basic Hornet commonality, was expected to be placed by Malaysia in August, after competitive evaluation against the Sukhoi Su-30/35 series.

First Super Lynx 300s completed

Production is now well advanced by AgustaWestland at Yeovil of the six

Flashman trials

In a trial known as Flashman, the RAF's SAOEU (Strike Attack Operational Evaluation Unit) painted a Tornado GR.Mk 4 (ZD792, below) and Jaguar GR.Mk 3 (XZ103, right, landing after its first flight in the new scheme at Coltishall) in an experimental grey scheme with sand uppersides, to test its effectiveness in a desert environment during a deployment to China Lake. The Tornado exhibits three shades of grey: the Flashman grey; F.Mk 3-style grey on the fuel tanks; and standard GR.Mk 4 grey on the Sky Shadow pod.

LHTEC T800-engined Super Lynx 300s ordered by the Royal Malaysian Navy, with the first (ZJ904) having started flight development in April. A second RMN example (ZJ905) was shown statically by AgustaWestland at Farnborough, and initial deliveries are planned to Malaysia early next year, following Super Lynx 300, T800 and 'glass' cockpit certification.

Mexico

New naval aircraft deliveries

Six armed twin-turboshaft NOTAR (no tail-rotor)-equipped MD Combat Explorer helicopters delivered to Mexican Naval Aviation (AAM) between 25 January and late March are now operating on drug interdiction and anti-smuggling missions. Each is armed with a 12.7-mm (0.5-in) General Dynamics GAU-19/A multi-barrel Gatling machine-gun and seven 70-mm (2.75-in) podded rockets. Alternative weapons include M134 7.62-mm (0.3-in) Miniguns, M2 12.7-mm machine-guns, Hellfire laser-guided attack missiles, and Rafael N-TD dual-operation missiles.

In September, the AAM was due to receive a Bombardier Dash 8-Q200 twin-turboprop transport, for which it had an additional option. The AAM also recently accepted its 10th and last Zlin Z242L primary trainer from a CKr80 million ($2.57 million) Czech Moravan contract.

Namibia

Libyan helicopters transferred

Continuing Libyan military aid to the Namibian Defence Force, which operates a small number of support and training aircraft, has recently included two surplus Mil Mi-8 'Hip' transport helicopters and two Mil Mi-24 'Hind' gunships. Four Nanchang K-8 turbofan trainers are also reportedly on order for the NDF.

Nepal

Helicopter reinforcements

Further expansion was reported earlier this year of the Royal Nepalese Army Air Service, from acquisition of two more Mil Mi-17 transport helicopters. These were reportedly obtained via Kyrgyzstan for some $5.8 million, supplementing a similar 'Hip' variant delivered earlier to the RNAAS. Two HAL-built Cheetah versions of the high-altitude SA 315B Lama/Alouette II utility helicopter were also delivered from Indian military aid, joining three previously-received RNAAS SA 316B Alouette IIIs, for operations against Communist rebel forces.

Peru

More transport aircraft

Recent Peruvian Army Aviation receipts have included two Antonov An-28 and a used Let L-410UVP twin-turboprop transports from eastern Europe, supplementing three FAP An-74 twin-turbofan STOL transports delivered in 1993. The FAP also acquired nine Harbin Y-12 light turboprop transports from China.

Poland

Fighter RFPs finalised

A decision on the Polish air force (PWLOP) requirement for up to 48 new multi-role combat aircraft, including 12 two-seat combat trainer versions, is now expected by 28 December, following the 8 July issue from Warsaw of final RFPs. Poland's planned $3.5 billion next-generation fighter procurement is being conducted at government levels, through RFPs circulated to official agencies in France, Sweden, UK and the US.

Three best and final offer submissions, including an extensive equipment and weapons package, were expected by the 10 November deadline. Apart from Sweden's Gripen International JAS 39C, the French government is offering the Dassault Mirage 2000-5 Mk 2, while the US is proposing Lockheed Martin Block 50/52+ F-16C/Ds.

All the companies concerned were negotiating with Polish industrial companies to meet stringent 100 per cent minimum offset contract requirements in the new fighter programme, including local assembly by PZL Mielec. From 2005-06, the first 16 aircraft, including nine two-seat versions, will equip the 3rd Tactical Air Squadron at Poznan-Krzesiny; then the 10th TAS at Lask in 2006-07; and 6th TAS at Powidz a year later.

New support aircraft

Six new business jets are being sought by the PWLOP, to replace nine Yakovlev Yak-40 and two Tupolev Tu-154M tri-turbofan transports, for government and VIP roles. Competitive bids for the $250 million programme will be invited for the Bombardier Challenger 604, Dassault Falcon 900 and Gulfstream GIV/V.

PWLOP orders are also planned for 47 Rolls-Royce 250-C20 turboshaft-powered PZL-Swidnik SW-4 light four/five-seat helicopters, now flying in prototype form, for pilot training and utility roles.

315 Sqn, KLu, celebrated its 50th anniversary on 1 July 2002. One of the unit's F-16AMs was painted with the squadron's lion badge on the fin, together with silhouettes of the aircraft types it has operated: F-84G, F-84F, NF-5 and F-16.

South Korea

F-15K selected

On 19 April the Republic of Korea announced that it had selected the Boeing F-15K as the winner of its $4 billion F-X fighter competition. The award covers 40 F-15Ks and associated weapons and support and ensures that the F-15 production line in St Louis will remain open through 2008, when the last F-15K is delivered. Based upon the USAF's F-15E, the new variant will feature AN/APG-63(V)-1 radar, a third-generation forward looking infra-red (FLIR), helmet-mounted cueing system (HMCS), and its weapons-control systems will support advanced weapons including the joint direct attack munition (JDAM), stand-off land attack missile-expanded response (SLAM-ER), AIM-120 AMRAAM and AIM-9X Sidewinder. Selected over Dassault Rafale, the first F-15K will be delivered during 2005.

Sudan

More Russian aircraft

The Sudanese air force has reportedly become one of the latest customers for the MiG-29, with imminent deliveries of a dozen or so aircraft, costing $120 million. These are believed to be refurbished surplus 'Fulcrum-A/Bs' from Russian storage, although accompanied by additional new-build Klimov RD-33 turbofans from Moscow's Chernyschev factory. They follow six Mi-24V 'Hind-E' attack helicopters acquired from Belarus around 1998, and a reported contract for Mil Mi-35s signed with Rostvertol in 2000.

Turkey

Boeing B737 AEW&C ordered

A $1 billion-plus contract was signed with the Boeing Company on 4 June by the Turkish Defence Industries Under-Secretariat for procurement of four B737 Airborne Early Warning & Control (AEW&C) aircraft. Northrop Grumman Electronic Systems and several major Turkish companies, including TAI and Havelsan, are represented in the international Boeing-led team undertaking the programme. This is based on a similar aircraft being developed for Australia's Project Wedgetail contract, signed with Boeing in December 2000, for four B737 AEW&C systems.

The Boeing Space and Communications (S&C) AEW&C system combines the B737-700 increased gross weight aircraft with Northrop Grumman's Multi-role Electronically Scanned Array (MESA) radar, incorporating integrated IFF capabilities. Turkey's B737 AEW&C also contains a flexible, open-architecture mission computing sub-system, and includes Turkish-built electronic support measures (ESM) and an extensive communications suite.

Morocco's 'Marche Verte' ('Green March') aerobatic team is in the process of replacing its Mudry CAP 231s with uprated CAP 232s.

In use for training at Kanoya AB with Dai 211 Kyoiko Koku-tai of the JMSDF is the Hughes 369E, known locally as the OH-6DE. It is distinguished from the earlier OH-6D by its pointed nose profile.

On 24 April 2002 the French Sécurité Civile (part of the Ministère de l'Interieur) received its first Eurocopter EC145 (previously BK 117C-2). The primary roles are coastal and mountain SAR, and fire-spotting.

New heavy-lift helicopter
The Turkish government has once again announced plans to purchase eight heavy-lift helicopters and will launch a new international tender in late 2002 or early 2003. The cost of the helicopters, which will be operated by the Army, is estimated to be $250 million. The leading contenders are expected to the Sikorsky CH-53 Super Stallion and the Boeing CH-47 Chinook. Turkey had previously planned to purchase eight CH-53Es but cancelled the contract due to a lack of funds.

United States

AIM-9X delivered
Raytheon delivered the first production AIM-9X Sidewinder missile to the US Navy at its Tucson, Arizona, facility during May 2002. The missiles, which will initially equip the F/A-18E/F Super Hornet, will be used for training but should achieve initial operational capability (IOC) during the summer of 2003. Over the past two years the AIM-9X has achieved 18 successes in 19 guided flights and 37 successes in 39 launches.

2,500th Blackhawk delivered
On 26 April 2002 Sikorsky marked delivery of the company's 2,500th production Blackhawk helicopter during a ceremony conducted at the Hawaii Army National Guard facility in Hilo, Hawaii. The aircraft is one four new UH-60Ls that were delivered to Company D 1-207th Aviation Regiment at the end of 2001. The contractor delivered the first production UH-60A to the US Army in 1978, and more than 2,000 H-60s and their derivatives currently serve with five US air arms. A further 600 S-70s have been delivered or are on order with 25 other nations worldwide.

T-6A order increased
The US Navy increased its orders for the Raytheon T-6A Texan II primary trainer to 43 aircraft when it purchased seven additional aircraft recently. Raytheon will deliver the first T-6A to the Navy late in 2002. The aircraft will initially be assigned to Training Air Wing Six at NAS Pensacola, Florida.

New aircraft to USAF Academy
The USAF has ordered three new glider types for use by the Air Force Academy in Colorado Springs, Colorado, and examples of each model arrived at Edwards AFB, California, on 9 April 2002. The gliders are undergoing flight tests at the hands of students from the USAF Test Pilot School and were ordered from Letecke Zavody a.s. via its distributor, Blanik America, in October 2001. Although the new gliders include three different models, they have been designated as if they were variants of the same mission design series (MDS). The aircraft, which comprise the L23 Super Blanik, L13AC Blanik and L33 Solo, have respectively been assigned the designations TG-10B, TG-10C and TG-10D. The USAF had previously purchased an L23, which had been operated by the USAF Test Pilot School as a TG-10A. USAF personnel had previously evaluated the three Blanik variants near Edwards AFB, California, in June 2000. The service hopes to order as many as 21 gliders to replace those currently in service with the 94th Airmanship Training Squadron at the US Air Force Academy.

New F-16Cs delivered
Lockheed Martin recently delivered the first of 10 new F-16Cs ordered in Fiscal Year 2000 to the USAF. The fighters, along with four ordered in Fiscal Year 2001, will be delivered at a rate of one-two per month through December 2002. The Block 50 single-seat versions are equipped with General Electric F110-GE-129 engines, a modular mission computer, colour cockpit displays and recording equipment, on-board oxygen generating system, and the AN/APX-113 advanced electronic interrogator/transponder. They are the first new F-16s delivered to the USAF since April 2001. The aircraft share a common configuration with the Block 50 F-16C being delivered under the first phase of the USAF F-16 Common Configuration Implementation Program (CCIP). The first CCIP-modified aircraft was delivered in January 2002.

Global Hawk improvements
Two contracts totalling nearly $300 million were recently awarded to Northrop Grumman to enhance the overall performance of the RQ-4A Global Hawk unmanned reconnaissance system. Specifically, the upgrades will include improvements to the integrated sensor suite, development of more powerful generators, increase of the air vehicle's gross take-off weight and improvement of its constant altitude flight mode. In related news, Northrop Grumman delivered the newest Global Hawk to the USAF at Edwards AFB, California at the conclusion of its 4.5-hour maiden flight on 23 April 2002.

Yemen

MiG-29 come-back
As a former MiG-29 operator, the Yemeni government is receiving about 10 more 'Fulcrums' from a $300 million contract reportedly signed with RSK MiG in September 2001. Eight MiG-29s, plus two two-seat MiG-29UB combat trainers, acquired from Moldova in 1994 by the former northern Islamic Yemen Arab Republic, were returned after three or four years' operation.

The new order, from the current Republic of Yemen air force, is unofficially reported to comprise eight or more refurbished MiG-29s and two MiG-29UB combat trainers, plus associated Vympel R-27 (AA-10 'Alamo') medium-range and Vympel R-73 (AA-11 'Archer') close-combat AAMs. The RYAF MiG-29s were probably drawn from about 50 undelivered 'Fulcrums' in long-term RSK MiG storage in Moscow. Options worth about $100 million were also reportedly included in the RYAF contract for another half-dozen MiG-29s. RSK MiG recorded some $977.3 million of business in 2001, including about $200 million from MiG-29 sales to Eritrea and Myanmar.

Open Skies certification

Between 15 and 28 April 2002 Nordholz naval air station in northern Germany played host to the Open Skies Joint Certification 2002 programme, which certified observation aircraft from Hungary, Russia and the Ukraine. The main aim of the exercise was to calibrate the minimum operating height for the aircraft, and involved an intense two-week assessment by inspectors from signatory nations. Sorties were flown over a calibrated target area at Nordholz, which included light beams and black/white boards. The aircraft used various cameras, film types and filters, and the results were analysed to establish the minimum altitude necessary to achieve the maximum 30-cm (1-ft) resolution as specified by the Open Skies treaty. This rigorous assessment is needed as the difference in quality and optical specifications between various sensor systems can result in an altitude difference of up to 1000 m (3,280 ft).

The Treaty itself finally came into effect during 2002, when Russia and Belarus signed, joining Benelux (Belgium, Netherlands, Luxembourg), Bulgaria, Canada, Czech Republic, Denmark, Finland, France, Georgia, Greece, Italy, Kyrgyzstan, Norway, Poland, Portugal, Romania, Slovakia, Spain, Sweden, Turkey, United Kingdom and the United States. As well as Hungary and the Ukraine, nations expected to sign up in the near future include Croatia and the Baltic States of Estonia, Latvia and Lithuania.

Paul Kievit and Carlo Kuit

The dedicated An-30 photo-survey aircraft is an obvious platform for Open Skies verification missions, and has been chosen by Russia (below) and Ukraine (right). The Russian aircraft is based at Kubinka, near Moscow, while the Ukrainian 'Clank' is based at Borispol with the 10th OAE. Hungary (below right) has modified an An-26 'Curl' of the 89th VSE at Szolnok to perform Open Skies missions.

Air Arm Review

Australia

JSF decision

The Australian government's JSF participation decision, to meet the RAAF's $A12 billion Air 6000 F/A-18 and F-111 replacement combat aircraft requirement from about 2015, caused some surprise among other industry bidders. Speculation also resulted that JSF time-scales might also necessitate procurement of interim combat aircraft, or additional measures and upgrades to maintain the viability of the RAAF's current operational types.

Other procurement programmes in the Australian Defence Force's (ADF) $A47 billion ($25.5 billion) 10-year Defence Capability Plan (DCP) are claiming priority, however, including the proposed acquisition of up to five air-refuelling tanker/transports – probably Boeing KC-767s – to replace the RAAF's current four Boeing 707-338Cs from around 2006. $A1.7 billion ($925 million) was allocated in the current defence budget for this requirement, apparently superseding tenders invited last year for the $A150 million five-year Private Finance Initiative (PFI) lease of five tankers.

New ADF transport aircraft requirements are also being proposed in an Air 9000 programme, to acquire a few Future Heavy Airlifters in the Boeing C-17 class for strategic operations, C-130H upgrades, and light tactical airlift projects.

A further $A400 million has been earmarked for 12 utility helicopters for Australian Army Aviation, together with studies for combining joint army and naval training operations with out-sourced modern IFR-equipped helicopters, through a PFI programme. At the moment, the RAN has 12 Eurocopter AS350BA Squirrels at Nowra, near Sydney, and four Bell 206B-1 JetRangers, while the AAA has 18 AS350BAs, plus about 40 Bell 206B-1 JetRanger scout helicopters at Oakey, near Brisbane. Most are due for replacement to provide representative training on next-generation equipment, represented by the AAC's new Eurocopter Tiger attack helicopters, due for delivery from 2006-07. A decision on future ADF helicopter training is expected by December 2003.

Bangladesh

MiG-29s to go?

Eight MiG-29s acquired by the Bangladesh air force for around $124 million only about three years ago are being withdrawn from service because of their high operational and support costs. In a leading national newspaper, Prime Minister Begum Haleda Zia was quoted as saying: "The MiG-29 purchase was an incorrect financial decision", although he refuted suggestions of corruption in their procurement by a previous Bangladesh administration.

Belgium

Service unification progress

Radical armed forces rationalisation from a government decision in January resulted in a single command structure for the three services. These are also being scaled-down from some 44,000 personnel to around 39,500 by 2015. Reduced to six squadrons operating only 72 of 107 F-16As and 19 F-16Bs, with mid-life upgrades for continued service until 2015, the air force has become a tactical air component known as Command Operations Air (COMOPSAIR).

With an increased emphasis on humanitarian support missions, it will operate alongside COMOPSLAND and COMOPSNAV, as the remaining armed forces' elements. Some 26 military bases and installations are also being closed, and military aircraft purchases restricted mainly to the seven Airbus A400Ms from Belgium's programme partnership.

Hungary

Newest NFTC partner

Hungary has opted to sign on for the NATO Flying Training in Canada (NFTC) programme. Beginning in 2003, Hungary will send seven students to Canadian Forces Base Moose Jaw, Canada, for the basic phase of the programme. At the conclusion of this phase the students will remain in Canada and either go on to the tactical fighter lead-in jet training phase at CFB Cold Lake or to rotary-wing training at the Southport Aerospace Center in Manitoba. Under the terms of the 17-year arrangement, the Hungarian Air Force will also provide seven instructors to the three phases of the NFTC programme. As well, some experienced Hungarian pilots will undertake the advanced and tactical fighter lead-in phases of NFTC over the next three years. The increased training requirements mean that the NFTC will require two additional Raytheon CT-156 Harvard IIs and two further BAE Systems CT-155 Hawk trainers, bringing the totals for these aircraft to 24 and 20, respectively.

The Georgia Air National Guard's 116th Bomb Wing has already begun ferrying out its B-1Bs in preparation for its transition to the E-8C Joint STARS mission. ACC intends to retain just 60 of the bombers, including 36 combat-coded aircraft, 16 for training and eight for test and as an attrition reserve.

This EC 2/30 Mirage F1CT was given a special scheme to commemorate the 60th anniversary of the 'Normandie-Niemen' squadron. The unit's badge consists of the two lions of Normandy and the white arrow flash of the 303rd Fighter Division, Soviet air force, to which the squadron was assigned during World War II. Formed in 1942 in Russia with French volunteers as Groupe de Chasse 3 'Normandie', the unit distinguished itself during the battle to cross the river Niemen in 1944, Stalin himself bestowing the name of the river as an honour title.

India

New fighter plans change

In a surprise move in July, the Indian government reportedly issued urgent requests for information for new single-seat multi-role combat aircraft. Some 125 aircraft are reportedly being sought to replace early series MiG-21s from around 350 still remaining in IAF service. These include 125 of the last-production HAL MiG-21bis series, now undergoing major avionics upgrades by Hindustan Aeronautics Ltd, from RSK MiG kits. After two or three years, however, only four MiG-21bis have reportedly so far been upgraded, and the IAF continues to lose one or more per month of its older 'Fishbeds' in accidents, or well over 100 in the past decade.

The IAF has apparently therefore felt compelled to seek additional combat aircraft, for early delivery. The new IAF RFIs are understood to have been invited for the Mirage 2000-5, JAS 39 Gripen, Sukhoi Su-35 and the new MiG-29M2, with requirements for early down-selection and delivery availability.

A second technology demonstrator prototype (TD2) of India's new GE F404-powered Light Combat Aircraft made its initial flight on 6 June, and a third LCA has also been rolled out from seven pre-production aircraft planned, including a two-seat trainer. This now appears to be regarded more as a long-term research project, however, rather than a prospective weapons system. Nevertheless, earlier this year, the Indian government released a further $625 million in LCA funding, for construction of eight IAF production aircraft for delivery from 2006, and purchasing 40 more GE 404 engines.

LCA R&D, including the first supersonic trials by TD2, is expected to total some Rs48 billion (around $1 billion), about half of which has already been spent in the past 15 years. Some Indian defence officials are still hoping to see the LCA in IAF service by 2010, with a fly-away unit cost of around $17 million.

Deliveries of the first two IAPO-built Sukhoi Su-30MKI two-seat long-range multi-role fighters, for which the IAF placed Rs 6,300-crore ($3.06 billion) orders for 32, in November 1998, plus 140 from 15-year HAL licensed production in Nasik, took place on 25 June. The two Su-30MKIs were flown partly-disassembled in an Antonov An-124 to the IAF base at Lohegaon, Pune. They were accompanied by 15 Russian technicians, and were to be followed by another eight Su-30MKIs in the next few months. The remaining 22 will be delivered in 2003, and the IAF's 18 earlier-delivered IAPO-built air-defence Su-30Ks, equipping No. 24 Squadron at Lohegaon, may also be upgraded by HAL to MKI multi-role standards.

Indian Navy interest was also reported in May in acquiring 17 of nearly 50 RN BAE Sea Harrier FA.Mk 2 advanced maritime air defence fighters from the UK, following their planned 2006 withdrawal. They would replace the IN's 16 earlier and less capable Sea Harrier FRS.Mk 51s, with Blue Fox instead of Blue Vixen radar. In June, the IN also received the first two of nine Kamov Ka-31 ship-borne AEW helicopters, from a $200 million contract.

Moldova

Fulcrums on the market

The Republic of Moldova has announced plans to sell six remaining MiG-29 fighters in the Moldovan National Army inventory. Funds received as a result of the sale will reportedly be used to purchase transport and cargo helicopters. Moldava previously sold 21 MiG-29 to the United States during 1997.

The US Navy's test and evaluation fleet has adopted new squadron numbers in the VX-range. This QF-4S, previously allocated to the NWTSPM (Naval Weapons Test Squadron Point Mugu), proudly wears its new VX-30 designation.

A rare and welcome visitor to Europe was this SA 330H Puma of the Moroccan Royal Gendarmerie, seen at the open day at Dax in France.

New Zealand

Long term defence plan details

A new 10-year $NZ3 billion ($1.41 billion) Long Term Development Plan (LTDP) announced in June pegs total defence expenditure in the current year to the same $NZ1.42 billion ($668 million) in 2001, and includes only $NZ212.4 million ($99.91 million) for capital equipment. Little new RNZAF procurement is in prospect, but some elements of the original $NZ445 million Project Sirius avionics upgrades of the six Lockheed P-3K Orions, plus new avionics and life extension of the five C-130s are receiving priority in $NZ1.47 billion LTDP equipment allocations. Replacements will also be sought for the RNZAF's five Bell 47G, costing $NZ11 million, and 14 Bell UH-1H helicopters ($NZ500 million).

Sweden

Fourth Gripen wing

Inauguration of the Swedish air force's F17 Wing with the Gripen lightweight multi-role fighter at Ronneby on 14 June marked a further advance in the SAF's re-equipment programme with this advanced combat aircraft. Under the SAF reorganisation plans approved in early 2001, two squadrons at F17 will comprise the fourth and last Swedish Wing to receive the Gripen, which will equip a total of eight squadrons. From 2004, F17 will undertake the roles allocated to the Swedish air force's Rapid Reaction Unit (SWAFRAP).

Since the first JAS 39A/B single- and two-seat Gripens entered SAF service in June 1996, more than 120 of the 204 ordered for Sweden have been delivered to date. These equip five squadrons, the first of which became operational with F7 at Såtenäs in late 1997, followed by F10 at Ängelholm and F21 at Luleå.

On 6 September, the SAF received its first upgraded JAS 39C Gripen, from the last 20 Batch 2 aircraft of 110 on order which are being completed to NATO-compatible Export Baseline standards. These will also feature in the SAF's third and last batch of 64 JAS 39C/Ds, some of which may be transferred to export customers. The remainder will replace all the SAF's current SAAB Viggen combat and reconnaissance aircraft by 2007.

United Kingdom

Eurofighter service delayed

Delays of about six months with RAF service entry of its first Eurofighters – previously acknowledged by BAE Systems – were confirmed on 23 May in Parliament by Defence Procurement Minister Lord Bach. He said: "We made clear in February that the June 2002 Eurofighter in-service date was becoming increasingly difficult to achieve. Although industry has made good progress with the first three instrumented production aircraft (IPAs) from successful first flights in early April, delays in achieving full maturity with the detailed design in some areas prevented flight-testing from starting on time.

"Following a thorough programme review by the partner nations, the NATO procurement agency, and industry, we have jointly concluded that acceptance of the first aircraft into (RAF) service should be delayed to the year-end. We'll now be working hard with industry to recover lost time, and achieve the planned Operational Employment date in the second half of the decade, thus avoiding any gap in front-line capability, or extra MoD costs".

BAE Systems plans to accelerate final assembly on its Warton Eurofighter line to meet original commitments of the December 1998 Contractor Support Service (CSS) agreement. This should speed delivery by mid-2003 of all 13 two-seat Eurofighters due to equip the RAF's initial Operational Evaluation Unit (OEU) at Warton.

Apache service delay

Restrictions on Boeing/Rockwell AGM-114K/M Hellfire anti-tank missile launches from the Army Air Corp's WAH-64 Longbow Apache attack helicopters, because of potential rotor damage, have delayed their AAC service entry. The MoD admitted in March that "a problem has been identified with the Hellfire rocket, involving damage to the aircraft from motor debris on firing. We're now working with the contractors to find solutions".

The MoD was unable to assess the degree of likely damage, or how long it would take to rectify the problem. Press reports, however, claimed that US experience had indicated that AH-64D missile launches were currently possible only from the starboard wing pylons, to avoid rocket efflux debris fouling the port-mounted tail-rotor blades. This, of course, effectively halves some of the Apache's weapon-load capabilities.

AgustaWestland is now well-advanced with co-production at Yeovil of 59 partially-assembled AAC WAH-64 Apaches ordered from a $2.57 billion March 1996 contract. This also included the first eight WAH-64s built entirely by Boeing for the UK, and delivered from August 1999. Initial AAC Apache weapons clearance, originally due in July 2001, and to be followed by full Military Aircraft Release (MAR 5) clearance this August, were then still awaited.

United States

'Diamondbacks' transition

On 1 May 2002 the 'Diamondbacks' of VF-102 were officially redesignated VFA-102 and began transitioning from the F-14B to the F/A-18F Super Hornet at NAS Lemoore, California.

New Hawkeye to deploy

The US Navy has announced that the 'Wallbangers' of VAW-117, stationed at Naval Base Ventura County/Point Mugu, California, will be the first operational squadron to receive the Hawkeye 2000 variant of the Northrop Grumman E-2C. The squadron will deploy as part of CVW-9 aboard USS *Nimitz* (CVN 68) later this year.

Clipper delivered

The first of four C-40As Clippers destined for the 'Sunseekers' of VR-58 was delivered to the unit at NAS Jacksonville, Florida, on 28 March 2002. The unit had transferred its last C-9B to VR-56 at Naval Station Norfolk/Chambers Field on 13 December 2001.

Knighthawks deployed

Stationed at Andersen AFB, Guam, the 'Providers' of HC-5 became the first operational MH-60S Knighthawk squadron on 31 March 2002 when the unit accepted three aircraft. The 'Chargers' of HC-6 at Naval Station Norfolk, also received their initial examples in early 2002.

New Hercs for California Guard

The first of eight new CC-130Js destined for the California Air National Guard's 146th Airlift Wing arrived at Channel Islands Air National Guard Station (ANGS) on 2 May 2002. The initial aircraft, which wears the serial 01-1461, is one of two aircraft purchased with Fiscal Year 2001 funding. The CC-130Js, which were previously known as C-130J-30s, will replace the wing's elderly C-130Es. The ANGS is located on Naval Base Ventura County/Point Mugu in Oxnard, California.

New colours for Japanese Eagles: above is an F-15DJ from the Nyutabaru-based Hiko Kyodo-tai (aggressor squadron) in a new sand/brown scheme, while below a Dai 303 Hiko-tai F-15J displays special 'wave' markings applied for the 2002 Air Combat Meet.

JSTARS delivery

Northrop Grumman delivered the 13th E-8C joint surveillance target attack radar system (JSTARS) aircraft to the 93d Air Control Wing at Robins AFB, Georgia, on 25 April 2002. The contractor is also upgrading the 10 earlier aircraft to the latest configuration, and the first of these has been redelivered to the USAF. In related news, Northrop Grumman has received a $7.9 million contract modification covering advanced procurement of for one Lot XI E-8C joint surveillance target attack radar system aircraft, referred to as aircraft P-17.

Test squadrons renamed

On 1 May 2002 five US Navy flight test squadrons assigned to Naval Air Systems Command were officially redesignated. The units, which are attached to the Naval Air Warfare Center Aircraft Division and Naval Air Warfare Center Weapons Divisions had previously carried names but were not numbered. Previously identified as Naval Aircraft Test Squadrons and Naval Weapons Test Squadrons, all five units were all redesignated as Air Test & Evaluation Squadron and assigned a number. The new activity titles are intended to increase the visibility of the squadrons.

Naval Air Warfare Center Aircraft Division – NAS Patuxent River, Maryland

VX-20 (prev. Naval Force Aircraft Test Squadron)
HX-21 (prev. Naval Rotary Wing Aircraft Test Sqn)
VX-23 (prev. Naval Strike Aircraft Test Squadron)

Naval Air Warfare Center Weapons Division – NAWS China Lake, California

VX-30 (prev. Naval Weapons Test Sqn Point Mugu)
VX-31 (prev. Naval Weapons Test Sqn China Lake)

Global Mobility Roadmap

The USAF has announced plans to acquire 60 additional C-17As beyond the 120 that were already planned. The additional aircraft will allow Air Mobility Command to transition more than half of its Lockheed C-5As to the reserve component, replacing Lockheed C-141Cs and older C-130Es. As part of the plan the service intends to assign 12 C-17As to the 305th Air Mobility Wing (AMW) at McGuire AFB, New Jersey, the 436th Airlift Wing (AW) at Dover AFB, Delaware and the 60th AMW at Travis AFB, California. The aircraft would also be assigned to the 452d AMW, an Air Force Reserve Command unit at March Air Reserve Base, California, the 517th Airlift Squadron (AS) at Elmendorf AFB, Alaska and the Hawaii Air National Guard's 204th AS at Hickam AFB, Hawaii.

Under the plan, each of these units would receive eight aircraft. The 97th AMW at Altus AFB, Oklahoma, which serves as the C-17A training unit, would receive six additional aircraft, while the wing would lose six C-5As that would in turn be assigned to the Air Force Reserve Command's (AFRC) 433d AW at Kelly Field, Lackland AFB, Texas, which already operates the Galaxy and would become the formal training unit (FTU). The 436th AW and 60th AMW would collectively retire 36 C-5A/Bs.

The 445th AW at Wright Patterson AFB, Ohio, also assigned to the AFRC, would retire its C-141Cs and receive 16 of the Galaxies, while a further 16 C-5As would be divided between the Air National Guard's 164th AW in Memphis, Tennessee, and the 167th AW in Martinsburgh, West Virginia. These units respectively operate the C-141C and the C-130H.

While all of the C-141Cs would be retired, the C-130Hs at Elmendorf – along with those operated by the two Air National Guard squadrons – would be distributed to other units, allowing older C-130Es to be retired. The service plans to retire 56 older C-130Es; equip 280 C-130E/Hs with a common avionics configuration, and acquire 150 C-130Js and stretched CC-130Js.

This hulk of a Mitsubishi/Sikorsky HSS-2B is in use at Omura AB, Japan, as an emergency egress trainer.

Unit restructuring

The USAF recently announced a series of changes that will take place within its Active and Reserve Component units in Fiscal Year 2002/2003. They are listed by command.

Air Combat Command (ACC)

Beale AFB, California: The 12th Reconnaissance Squadron (RS) will receive two RQ-4A Global Hawk unmanned aerial vehicles. The squadron is part of the 9th Reconnaissance Wing (RW).

Cannon AFB, New Mexico: The 522d Fighter Squadron (FS) will transition from the Block 30 F-16C to the latest Block 50 variant. The squadron is part of the 27th Fighter Wing (FW)

Davis Monthan AFB, Arizona: As a result of the migration of the airborne command and control mission to the E-8C Joint STARS aircraft, the 42d Airborne Command and Control Squadron (ACCS), assigned to the 355th Wing at Davis Monthan AFB, will be inactivated and its six EC-130Es will be retired.

Nellis AFB, Nevada: The 422d Test and Evaluation Squadron (TES) will receive four F-22A Raptors. Although based at Nellis, the squadron is assigned to the 53d Wing at Eglin AFB, Florida.

Shaw AFB, South Carolina: One of the four F-16C squadrons assigned to the 20th FW at Shaw will be inactivated.

Air Mobility Command (AMC)

Andrews AFB, Maryland: The 99th Airlift Squadron (AS), part of the 89th Airlift Wing (AW), will transfer two of its C-20Bs to United States Air Forces in Europe (USAFE).

Charleston AFB, South Carolina: The 16th AS will be reactivated under the 437th AW and will be equipped with nine C-17As.

Air National Guard (ANG)

The Iowa Air National Guard's (ANG) 185th FW/174th FS in Sioux City will transition from the F-16C to the KC-135E in late 2003. A total of eight Stratotankers will be assigned.

North Dakota's 119th FW/178th FS in Fargo will transition from the F-16A(ADF) to the F-16C in late 2003. The Fighting Falcons will likely be acquired from the Iowa ANG.

Pennsylvania's 193d Special Operations Wing (SOW)/193d Special Operations Squadron (SOS) in Harrisburg, will receive three EC-130Js in 2003. The aircraft are currently undergoing modification for the special operations role.

The Washington ANG's 141st Air Refueling Wing (ARW)/116th Air Refueling Squadron (ARS) will transition from the KC-135E to the KC-135R beginning in late 2003.

New UAV designations

Two new military design series (MDS) were recently assigned to unmanned aerial vehicles. The designations include MQ-1B, which has been assigned to the AGM-114 Hellfire missile-equipped variants of the USAF's RQ-1B Predator. Similarly, the turboprop-powered Predator B, which is being developed by General Atomics, has been assigned the MDS MQ-9A, indicating its planned multi-mission capability.

Operations and Deployments

NATO

Patrol flights end

Following a review of US homeland security efforts, NATO Airborne Warning and Control System (AWACS) crews ended their flights over the continental US in support of Operation Eagle Assist on 16 May 2002. NATO deployed five E-3As from the NAEWF at Geilenkirchen to Tinker AFB, Oklahoma, for the operation, which began on 9 October 2001. Over the following seven months NATO crews flew more than 350 sorties tasked by North American Aerospace Defense Command (NORAD). The historic deployment, a result of the September 2001 terrorist attacks, marked the first time NATO forces had deployed from Europe to the US in an operational capacity.

Left: Manas, in Kyrgyzstan, is one of the main bases for aircraft operating over Afghanistan. Here a VMFA(AW)-121 F/A-18D taxis past a row of Mirage 2000Ds.

Right: In Afghan operations the Chinook has proved its great value, being able to operate at high altitude. This is an RAF aircraft, deployed to support the Royal Marines in the mountains on Operation Snipe.

Rapid Terrain Visualization

Advanced terrain mapping with the DHC-7 Dash 7

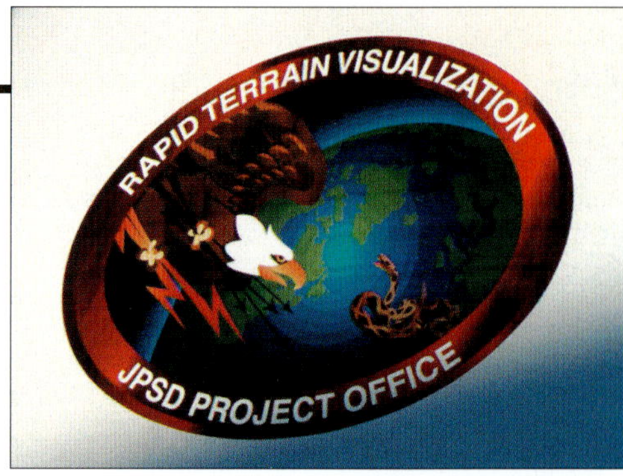

Right: N176RA wears the programme badge on the fin. The RTV effort is aimed at producing a means of terrain-mapping at various levels of resolution. The end result is a three-dimensional map which greatly aids the planning and execution of ground manoeuvres.

Over the years the US military has used a variety of aircraft that have not received a mission design series (MDS) designation. Often used for research and development (R&D) programmes, these 'non-standard' aircraft – which may or may not carry military serial numbers – have included either aircraft that have been confiscated by other government agencies or been purchased on the commercial market. 'Non-standard' aircraft that have recently supported such efforts include a CASA 212-300 used by the US Army's Grisly Hunter intelligence programme, and a Boeing 707-321B that is currently operating as a testbed for the USAF's Multisensor Command and Control Aircraft (MC2A) programme.

Although the US Army operates a small number of de Havilland Canada DHC-7 (Dash 7) airliners as reconnaissance aircraft, under the designation RC-7B, it recently placed another example of the former airliner in service as a testbed for new technologies. This particular DHC-7-102 carries the civil registration N176RA (construction number 76) and was modified by a civilian contractor to support an Advanced Concept Technology Demonstration (ACTD) known as Rapid Terrain Visualization (RTV). The Joint Precision Strike Demonstration (JPSD) Project Office at Fort Belvoir, Virginia, manages this ACTD. JPSD is part of the Army Material Command's Communications Electronics Command (CECOM) and reports to the Program Executive Office Intelligence, Electronic Warfare & Sensors (PEO IEWS). This office is also responsible for a number of other Army programmes, including the Guardrail Common Sensor (GRCS), Airborne Reconnaissance Low (ARL) and Aerial Common Sensor (ACS).

Before entering the US Army inventory in 1999, N176RA saw several years of service as a commuter airliner. The aircraft was originally delivered to Ransome Airlines in 1982 and was later operated by Pan Am Express and Trans World Express, before being registered to the Department of the Army.

Developed for the Army XVIII Airborne Corps at Fort Bragg, North Carolina, the RTV programme was initiated in 1995 and is demonstrating technologies associated with generating digital topographical elevation data (DTED), feature data and imagery. The resulting digital topographical data allows troops on the ground to better visualise complex battlefield situations. RTV comprises an interferometric synthetic aperture radar (IFSAR), developed by the Sandia National Laboratories, and light detecting and ranging (LIDAR) sensors built by the Canadian company Optech Inc. The RTV system is designed to provide high-resolution elevation, feature data and imagery of a 20 x 20-km (12.4 x 12.4-mile) area in 18 hours, a 90 x 90-km (56 x 56-mile) area in 72 hours, and a 300 x 300-km (186 x 186-mile) area in 12 days. The higher resolution terrain information, elevation and feature data allows field commanders and their staffs to more effectively plan and rehearse missions, evaluate courses of action and maintain situational awareness using two-dimensional and dynamic three-dimensional displays.

The IFSAR, which is installed on the DHC-7's fuselage centreline, is basically a SAR that collects imagery (echoes) from the same scene using two spatially separated antennas. The returns from the two antennas are used to calculate the height of the surface and build a three-dimensional digital elevation model (DEM). The primary advantage offered by the system is its ability to acquire terrain data through clouds and fog, during daylight or at night, using two different resolution modes. The first is a Level III mode, which can sweep 1.6-km (1-mile) wide swaths over a mission area. Level III data, which provides 10-m (33-ft) resolution, allows a soldier to see roads, trees, hills and waterways to support manoeuvre planning. The second is a Level IV mode, with 1.3-km (0.8-mile) wide swaths. Targeting data can be developed from Level IV returns, which

The US Army has operated the DHC-7 since 1990, when it entered service as a transport. By 1993 the Dash 7 was in use as a surveillance platform under the Grisly Hunter/ARL (Airborne Reconnaissance Low) programme. RTV uses the DHC-7 as a testbed only: any resulting operational system would probably be carried by UAVs.

Externally the RTV testbed is easily identifiable thanks to the IFSAR installation on the centreline, complete with sawtooth surround to avoid spurious returns. Aft of the dome are sliding windows covering the LIDAR equipment. There are also numerous additional communications/navigation antennas, including accurate GPS receivers.

offer 3-m (9.8-ft) resolution. The system is capable of transmitting over a swath width of 500-2000 m (0.31-1.24 miles).

LIDAR equipment combines a laser scanner with Global Positioning System (GPS) and inertial measurements to accurately measure terrain heights, and is also installed on the fuselage centreline. It uses the same principle as radar, but transmits light rather than radio waves. The heart of the LIDAR is an Airborne Laser Terrain Mapper (ALTM), which emits 33,000 laser pulses per second. The high-accuracy laser rangefinder, which is mounted on the aircraft centreline, scans a 160-m (525-ft) swath beneath the aircraft. The angle at which the laser is scanned is measured, an inertial reference system (IRS) records the motion of the aircraft for use during post-mission processing, and a GPS receiver aboard the aircraft records its position at fixed intervals. A second, ground-based GPS receiver provides differential correction for a more accurate position estimate. The laser range, scan angle, GPS data and IRS data are combined to accurately create DEMs and imagery at DTED Level V (1-m/3.3-ft resolution). This extremely high-resolution Level V data is useful in developing DTED that will support urban warfare. The data from both of these systems is processed and displayed in real time on work-stations aboard the DHC-7.

Although the modified DHC-7 and the RTV have proven to be extremely capable, the aircraft's vulnerability will preclude it from operating over hostile territory. The Army plans to continue development of the system with the hope of deploying a version capable of being carried by unmanned aerial vehicles (UAVs) or a space-based vehicle.

Tom Kaminski

Israeli King Airs keeping watch

Raytheon (Beech) King Air B200/B200T in IDF/AF service

The Israel Defence Force/Air Force (IDF/AF) is gearing up towards the planned standardisation of its light transport/surveillance fleet on the Raytheon (Beech) King Air B200. The Beech Queen Air B80, in service since 1974 as the Zamir (Nightingale), has been withdrawn from front-line service. The Zamir fleet utilisation as a navigation trainer by the IDF/AF Academy will continue, but the staff transport role will be contracted to commercial lightplane operators pending the delivery of a second batch of eight King Air B200T aircraft from November 2002, at a rate of one aircraft per month.

The first variant of the King Air to enter IDF/AF service was the RC-12D Kokiya (Cuckoo) Comint platform, of which five entered service in 1984, with a sixth augmenting the fleet at a later date. Four King Air B200 Zufits followed in 1990, and the first batch of five B200T multi-mission light transport aircraft was delivered in late 2000.

Almost two years after delivery of the first Beech B200T Zufit (Thrush) to the Israel Defence Force/Air Force (IDF/AF) in September 2000, the promised "special mission equipment to be installed in Israel" has appeared in public. Zufit 625 (the second of the five aircraft delivered: 622, 625, 629, 633 and 636), equipped with a side-looking underbelly electro-optical payload and covered windows, participated in the IDF/AF Day flypast over Hatzerim air base on 26 June 2002.

Plans to purchase an additional 14 B200Ts were scaled-down to the present contract for eight, possibly as a result of the Israeli Navy reluctance to replace its small fleet of three IAI 1124N Shahaf (Seagull) maritime surveillance aircraft with an MPA (maritime patrol aircraft) version of the Zufit. This had been proposed by the IDF/AF for commonality reasons, as the air force operates the MPA fleet on behalf of the Navy. On its part, the Navy is currently upgrading its three Shahaf MPAs with the Elta EL/M-2022 Maritime Surveillance Radar, and the jet performance of the Shahaf is cited as essential for the MPA mission. However, the Shahaf is certainly in need of replacement, having entered service in 1977.

The third batch of Zufits is planned to replace the long-serving Dornier Do 28 Agur (Crane) observation aircraft which, on service entry in 1971, was equipped with a stabilised 20 x 120 binocular set. From 1975 this was replaced by the Shelef (Stubble) system, a stabilised long-range TV observation system. Original planning called for the retirement of the Agur in 2003, about the time of completion of the delivery of the second batch of eight B200Ts. However, recent hostilities between Israel and the Palestinian Authority have led to a revision to these plans and the last eight Agur observation aircraft will continue to serve for as long as it will take to decide otherwise.

Shlomo Aloni

Above is a B200T Zufit carrying a long-range electro-optical sensor package under the fuselage, while at right – in a rare public appearance – a Comint-dedicated RC-12D Kokiya leads a utility transport B200 Zufit. The second batch of Zufits, delivered in 2000, were to B200T standard with wingtip fuel tanks.

21st Century Pigs

F-111 in RAAF service

The Royal Australian Air Force is now the sole operator of the General Dynamics F-111 swing-wing strike aircraft, and the two operational squadrons of Amberley's No. 82 Wing remain the supreme strike force in the Southeast Asia/Pacific region.

It is a firm belief of the RAAF that the combination of 'hardware' (platform and weapons), 'software' (in the form of recent upgrades) and 'fleshware' (the men and women of No. 82 Wing and the supporting Amberley Weapons Systems Business Unit) will prevail against all comers if the force is ever called upon to go to war. Apart from deployed United States Navy carrier air wings, Nos 1 and 6 Squadrons have long been the only night-capable strike force south of Japan and the Korean peninsula, and even the introduction of more modern types such as the F/A-18D Hornet, MiG-29N and Su-27/30 series have not diminished their supremacy.

Australia is a vast country and, as an island, has maritime approaches to the North, South, East and West that have to be safeguarded. Therefore, any strike aircraft that Australia operates has to have long legs, and there are none better than the F-111. Thirty years ago, the F-111C provided the ideal solution to these 'tyrannies of distance', and the RAAF hopes that it will continue to do so for at least another fifteen.

The long-awaited Defence White Paper, titled 'Defence 2000, Our Future Defence Force' (released in December 2000) reaffirmed the aircraft's future until the planned retirement date of 2020, and committed the Australian taxpayer to an average A$500 million per annum over the next decade to retain the current strike capability. Such is the regard for the F-111's capabilities, the White Paper states that, even during the 2015-2020 planned replacement window, 'it is unlikely that there will be any comparable specialised strike aircraft suited to our needs available at that time'. Known affectionately as the 'Pig' by the RAAF, for the same reasons that the aircraft earned the epithet 'Aardvark' in the USAF, it is worth noting that, in 2020, the F-111Cs will be 52 years old, and will have been in continuous operational service for 47 years!

The 'dump and burn' is the F-111's most famous party trick, and has been a feature of RAAF 'Pig' displays for 30 years, at its most spectacular during night-time shows. 'Torching' has been sustained for over 90 seconds, with the aircraft climbing in front of the crowd.

In order to ensure that the aircraft maintains its viability in the modern arena, the F-111C fleet has recently completed an Avionics Upgrade Programme (AUP), and has now begun a series of weapons and systems upgrades collectively known as the Block Upgrade Programme (BUP). Of prime importance to the future of the type are upgrades to the Electronic Warfare Self Protection (EWSP) suite, and the acquisition of stand-off weapons designed to allow the aircraft to launch an attack from outside the range of enemy air defence systems.

Right: 'F-111 day' – three F-11Cs sweep in over the Queensland coast on 1 June 1973, the day the first six aircraft arrived in Australia, amid great media interest. Their arrival signified the end of a procurement saga which had begun in the mid-1950s as a routine military requirements study, but which turned into a major national debate from 1963 – the year in which they were controversially ordered. For the next 40 years, the F-111 force has never been far from Australian headlines. Cost-cutting politicians and military chiefs envious of its funding levels have offered the 'Pig' up for roasting on the barbecue on many occasions. Each time it has survived, and today remains at the very heart of Australia's offensive capability.

Left: Thanks to its variable-geometry wings, highly accurate navigation system and a terrain-following radar, the F-111 can fly long night/bad weather sectors at very low level in safety and comfort. This capability is regularly practised by 82 Wing crews: the Queensland hinterland offers a sparsely populated area of varying terrain, perfect for regular low-level flying.

Below: Another low-level pass, from an earlier era. The Vietnam-era 'SEA' camouflage was effective when viewed from above, but the fleet has now moved to an all-over Gunship Grey which is better against a sky background and is highly effective over the sea. Pulling hard in the moist atmosphere of sub-tropical southern Queensland, this aircraft carries Harpoons.

However, the fleet has recently experienced some severe fatigue problems, requiring drastic remedial action. It would appear that a solution to these latest problems has been found, but it remains to be seen whether the type can remain free of any further problems between now and the planned retirement date. Given the high regard that the Amberley strike force enjoys within the region, these problems are the cause of much concern among Australian Defence Planners.

History

Just how the F-111 came to wear kangaroo roundels in the first place is a long and drawn-out story, involving many contributing factors and a good dose of political interference – from all the major players! The mid-1950s had seen the delivery of the bulk of the RAAF's Canberra Mk 20s from the Government Aircraft Factories' airfield at Avalon, and thoughts were turning to the eventual replacement of the type. Air Marshal Sir Frederick Scherger, the Chief of Air Staff at the time, favoured the Avro Vulcan, equipped with nuclear weapons, and later became a proponent of the BAC TSR-2 – though by 1959 held doubts that the programme was sufficiently advanced to meet a future RAAF requirement.

By the early part of the 1960s it was becoming obvious that the clear-weather/daylight Canberra was fast reaching obsolescence. This alone was regarded as a problem to be managed with some urgency by the RAAF, and when viewed against a backcloth of increasing unrest in Vietnam, and Indonesian aspirations in Southeast Asia, caused concern on both sides of Australian politics. This last fact initiated opposition pressure to replace the Canberra as a matter of utmost importance and, in the face of a looming election, the government of Robert Menzies was stirred enough to dispatch an overseas mission to evaluate then-current and projected long-range, all-weather (day or night) strike aircraft. The requirement was by now defined, and known as Air Staff Requirement (ASR) 36.

Led by the Chief of Air Staff Air Marshal Sir Valston Hancock, the mission departed Australia in June 1963 to examine the Dassault Mirage IV, BAC TSR-2, McDonnell F-4C Phantom, North American RA-5C Vigilante and the General Dynamics TFX. Its brief was to suggest an aircraft that was not only a long-range/all-weather strike platform, but also one that had an inherent reconnaissance capability and the ability to carry (and deliver) 'special stores'. This latter requirement stemmed from the belief of many senior RAAF officers that the only way that Australia could afford to 'punch above its weight' was to employ nuclear weapons. In the event, no Australian military aircraft has been equipped for the carriage and release of such weapons.

Hancock noted that, although the TFX offered the best solution, it was still very much a 'paper aeroplane' at the time (as was the TSR-2). Taking this in to account, he concluded that the Vigilante offered the best compromise – as both the Mirage and Phantom failed to meet the ASR 36 requirements. Accordingly, the recommendation to acquire 36 aircraft was subsequently approved by the Minister for Air, David Fairbairn, on 24 August 1963.

It was at this point that politics intervened as, with an election due in November, Minister for Defence Athol Townley travelled to the United States in October to meet with his American counterpart, Robert S. McNamara. The reason

Seen in the days before unit markings were applied, an F-111C taxis from the Amberley car-port complex with the Mirage IIIOs of No. 2 OCU in the background. F-111s routinely operate from Butterworth, in Malaysia, where the RAAF maintained a permanent presence (with Sabres and Mirage IIIs) until April 1988.

behind this trip was the fact that McNamara had offered to sell Australia the TFX (by now known as the F-111A) on 'extremely favourable terms'. Townley did not consult with Hancock, or anyone else within the RAAF, and entered in to a formal agreement to purchase 18 strike and 6 reconnaissance F-111As (plus an unspecified number of attrition replacements), for delivery commencing in July 1968. Total cost of the deal was reported as US $125 million, though in reality was somewhat 'open-ended', as the final unit cost would be calculated as a fraction of the total production run. As an interim measure, this deal also provided for the lease of 24 B-47E/RB-47E Stratojets, with Australia being only responsible for equipment and training costs during the period of operation.

After receiving news of the American offer, Menzies hastily convened a cabinet meeting on 23 October 1963 to 'rubber stamp' the deal, and details appeared in the following day's press – just five weeks before the election was to take place. The perception of the F-111A acquisition as a 'done deal' not only answered opposition criticism of the Menzies Government's perceived lethargy in the face of Indonesian hostility, but also went a long way towards placating a nervous public. Immediately prior to the election, three USAF B-47s toured Australian air bases, but the RAAF were, at best, lukewarm to the idea of operating such a large aircraft, and the offer was left to lapse.

The 'Pig' takes flight

On 21st December 1964 the first F-111 took to the air, and subsequent flight trials soon highlighted a range deficiency with the F-111A variant, meaning that the aircraft would obviously now no longer meet the original RAAF requirement. To rectify the problem, it was decided to modify the Australian order to include the longer-span wings and heavier landing gear of the F-111B (and later FB-111A). These new wings conferred a 5 per cent increase in range and, coupled with the fact that the aircraft were to retain the original avionics, Triple Plow I engine intakes and TF30-P-3 engines of the F-111A, created the unique F-111C.

As the USAF production programme progressed, the F-111 began to earn an unenviable reputation, with several being lost in training accidents in the USA and on operations in Southeast Asia. In response, the RAAF sent a technical team to America to learn more about the problems and, among other shortcomings, identified technical deficiencies in the horizontal stabiliser control system. The subsequent modifications put delivery back from July to September but, before this could occur, a fatigue specimen at General Dynamics suffered a failure of the wing carry-through box and it became obvious that major structural modifications would be required if the F-111 was to meet its planned fatigue life.

After undergoing rectification, a hand-over date of 4 September 1968 was set and eagerly awaited by all involved with the project. A further setback occurred just two weeks prior to the ceremony when the wing carry-through box of a test specimen failed much earlier in the fatigue trial process than had been anticipated. In the event, the ceremony went ahead, though any euphoria was to be short-lived, as an F-111B was lost barely a week later, and was followed by the (unrelated) loss of a USAF F-111A less than two weeks after that.

By now the whole Australian F-111 acquisition project was threatening to degenerate to a farcical level, with much scathing criticism regularly aired in the pages of local newspapers – once again the government found itself under pressure from the opposition. No doubt due to their own internal problems with the F-111, American pressure was brought to bear on the RAAF to accept delivery – albeit with severe operational restrictions – until such time as the aircraft could be returned to General Dynamics for permanent rectification. Resisting the pressure, the RAAF insisted that all problems be rectified before delivery could take place.

History would later record that the RAAF's refusal to accept its aircraft until all of the problems were understood and rectified was justi-

RAAF operating units

No. 82 Wing

No. 82 Wing was formed at Ballarat on 24 March 1944 as part of a restructuring of the RAAF to allow command units to be more mobile. During October 1944 the wing moved to Fenton in the Northern Territory, and assumed responsibility for three Liberator units: Nos 21, 23 and 24 Squadrons. The wing moved to Amberley in 1946 to become the peacetime RAAF's sole heavy bomber wing. The three subordinate squadrons were by then Nos 12, 21 and 23 Squadrons, and the Liberators began to be replaced by the new GAF-built Avro Lincolns.

No. 12 Squadron was disbanded in early 1948, and replaced by the re-formation of No. 1 Squadron on 23 February of that year. As Nos 21 and 23 Squadrons were Citizen's Air Force Units, they were renumbered as No. 2 and No. 6 on the same date, allowing their numberplates to be re-issued to Mustang-equipped reserve units.

After the F-111C entered service, the operational arm of the RAAF was divided into five Force Element Groups (Strike & Reconnaissance, Tactical Fighter, Maritime Patrol, Airlift and Combat Support), with No. 82 Wing therefore reporting to Headquarters Air Command via the former. In February 2002, the Strike & Reconnaissance and Tactical Fighter Groups were merged in to the newly-created Air Combat Group.

No. 1 Squadron

Tracing its lineage back to No. 1 Squadron, AFC, of World War I, No. 1 Squadron is also the equal oldest unit in the RAAF – an honour shared with No. 3 Squadron. Formed as a composite unit at Point Cook on 1 July 1925, the unit was initially equipped with the Airco DH.9, DH.9a and Royal Aircraft Factory SE.5a.

By then operating the Lockheed Hudson, No. 1 Squadron was dispatched to Singapore and Malaya during 1941, and began operations against the Japanese invasion fleet on 6 December 1941. After the British and Commonwealth forces were overwhelmed by the marauding Japanese, the remnants of the squadron were withdrawn to Australia to regroup. Reformed with the Beaufort in 1943, No. 1 Squadron re-entered the fray during March 1944. VJ Day saw the unit operating the Mosquito FB.Mk VI before being disbanded in August 1946.

F-111s operate from 'car-ports' on the Amberley flightline, each capable of housing six jets. The shelters provide necessary protection to both aircraft and crew from a mix of fierce Queensland sun and tropical downpours.

Above: When they first arrived, the F-111Cs were only capable of free-fall bombing, for which a typical load was 24 Mk 82 500-lb (227-kg) bombs held on four BRU-3 multiple ejector racks. An alternative was a load of 20 Karinga cluster bombs.

Below: The use of forward operating bases significantly enhances the reach of the F-111 fleet to the north. This 'Pig' rests in a shelter at Darwin, one of Australia's two northerly bases and the closest to the troubled island of Timor.

fied by later events. The delivery date had by now slipped to June 1969, and General Dynamics' confidence that this could be achieved suffered yet another blow in February, when another wing carry-through box failed during static testing.

The bulk of Australian personnel in the United States were sent home while General Dynamics carried out further testing with a strengthened wing carry-through box. This also failed short of its target in June 1969, though in excess of the proposed 4,000-hour service life of the aircraft, and it became apparent that a structure with greater strength would have to be designed. Another crash occurred during December 1969 due to the same cause, though a breakthrough of sorts was to occur during the investigation into this accident: it was discovered that the type of steel used in the manufacture of wing carry-through boxes, known as D6ac, could only be adequately tested for flaws during cold proof loading.

This revelation was still not fully understood in 1970, when Australian Defence Minister Malcolm Fraser (under pressure to cancel the entire F-111 order) announced that a deal had been struck with the US Government for the interim lease of 24 F-4E Phantoms. The F-111Cs were put into long-term storage at Carswell AFB pending modification, and the Phantoms were ferried to Amberley in September and October 1970 – becoming an immediate success in the process. After seven years of waiting, No. 82 (Bomber) Wing finally had a modern replacement for the Canberra, albeit not the one it had ordered.

Following the successful modification of the USAF F-111 fleet, further structural test results and the operational experience of US squadrons (who had amassed many hours flying unmodified aircraft throughout the gestation period of the Australian project), Cabinet finally endorsed a proposal to reactivate the F-111Cs, and they duly entered the modification programme in January 1972. Though understandably loathe to part with their Phantoms, No. 82 Wing crews began returning to America in the following month, with the intention of ferrying the F-111Cs back to Amberley in June 1973.

Entry into service

Australia had taken a single F-111C on strength on 6 September 1968, when A8-126 was flown to Edwards AFB for inspection and acceptance (though for the ill-fated ceremony two days before, A8-127 – painted as A8-125 for the occasion due to the external condition of the latter – fulfilled the 'official' hand-over duties), but returned it to the USAF for storage shortly afterwards. The RAAF ultimately gained possession of its aircraft on 15 March 1973. By now though, the cost had escalated dramatically, and as the unit price in the original contract was dependent upon the total production run (now much less than envisaged), the aircraft were now no longer a bargain – though this price structure would work to the RAAF's advantage in years to come.

Reformed again on 23 February 1948, as a component of No. 82 (Bomber) Wing at Amberley, No. 1 Squadron began operating with the Australian-built Lincoln. Deploying to Tengah in July 1950, operations were flown against communist terrorists during the Malayan Emergency until a return to Amberley in July 1958. After returning from Malaya, No. 1 Squadron re-equipped with the GAF Canberra and has remained at Amberley to the present day. F-4E Phantoms were briefly operated between 1970 and 1973, as a stop-gap measure, prior to the arrival of the F-111C in June 1973.

As the operational F-111 unit, No. 1 Squadron received the RF-111Cs from No. 6 Squadron in July 1996, and continues to operate both F-111C and RF-111C in the strike and reconnaissance roles.

No. 6 Squadron

Also able to trace its origins back to the AFC, No. 6 Squadron, RAAF, was formed at Richmond on 1 January 1939, by way of the re-numbering of No. 4 Squadron, and was equipped with the Avro Anson.

The unit re-equipped with Lockheed Hudsons in April 1940, and was to later distinguish itself at Milne Bay during August 1942, where (alongside the Kittyhawk-equipped Nos 75 and 76 Squadrons) it was instrumental in the ultimate repulsion of a large Japanese invasion force. No. 6 Squadron later re-equipped with Beauforts and operated that type to the end of the war, before being disbanded on 31 October 1945.

On 23 February 1948 No. 23 Squadron, based at Amberley and equipped with the GAF Lincoln, was renumbered as No. 6 Squadron, and began operations as the operational training and conversion unit within No. 82 Wing – a role which is continued to this day.

Canberras replaced Lincolns during 1955 and, as with No. 1 Squadron, the F-4E Phantom was flown between 1970 and 1973 pending the delivery of the first F-111Cs. Following the withdrawal of the Canberra from service and the disbandment of No. 2 Squadron in 1982, a Survey Flight was formed within No. 6 Squadron to operate civil-leased Learjets in the photo-mapping role. The Learjets were operated between 1985 and 1987.

As the F-111 operational training and conversion unit, No. 6 Squadron today operates the F-111C and F-111G.

ARDU

Formed as No. 1 Air Performance Unit at Laverton on 1 December 1943, as the RAAF's Testing and Trials unit, the Aircraft Research and Development Unit has operated many and varied types of aircraft over the years. ARDU moved from Laverton to Edinburgh in February 1977, and continues to support all three services of the ADF.

Numerous F-111s have been loaned to ARDU for various trials since 1973, but for a period during the 1980s the unit actually had an aircraft on its books. A8-132 was transferred from 82 Wing in 1981 and was specially instrumented for trials, including the AGM-84 Harpoon anti-shipping missile and High Speed Anti-Radiation missile, before eventually becoming the prototype Avionics Upgrade aircraft. After AUP conversion and testing, the aircraft passed back to No. 82 Wing, but ARDU's involvement with loaned F-111s continues.

A8-132, ARDU's 'own' F-111C, is seen at Edinburgh in 1985 during trials of the GBU-15 electro-optical guided bomb. The white undersides aided calibration of stores releases.

RAAF F-111 serials

Serial	Type	ex USAF	Remarks
A8-106	F-111A	67-0106	NTU. Failed fatigue tests. AMRL 11/99
A8-108	F-111A	67-0108	NTU. Failed fatigue tests.
A8-109	F-111C	67-0109	converted F-111A. Current
A8-112	F-111C	67-0112	converted F-111A. Current
A8-113	F-111C	67-0113	converted F-111A. Current
A8-114	F-111C	67-0114	converted F-111A. Current
A8-125	F-111C	67-0125*	Current
A8-126	RF-111C	67-0126*	converted F-111C. Current
A8-127	F-111C	67-0127*	w/o Guyra NSW 13/9/93
A8-128	F-111C	67-0128*	w/o Tenterfield NSW 02/04/87
A8-129	F-111C	67-0129*	Current
A8-130	F-111C	67-0130*	Current
A8-131	F-111C	67-0131*	Current
A8-132	F-111C	67-0132*	Current
A8-133	F-111C	67-0133*	w/o Evans Head NSW 29/9/77
A8-134	RF-111C	67-0134*	converted F-111C. Current
A8-135	F-111C	67-0135*	Current
A8-136	F-111C	67-0136*	w/o Armidale NSW 28/4/77
A8-137	F-111C	67-0137*	w/o Ohakea NZ 24/08/79
A8-138	F-111C	67-0138*	Current
A8-139	F-111C	67-0139*	w/o Batemans Bay NSW 28/1/86
A8-140	F-111C	67-0140*	Current
A8-141	F-111C	67-0141*	w/o Hauraki Gulf NZ 25/10/78
A8-142	F-111C	67-0142*	Current
A8-143	RF-111C	67-0143*	converted F-111C. Current
A8-144	F-111C	67-0144*	Current
A8-145	F-111C	67-0145*	Current
A8-146	RF-111C	67-0146*	converted F-111C. Current
A8-147	F-111C	67-0147*	Current
A8-148	F-111C	67-0148*	Current
A8-259	F-111G	68-0259	converted FB-111A. Stored
A8-264	F-111G	68-0264	converted FB-111A. Stored
A8-265	F-111G	68-0265	converted FB-111A. Stored
A8-270	F-111G	68-0270	converted FB-111A. Stored
A8-271	F-111G	68-0271	converted FB-111A. Current
A8-272	F-111G	68-0272	converted FB-111A. Current
A8-274	F-111G	68-0274	converted FB-111A. Current
A8-277	F-111G	68-0277	converted FB-111A. Current
A8-278	F-111G	68-0278	converted FB-111A. Stored
A8-281	F-111G	68-0281	converted FB-111A. Stored
A8-282	F-111G	68-0282	converted FB-111A. Stored
A8-291	F-111G	68-0291	converted FB-111A. w/o South China Sea 18/4/99
A8-506	F-111G	69-6506	converted FB-111A. Stored
A8-512	F-111G	69-6512	converted FB-111A. Current
A8-514	F-111G	69-6514	converted FB-111A. Current

** USAF serials allocated*

Amid a wave of publicity, the first six aircraft finally touched down at Amberley on 1 June 1973, dubbed 'F-111 Day' by the RAAF publicity machine. An estimated 3,000 people were on hand to witness Group Captain 'Jake' Newham (OC 82 Wing) and Wing Commander Trevor Owen lead the formation in an impressive low-level pass, before individually breaking off for a stream landing. No time was lost introducing this initial batch of aircraft into operational service; No. 6 Squadron flew its first sorties on 6 June, and dropped live bombs on the Suamarez Reef range a few days later. The remainder of the fleet arrived in equal batches on 27 July, 28 September and 4 December of 1973 and, after a decade of upheaval, the RAAF finally had a long-range, all-weather strike capability. It is worth reiterating at this point, that 'Strike', in Australian terms, has no nuclear connotation.

A real Precision Strike capacity was lacking from the outset, however, as the aircraft had no designation capability for Precision Guided Munitions, and relied largely on radar-bombing using the ubiquitous Mk 82/84 GP bomb combination, on up to six external pylons (the outermost pair do not swivel, and are not normally fitted). In addition, an M61A1 Vulcan 20-mm gun occupied the weapons bay, for close air support and (a limited level of) self defence.

Precision strike capability

As delivered, the major roles for the aircraft were: Precision Strike, Maritime Strike and Battlefield Air Interdiction, so the purchase and integration of weapons to meet these requirements became an early priority. Ten Ford Aerospace AN/AVQ-26 Pave Tack pods were purchased in 1980, and all surviving F-111Cs were wired to accept the unit, albeit in lieu of the gun, as the Pave Tack system utilises the aircraft's weapons bay. The first aircraft to be modified (A8-138) was flown to General Dynamics for conversion in December 1983 and the remainder (commencing in March 1985 with A8-147) were modified at Amberley by No. 3 Aircraft Depot. At least now the F-111C had a Precision Strike capability, and this then allowed the use of the GBU-10/12 Paveway II laser-guided bombs. The GBU-15 was also acquired and tested, with the AN/AXQ-14 datalink pod being carried on a hardpoint under the rear fuselage, between the engines. After modification, the gun could still be carried if the Pave Tack pod was not fitted, and remained in use until the late 1980s or early 1990s. In the latter years it was seldom carried, and only regularly used on an annual basis by 6 Sqn Training Flight instructors, in order to retain some form of currency with the weapon.

The 'Supersonic Kookaburra' struts its stuff – a load of Harpoons and GBU-12 LGBs fail to prevent this F-111 crew from demonstrating the low-level manoeuvrability of the 'Pig'. These No. 1 Sqn markings were not around for long, being replaced by the earlier and more discreet yellow lightning bolt.

In 1982 the Maritime Strike role was addressed, when A8-132 (allocated to the Aircraft Research and Development Unit) was flown to the US Naval Warfare Center at China Lake to begin trials with the AGM-84 Harpoon missile. Following the successful completion of these trials, a quantity of AGM-84A Block 1C missiles, training rounds and test/support equipment was purchased, and the weapon (in the updated AGM-84D form) is still responsible for the superb anti-shipping capability of the force. Prior to this, Maritime Strike would have been performed by several jets (each typically loaded with 12 Mk 82 bombs) carrying out a multi-axis, simultaneous mass attack – similar to the tactics employed by RAF/RN Buccaneers, but with the added advantages of supersonic egress and more effective ECM.

Battlefield Air Interdiction has always been considered to be more suited to a smaller (arguably, lower-value) platform such as the Mirage IIIO(A) and later F/A-18A Hornet, but the use of the F-111 in this role remains an option. The weapons of choice are Mk 82 and Mk 84 GP bombs, although the indigenous Karinga cluster bomb had been cleared for use, prior to its deletion from the inventory on ideological grounds (Australia does not condone the use of sub-munitions). Today, with the introduction of precision weapons, the delineation between Battlefield Air Interdiction and Precision Strike is blurred, and in some respect the different doctrines are now quite similar. As an aside, the success of the USAF's 'tank-plinking' methods – using small LGBs against armour – in the Gulf War came as no surprise to the RAAF, as this doctrine had been understood for some four or five years by that time.

AIM-9B Sidewinders were also integrated for use, and began to be seen regularly in the mid-1980s. Primarily employed for self defence, the 'Winder also conferred the additional (and little-known) secondary role of Long Range Air Defence, being a useful means of countering maritime patrol or surveillance aircraft beyond the range of the Hornet – just so long as the target couldn't shoot back! Early model

Carrying a single AIM-9M each, these two F-111s fly in the hazy conditions prevalent in the sub-tropics. In both anti-ship (with Harpoon) and anti-maritime patrol aircraft roles the F-111 rules the northern approaches to Australian territory – and beyond.

In a region where range is everything, Australia chose the long wings of the FB-111A for its aircraft, conferring a significant increase in range over the baseline F-111A upon which the Aussie Cs are principally based. As well as the four moving inboard pylons, all F-111s can carry two or four additional non-swivelling pylons under the outer wings, but they are not used as the wing has to be maintained at the 26° sweep position.

Night capability is conferred by the integration of the TFR, INS and ARS (attack radar system). When engaged at medium altitude, the TFR noses the aircraft into a dive before levelling out at 1,000 ft (305 m). The crew then step down through six intermediate Set Clearance Planes (SCPs) to 200 ft (61 m).

Sidewinders have long been replaced by the AIM-9M model, though the secondary tasking remains.

Trials with the AGM-88 HARM missile were carried out at Woomera in the 1980s, but the weapon did not find its way into the inventory. An RAF offer of BAe ALARM missiles on a 'cost-only' basis was considered during the late 1990s, but rejected on the grounds of integration costs, and the aircraft lacks a dedicated Suppression of Enemy Air Defence (SEAD) capability to this day. A lack of indigenous air-to-air refuelling capability has hampered unrestricted operations since service entry, although RAAF crews do attempt to retain proficiency by using deployed USAF KC-135 and KC-10 tankers as often as possible. Four of the RAAF's six Boeing 707s were converted to tankers during the early 1990s, but equipped only with the probe and drogue podded system, and a replacement tanker that also utilises the boom and receptacle AAR system is still keenly awaited.

Reconnaissance jets

Under the original contract a reconnaissance capability was outlined, stating that a quarter of the fleet were to have been an RF-111 variant, then under development for the US Air Force. Unfortunately, the USAF cancelled its RF-111A project in 1968 (the entire RF-111 project followed suit two years later) and therefore, as delivered, the F-111Cs fell short of the original ASR 36 requirement. Australia decided to resurrect much of the work done on the RF-111A, and engaged General Dynamics to study the conversion of four aircraft to RF-111C standard in December 1974. Upon completion, funding was obtained and the prototype aircraft (A8-126) was flown to Fort Worth in December 1978, flying as an RF-111C for the first time in May 1979, before being delivered back to Amberley three months later. The aircraft was then forward-deployed to Darwin for tropical trials of the reconnaissance suite, prior to the conversion of the remaining three aircraft.

Reconnaissance equipment is pallet-mounted in the weapons bay and comprises two video cameras, a Honeywell (now Lockheed Martin) AN/AAD-5A infra-red line-scanner, Fairchild KA-56E low-altitude and KA-93A4 high-altitude panoramic cameras, and two CAI KS-87C split-vertical cameras. The system allows video photography of the AN/APQ-113 (AN/APQ-169 after AUP) attack radar display on the navigator's side of the cockpit. The aircraft retain the capability to perform strike missions, but have no self-designation capability, and were not initially compatible with the Harpoon.

The three subsequent conversions (A8-134, 143, 146) were performed by 3 AD at Amberley during the course of 1980, and began operations with No. 6 Sqn Reconnaissance Flight shortly thereafter. The four 'recce-jets' regularly deploy to the United States for exercises, firstly the Reconnaissance Air Meet (RAM) series and, after that competition's demise, to the Red and Green Flag series at Nellis AFB in Nevada. Following the introduction of the F-111G to 6 Sqn service in 1992, the four aircraft were transferred to No. 1 Sqn and remain to the present day. The 'wet film' recce system is now dated, and an upgrade to a modern digital system has now completed the project definition stage of its gestation, with a view to new equipment being installed later in the current Block Upgrade Programme.

Attrition replacements

In April 1977 the fleet incurred its first loss, when A8-136 suffered a bleed air duct failure and subsequent fuel vapour explosion, and a further three aircraft had been written off by 1980 – none, it should be noted, due to structural failure. A clause in the original contract signed in 1963 allowed for the purchase of an unspecified number of attrition replacements at the same unit cost as the original 24 aircraft, something that was turned to advantage when it came time to seek replacements for the lost aircraft. With the type long out of production by this stage (early 1980), General Dynamics had proposed to sell 10 early F-111s to the RAAF, but the aircraft in question were pre-production examples and deemed unsuitable due to the lack of compatibility with the F-111C. In the event, a deal covering the purchase of four F-111As was signed with the USAF and the 1963 price tag, even for a second-hand aircraft, was somewhat of a bargain. The deal was valued at A$60 million, and went some way towards compensating the Australian taxpayer for the large cost blow-out suffered in the 1960s.

Four aircraft which were then in service with the 366th TFW at Mountain Home AFB in Idaho were selected, but two (67-0106, -0108) failed the required cold proof load test prior to delivery and were replaced in the contract by two further machines from the same source. The two aircraft that were not accepted returned to USAF service before being retired to AMARC in 1991, and held for possible future use by Australia. The replacement jets were ferried to Amberley during May, July and August 1982 and, due to the differences between the two variants, were consolidated exclusively within No. 6 Sqn.

In later years these four aircraft were modified with the longer-span wings and heavier landing gear of the F-111C, though retained the lower-strength wing carry-through box. Apart from this, the only differences are now minor (engine fire detector loops as an example), and the aircraft are treated in the same way as the remainder of the fleet, swapping squadrons as the maintenance programme and operational requirements dictate. Despite reports of these aircraft being referred to as F-111A[C], to the RAAF they are simply F-111Cs – only the logistics computer system differentiates between the two models.

The main anti-ship weapon is the AGM-84 Harpoon, of which four can be carried. Here the weapon is seen while under clearance trials in the 1980s using ARDU's own F-111C. The F-111's radar can pick out surface targets at very long ranges.

Camouflage and markings

When completed, the F-111Cs were painted in the USAF's 'South East Asia' camouflage scheme of Dark Green (FS34079), Green (FS34102) and Tan (FS30219) disruptive camouflage over Light Grey (FS36622) undersides. However, by the time of delivery in 1973 the undersides had been painted Black (FS37038) in line with the USAF fleet. RF-111Cs retained this scheme upon conversion, as did the four F-111As. The only exceptions to this were the application of white fuselages sides and undersides to the ARDU-attached A8-132 in order to facilitate its varied trials, and the overall light grey scheme (with blue trim) applied to this same aircraft on its departure to the US as the prototype AUP aircraft. The F-111Gs arrived in the USAF overall-Gunship Grey (FS36118) scheme, and this was adopted by the rest of the fleet as they cycled through the AUP and/or depot maintenance. The original 'green jets' (as they had become known) had all been repainted by late 1999/early 2000. The Gunship Grey scheme is said to be particularly effective in the hazy tropical skies of Asia, the likely area of operation. Over the next couple of years it is planned to change the colour of the radomes from black to grey.

Squadron markings

Aircraft were originally 'pooled' within 82 Wing and were therefore fairly anonymous. From 1 February 1983 each squadron was allocated its own aircraft and squadron markings in the form of a yellow (No. 1 Sqn) or blue (No. 6 Sqn) lightning flash were painted on the fin – a tradition carried over from Canberra days. Later still, No. 6 Sqn added a boomerang in a white circle to the centre of the flash, while No. 1 Sqn changed to a large yellow '1' running parallel to the fin leading edge, with a diving Kookaburra superimposed. This was known as the 'Supersonic Kookaburra' scheme, but fell foul of higher authority, and markings reverted to the earlier lightning flash, though with a Kookaburra at the centre as per the No. 6 Sqn boomerang. Since the introduction of the Gunship Grey scheme, all squadron markings (with the exception of the specials noted below) have been toned-down to black.

Aircraft in special markings

A8-125 of No. 6 Sqn carried '25th Anniversary' (RAAF F-111) markings during 1998
A8-146 of No. 6 Sqn wore '20 Years' (of F-111 operations) markings during 1993
A8-??? of No. 1 Sqn wore '20 Years' (of F-111 operations) markings during 1993
A8-272 is 'The Boneyard Wrangler'
A8-274 No. 6 Sqn '60th Anniversary' marks during 1999
A8-281 of No. 6 Sqn wore RAAF '75th Anniversary' marks during 1996

(With the exception of A8-272, all special markings have been short-lived)

As originally completed, the F-111Cs sported light grey undersides, as demonstrated here by A8-126. This aircraft subsequently became the first RF-111C conversion.

The standard black undersides scheme is displayed by this Pave Tack-equipped No. 1 Squadron aircraft. The scheme remained unchanged for 20 years.

F-111G A8-272 displays the current all-over grey scheme, soon to be modified by the adoption of a grey radome. The aircraft carries its 'Boneyard Wrangler' markings on the fin, one of only a handful of 'Pigs' to carry special markings.

The lightning bolt is retained on the toned-down camouflage with the unit insignia at the centre. This is No. 1 Squadron's diving Kookaburra badge.

In 'green jet' days the lightning bolt reflected the squadron colour (here blue for No. 6 Sqn). Note the full-colour fin-flash and Australian ensign.

No. 6 Squadron celebrated its 60th anniversary by applying this scheme to an F-111G. The squadron's boomerang badge was to the fore.

Appropriately it was A8-125 (the first F-111C) which received 25th anniversary markings, which incorporated the yellow and blue of both squadrons.

During the F-111A delivery process it was announced that the remaining 20 F-111Cs would be flown back to the United States for further cold proof load testing at a reported cost of $259,000, and rekindled debate as to whether the force was any longer effective. Over the years of operation, several fatigue problems had become apparent, and the use of Boron adhesive patches, developed in Australia, were by now routine. It was also evident that close monitoring of the airframe would remain an expensive fact of life.

Throughout the F-111s long career with the RAAF, it has had to endure ongoing criticism – even open hostility – from both within and without the service, and would have been retired many times over if detractors had their way. In one infamous case during the mid-1980s, a senior defence official claimed that the F-111 force was no longer operationally effective, and that the money saved by withdrawing the type could be better spent elsewhere. In what was later to be shown as a master-stroke, a Pave Tack-equipped jet was flown non-stop from Amberley to the Defence Department offices at Russell Hill, in Canberra: the navigator placed cross-hairs of the aiming system on the official's office window and filmed the event with the video recorder. The aircraft then returned directly to Amberley and the person in question was presented with the video footage shortly afterwards. This effectively demonstrated the unrivalled long-range, all-weather, precision-attack capability of the F-111 force and one critic, at least, was silenced.

Nevertheless, the problems experienced with the interface between an analogue aircraft and a digital system during the Pave Tack modifica-

A store commonly seen on RAAF F-111s is the SUU-20 practice bomb dispenser, which can house six BDU-33 'blue bombs', the ballistics of which approximate to those of the Mk 82. The four 2.75-in rocket tubes are not used. Note the underfuselage bulge which signifies installation of the Pave Tack cradle in the weapons bay.

82 Wing is justifiably proud of its F-111 operations, and seized the opportunity in 1998 to celebrate its 25 years with the 'Pig'. The aircraft has its Pave Tack turret in the deployed position, with bay doors open.

tions, and the fairly poor Mean Time Between Failure (MTBF) of some of the older systems, underlined the need for a fairly wide-ranging electronics and cockpit upgrade.

AUP – Avionics Upgrade Programme

Conceived in 1989 as a necessary modification to replace the ageing analogue avionics, the Avionics Upgrade Programme (or Project Air 5225) contract was signed with the Rockwell/Hawker De Havilland team in August 1990. Design and development began at Palmdale during 1990 with the stated aim of increasing the MTBF, from an average 3.5 hours then being experienced, to 19.5 hours.

An improved stores management system (as fitted to the F/A-18A Hornet) was incorporated, as were dual-redundant MIL-STD-1553B digital data busses (with MIL-STD-1760 interface with the weapons pylons), twin fully-redundant IBM AP-102A (MIL-STD-1750A) mission computers, Honeywell H423 (AN/ASN-41) ring laser gyro (with embedded Rockwell-Collins 3M MAG-R GPS), a triple-redundant Lear Astronics digital flight control system, and KY-58 secure voice and Have Quick UHF radios. The attack radar was upgraded to AN/APQ-169 configuration, and the terrain-following radar was replaced by a Texas Instruments AN/APQ-171 unit.

The Harpoon-test aircraft (A8-132) was selected as the prototype and departed for the USA in 1992, the remainder of the fleet being modified at Amberley by Hawker De Havilland (later Boeing Australia), and the production phase commenced during 1995. In order to minimise fleet disruption, regular R3 servicing was carried out concurrently, and each conversion took an average 110 days to complete. The project soon fell behind schedule, due to the underestimation of the complexity of production and integration of equipment, but allowed the opportunity to incorporate some external projects, such as the Interim CMDS (ALE-40/ Project Air 5391) and RWR (BAE Systems ALR-2002A/Project Air 5416). In September 1999 the final aircraft was completed, within one month of the revised schedule, at a total project cost of A$474 million and hailed as a great success by all parties involved. During his speech at the official completion ceremony, the then-Chief of the Air Force, Air Marshal Erroll McCormack, noted that the MTBF had increased to an unprecedented 179 hours.

Concurrent with the AUP was a project (Air 5208) to replace the 1969-vintage simulator with one that would reflect the new standard of equipment. An ex-USAF F-111F simulator had been acquired from Cannon AFB in New Mexico in 1993, and the upgrade contract was awarded to Wormald Australia (later Thomson-CSF Pacific). It was commissioned in an interim capacity (to be used for training and AUP development as the aircraft modification programme progressed) in 1997, and the official opening was eventually timed to coincide with the AUP Completion Ceremony in November 1999.

Despite efforts to the contrary, the delays to the AUP programme caused some degree of fleet disruption, and for a while (amongst other measures) the RF-111Cs were operated as standard aircraft to cover the lack of on-line aircraft. One added bonus however, was that, post-AUP, the 'Recce Jets' became Harpoon-compatible.

The availability of the simulator was a direct result of the USAF F-111 fleet drawdown, beginning in 1992 and concluding with the retirement of the last EF-111A in 1998. This upheaval within the USAF was further advantageous to the RAAF in that it enabled the purchase of additional aircraft and engines, ensuring fleet viability to the planned retirement date of 2020. The downside was that Australia had to become much more self-sufficient when it came to supporting the F-111 force – even to the extent of constructing a Cold Proof Load Test Facility at Amberley, as the USAF facility at the Sacramento Air Logistics Center had been closed down.

F-111G purchase

The first manifestation of the advantages came on 15 October 1992 when the then-Defence Minister, Senator Robert Ray, announced that his government had accepted an American offer to purchase "up to" 18 F-111G aircraft at a cost of A$77 million (a

F-111C weapons

This weapon display includes the F-111's current repertoire, dominated by Mk 80 series free-fall weapons and Paveway II LGBs. Noteworthy are the GBU-15(V)1 EO-guided bombs at the back (no longer in use), AGM-84 and, in front of the Harpoons, the indigenous Karinga cluster bomb (no longer in use).

Bomb options include the Mk 80 series (inert Mk 82s shown above on a BRU-3 rack) and Paveway II series (GBU-10 shown below). Note also the AIM-9 acquisition round on the inboard pylon.

The AGM-84 Harpoon (above) gives the F-111 a powerful anti-shipping capability. On strength, but not yet in service, is the AGM-142 Popeye (below) and its wing-mounted datalink pod.

Inside the 'Pig'

Above: Significant changes, notably at the navigator's station, were introduced by the digital AUP jets. Two MFD screens were incorporated for nav/tactical displays. The F-111G cockpit is essentially similar.

Left: The pre-AUP F-111C had a 1960s vintage cockpit dominated by dials and gauges. Note the hood covering the navigator's radar/Pave Tack display.

further $68 million would be spent on maintenance, training and logistics). The aircraft involved were earmarked for retirement from the 27th TFW at Cannon AFB, and by the time that the deal was concluded, one aircraft (to become A8-272, appropriately 'The Boneyard Wrangler') had actually been retired to the AMARC facility in Arizona. The F-111G variant was chosen due to its commonality with the F-111C (wings, landing gear and, following AUP, digital avionics), cost and remaining fatigue life. In USAF service the aircraft had already undergone the Avionics Modernization Program (AMP) digital upgrade, which (among other improvements, such as a twin inertial Doppler mix) had installed an AYK-18 mission computer. Although not compatible with the AP-102A computers fitted to the post-AUP F-111Cs, it did provide a great deal of similarity and allowed crews to progress from one to the other without too much trouble.

Behind the purchase of extra aircraft lay the logic of diluting F-111C flying hours in order to extend the service life of the fleet, and the original intention was to hold the F-111Gs in rotational storage rather than forming an extra squadron. The announcement of the deal did not go down well in Indonesia, however, causing the ambassador to Australia at the time, Mr Sabam Siagian, to question whether the extra aircraft were 'necessary or appropriate in the current strategic climate'. Not for the first time, Australia's F-111 force was playing a role in relationships between the two countries.

Arrival of the G

Fifteen aircraft were eventually purchased and ferried to Amberley during 1993 and 1994, with the first being delivered by Squadron Leader Richard Owen (now Group Captain, Chief of Staff Air Combat Group). This event amply demonstrated the longevity of the design, as Richard Owen is the son of the navigator who delivered the very first F-111C some 20 years previous. Initially operated by G Flight/82 Wing until conversion was complete, the F-111Gs were eventually integrated into No. 6 Sqn. There are currently seven aircraft in service with the unit, and seven in storage at Amberley (one aircraft was lost in 1999).

In practice, the rotational storage idea has not happened, as each aircraft requires an R5 service before entering squadron use – rendering it uneconomical to give every aircraft an overhaul when approximately only six are required. Two of the stored aircraft, A8-265 and -278, will not enter service and will be broken down as a source of spares. With the F-111C AUP project getting under way in the mid-1990s, the four 'Recce Jets' were transferred to 1 Sqn, and as post-AUP jets were initially delivered to 6 Sqn, this latter unit could (for a while at least) lay claim to the fact that it was the 'all-digital' F-111 squadron.

Initially, the F-111G was operational in the Tactical Close Air Support role, but the 'Achilles heel' of the variant is its lack of self-designation capability for guided weapons. Medium-level dive-bombing trials were successfully carried out during the latter part of the 1990s, in order to increase the accuracy of weapons delivery, but the interoperability originally intended has never been fully realised. Today the aircraft are used in the Operational Conversion Support role within No. 6 Sqn. An F-111G upgrade programme (Air 5404), along the lines of the F-111C/AUP, had been planned, and by 1998 had progressed to the Project Definition stage. However, funding was delayed by the strains placed upon the budget by the RAN's 'Collins'-class submarines. Due to the lack of an operational role for the aircraft, it would appear that an upgrade will not now take place, and the F-111G will be the first variant to be retired.

In spite of this, in an effort to bring the F-111Gs into line with the remainder of the fleet, an advanced digital flight control system was installed by Lockheed Martin, with the final aircraft being handed back to 82 Wing on 30 October 2000. The $32 million contract also included the installation of a crash data recording system, data transfer system and a mission computer software upgrade.

For those who enjoy trivia, the delivery of F-111G A8-514 (formerly FB-111A 69-6514) means that the Amberley flight-line now boasts the last F-111A (A8-114), F-111C (A8-148) and FB-111A ever built.

New engines

A second major windfall that resulted from the USAF's retirement of its F-111 fleet was the availability of the higher-powered TF30-P-109 engines fitted to the F-111D and EF-111A. The RAAF purchased a relatively large quantity of these engines, and began incorporating a series of modifications that had been planned by Pratt & Whitney at the time of the USAF fleet withdrawal. These were mainly reliability enhancement improvements (for example, the use of some components from the TF30-P-414A fitted to the F-14A). Once modified, they began to be installed in the F-111C fleet, replacing the earlier TF30-P-103.

The P-109 engine confers an increase in thrust from 18,500 lb (82.32 kN), to 20,840 lb (92.74 kN), operates at higher exhaust gas temperatures and has a higher mass flow than

Above: Although the RAAF has no boom-equipped tankers, the F-111 fleet maintains inflight refuelling proficiency by taking whatever opportunities arise to practise with USAF tankers. F-111G A8-272 wears its 'Boneyard Wrangler' artwork on the fin.

Right: Despite not being able to carry Pave Tack, the F-111G is useful for other missions, and the purchase of 15 aircraft (six or seven are in use at any one time) relieves the hard-pressed F-111C fleet, especially on training duties and on general bombing tasks.

Above: An F-111G thunders into the Queensland sunset. Both C and G have received uprated engines in the form of the TF30-P-109, plundered from the USAF's retired F-111D/EF-111A fleet. In the G model the powerplant is slightly modified and designated P-108.

the earlier engine. Despite the original Triple Plow I intake of the F-111C, the later engine provides a greater operating margin, and installation in the aircraft was relatively straightforward. Following conversion of a prototype, test flying was completed in 1998, and the engines were fitted to the fleet between 1998 and the end of 1999. After modification, engines are RAAF-specific and are known as the P-109RA.

In the interests of commonality, studies were conducted with regard to fitting the P-109 engine to the F-111G, but a major engineering difficulty had to be overcome first. The P-107 of the G is a 'straight' engine, with no change of thrust line between core and afterburner, whereas both the P-103 and P-109 have a 3° upwards 'kink'. As an aside, the afterburner petals of the F-111G are therefore not only closer to the ground than those of the F-111C, but angle the exhaust efflux towards, rather than parallel, to it. This created problems on the engine run pads at Amberley, as the use of afterburner during engine tests began melting the asphalt surface. The solution was to construct new concrete pads that are (relatively) impervious to the efflux.

In conjunction with Pratt & Whitney, RAAF technicians developed a hybrid engine, mating the 'straight' afterburner section of the P-107 to the core of the P-109. Test flying with the hybrid engine, unofficially dubbed the 'P-108', began in August 1998. Since that time, the engine has received this designation officially and, though most F-111Gs still operate with the earlier engine, the reliability advantages should see the new engines fitted to all operational aircraft.

Self-sufficiency

With a view to maintaining the viability of the fleet through to 2020, a quantity of aircraft stored at AMARC were earmarked for the RAAF and held in 'inviolate storage' as a potential source of major airframe components should they be required. These are a mix of different variants, comprising one F-111A, four EF-111As, two F-111Fs and three F-111Gs. In addition, a single pre-production aircraft (GF-111A 63-9768) that had been in use as a ground training aid at Sheppard AFB was brought to Amberley, and is today in use as a BDR training airframe at 82 Wing. The final acquisition to date, is the fuselage of F-111A 67-0106 that was shipped to Melbourne for use as a static fatigue test specimen at the Aeronautical and Maritime Research Laboratories in November 1999. This aircraft was one of the aircraft initially rejected by the RAAF when they were seeking attrition replacements in the early 1980s, and has had to wait almost 20 years for its call to duty.

Above: 63-9768 was the third F-111 to be built, and is now at Amberley for use by 82 Wing in the battle damage repair training role.

Right: Unzipped for deep maintenance, this F-111C is shown with the crucial wing carry-through section exposed. Boeing Australia heads the effort to keep the 'Pig' flying, handling both ongoing maintenance and upgrade work.

With the withdrawal of the USAF F-111 fleet, the Cold Proof Load Test hangar at the Sacramento Air Logistics Centre was closed down, and it became immediately obvious that the RAAF would have to find an alternative. Each RAAF F-111 requires at least one further cold proof test during its remaining life (each test confers a further 2,000 flying hours), and it was decided to create an in-house capability at Amberley. Lockheed Martin Australia was engaged to construct the facility, and it was duly commissioned on 26 July 2001, combining some refurbished ex-USAF equipment with purpose-built items. Several aircraft have now successfully cycled through the facility.

After taking its F-111s out of service, the USAF had no further platform that was integrated with Pave Tack, and Australia once again has benefited from this with the purchase of sufficient ex-F-111F stocks to enable each F-111C to be fitted out with a pod.

No. 482 (Maintenance) Squadron and No. 3 Aircraft Depot had provided support and deep maintenance for the two operational squadrons since 1973, but a restructuring of the RAAF in late 1991 saw these two organisations merge to form No. 501 Wing. At the same time, much of No. 482 Squadron was merged into Nos 1 and 6 Squadrons, and this idea was mirrored at RAAF bases around the country. With an influx of maintenance and support personnel to the two squadrons, each became more independent and capable of deploying to one (or more) of the Bare Bases for longer periods in times of tension.

As with many other air arms around the world at this time, the RAAF was evolving in to a 'sharp end first' organisation, with many support functions being tendered out to industry. No. 501 Wing was identified as a candidate for this process, and the Strike & Reconnaissance Support Project Office (SRSPO) was formed to oversee the various facets of the project.

Unravelling the numbers

The development and numbering of the many RAAF projects involving the F-111 fleet have caused confusion in the past. The table below aims to provide a quick reference to these projects.

Project	Description
AIR.5136	F-111C/RF-111C Automatic Test Equipment Replacement (Harris Intermediate Level Automatic Tests Station) (complete)
AIR.5208	Phase 1: Acquisition of ex-USAF F-111F simulator (complete)
	Phase 2: Modification of simulator to AUP standard (complete)
AIR.5225	Phases 1 & 2: F-111C Avionics Upgrade Programme (AUP) (complete)
	Phase 3A&B: F-111C/G Commonality (complete)
	Phase 3C: F-111G Acquisition (complete)
AIR.5333	Link 16 Installation (was to be part of BUP C6, now suspended)
AIR.5391	Phase 2: Interim EWSP Capability, CMDS/ALE-40
	Phase 6: Interim EWSP Capability, RWR/ALR-62(V)6, Upgrade ALE-40 to ALE-47, Terma Elektronik A/S Warfare Management Unit (AN/ALQ-213[V]) (to AIR.5416)
	Phase 7: Interim EWSP Capability, Jammer Pod with DFRM (Elta EL/L-8222) (to AIR.5416)
AIR.5395	Phase ?: Cockpit Data Recorder
	Phase 3: Air Combat Training System (ACTS) Pod Integration
AIR.5398	Phase 1A: Stand Off Weapon, Long Range Air/Surface Missile (AGM-142E) (to AIR.5418)
	Phases 1B, 2, 3, 4, 6: Follow On Stand Off Weapon (FOSOW) (all to AIR.5418)
AIR.5404	Phase 1: F-111 Capability Optimisation, Fleet Commonality component purchase (ALR-2002 RWR)
	Phase 2: F-111 Capability Optimisation, F-111G Precision Weapon/Maritime Strike Capability Modification (incorporation of a Laser IR Designation System, upgrade of Mission Computer and Stores Management System, Programmable Display Generator, GPS Installation)
AIR.5408	Phase 2: GPS Enhancement
AIR.5409	Phase 1: Bomb Improvement Programme (GBU-31/32/35 and/or JDAM-ER?)
AIR.5412	WVR Missile and Helmet-Mounted Cueing System (AIM-9M replacement) (was to be part of BUP C5, now suspended)
AIR.5416	Interim EWSP Capability (Project Echidna) (replaces AIR 5391/5394)
	Phase 1: ALR-2002A FSED (June 2002)
	Phase 2: Elta EL/L-8222
	Phase 3: ALR-2002A Integration, Missile Approach Warning System (MAWS) (2004)
AIR.5418	Anti-Radiation/Littoral Weapon Acquisition (Follow-On SOW capability) (ex AIR.5398 Phases 1B/2/3/4/6)
	Phase 1: Anti-Radar, Aerial Denial, Maritime Strike Weapon (100+ nm range)
	Phase 2: Hardened Target Weapon
AIR.5421	Tactical Reconnaissance and Strike Support (RF-111C Upgrade) – Synthetic Aperture Radar, LOROP System
AIR.5422	F-111 Block Upgrade Programme
AIR.6000	F-111/Hornet Replacement Project
JP.129	(Risk Mitigation Phase) DB-110 Raptor Reconnaissance Pod Trials
JP.2008	MILSATCOM (was to be part of BUP C5, now suspended)
JP.2036	Phase 2: Advanced Narrow Band Digital Voice Terminal
JP.2043	HF Modernisation Programme (was to be part of BUP C5, now suspended)
JP.2045	Phase 1A: Sea Mining Capability (was to be part of BUP C5, now suspended)
	Phase 1B: Stand-Off Mining
	Phase 2: Stand-Off Mining
JP.5406	GPS Enhancement

Boeing Australia had by then recently completed the Avionics Upgrade at Amberley, working closely with No. 501 Wing, and had already established the Amberley Weapons System Support Facility (WSSF) for that project. In August 2000 the Minister for Defence, John Moore, announced that Boeing had been selected as the preferred tenderer for the establishment of the Amberley Weapons Systems Business Unit (WSBU) and complementary Block Upgrade Programme. The WSBU contract is for a 10-year period, with two five-year options, and will encompass the entire Block Upgrade period. Subsequently, Boeing Australia has been awarded a further 10-year contract for the ongoing maintenance and logistics support for the fleet (including the operation of the Cold Proof Load Test Facility) under the auspices of the WSBU, and has become the major civilian partner in the future of the F-111.

Other areas have also been outsourced, and today the SRSPO oversees not only the WSBU, but also the Engine Business Unit (run in-house by the Defence Materiel Organisation), the Avionics Business Unit (Honeywell), Workshop Business Unit (Tasman Engineering Australia, a subsidiary of Air New Zealand), and Rosebank Engineering, which had a pre-existing contract for the overhaul of select hydraulic components. This integrated approach to the support of the F-111 will allow the RAAF to concentrate on operational matters, but retain the 'corporate knowledge' required to support the fleet at Amberley. It is seen as an innovative way of attaining self-sufficiency.

Operational use

The completion in August 1998 of RAAF Scherger (near Weipa in North Queensland), the final 'Bare Base', allowed the RAAF the use of a string of forward operating locations, and the aforementioned restructure of the service was designed to take advantage of this fact. In times of tension, it was planned that the RAAF would restructure into two flexible combat forces known as No. 95 Contingency Strike and No. 96 Contingency Air Defence Wings, and therefore the two F-111 squadrons could expect to be operating for the former unit, from any or all of the Bare Bases.

This concept had been exercised throughout the latter part of the 1990s, but was ultimately put in to practice during the period of unrest in East Timor from September 1999. Aircraft from both Strike and Reconnaissance F-111 squadrons were forward deployed to Tindal (near Katherine in the Northern Territory), and requests were made to the Indonesian Government to allow reconnaissance over-flights of East Timor in support of the United Nations-mandated INTERFET force. Initially, the TNI-AU Operational Commander for Eastern Indonesia, Air Marshal Santoso, rejected these Australian requests and threatened to shoot down any reconnaissance aircraft encountered over East Timor. Ultimately, Indonesian forces withdrew from East Timor on 30 October 1999, and reconnaissance over-flights officially began six days later, on 6 November.

This was the first and, at the time of writing, only operational use of the aircraft in RAAF service, and the scale of the effort at Tindal was such that the remainder of the two squadrons still at Amberley operated together under the umbrella of No. 82 Wing, and jokingly referred to themselves as No. 7 (i.e. 1+6) Squadron.

During this period, trials were being conducted by the Aircraft Research and Development Unit with a Raytheon DB-110 podded reconnaissance system, attached to A8-134. Part of a project to define future ADF reconnaissance requirements (JP.129), the presence of the aircraft in the Northern Territory led to much speculation that the system was used operationally over East and West Timor during August 1999. This has been denied by the RAAF, which states that the aircraft was undergoing trials under northern climatic conditions from Darwin at the time.

The last operational sortie was flown on 9 December 1999 and all aircraft returned to Amberley and normal operations. Further history was soon to be written, however, with

Above: RAAF F-111 crews test their tactical skills during occasional deployments to the USAF's Red and Green Flag exercises. Here six Pave Tack-equipped F-111Cs from No. 1 Sqn line the Nellis ramp.

Right: Amberley is now the last refuge of the F-111, and the 82 Wing roster includes a number of ex-USAF personnel. The latest round of wing-crack problems have cast a shadow over the future of the type.

Right: The four 'Recce Jets' also underwent AUP, retaining their multi-sensor reconnaissance equipment but adding new capabilities, such as Harpoon. Having already evaluated the DB-110 sensor, the RAAF is studying a podded stand-off sensor system for the F-111.

This post-AUP F-111C carries the boomerang badge of No. 6 Squadron. AUP revitalised the F-111 fleet in terms of avionics systems, and was broadly similar in capabilities to the upgraded F-111G AMP version arriving from USAF stocks.

the graduation of the RAAF's first female fast-jet aircrew at No. 6 Sqn. Pilot Officer Aroha Fifield and Flying Officer Brooke Chivers graduated as navigators in March 2000, and their achievement was hailed as a 'milestone event' for the RAAF. At the time of writing there are no female F-111 pilots.

BUP – Block Upgrade Programme

Following on from the completion of the Avionics Upgrade Programme, the Block Upgrade Programme (BUP, Project Air 5422) commenced at Amberley, with Boeing Australia as the prime contractor. The BUP is designed (by and large) to introduce a new stand-off weapons capability and increase the level of electronic warfare self protection and, as its name suggests, is a series of incremental modifications that will see the aircraft cycling through the Amberley facility several times each over the next decade.

Several diverse projects were under way at the time, and the BUP was seen as a way of harnessing them all under the same roof, with all the synergy that brings. Boeing Australia was awarded an interim contract to perform modification integration tasks, as the 'commercialisation' of 501 Wing had yet to be completed, with the definitive contract following after that event had taken place. There are currently six components to the programme, though the content of each tends to be fluid, as component projects stretch out or are redefined, and the reader is directed to the accompanying tables for the current state of play.

As the BUP was designed to build on the success of the AUP, Phase C1 (referring to the modifications specific to the F-111C) was deemed to be the AUP, and (the concurrent) C2 has seen the introduction of the ALE-40 CMDS as an interim step (the F-111Gs were delivered with the ALE-40, and this will standardise the fleet prior to the definitive ALE-47 being installed in the future). Block C2 also saw the RWR upgraded from the original Dalmo-Victor ALR-62 to an interim ALR-62(V)6 standard, and the full-scale engineering development of the definitive Australian-developed BAE Systems ALR-2002A. At the time of writing Blocks C1, C2, C2A and G1 have been completed, Block C3 is in production, and Blocks C3A and C4 are in the design development stage.

One of the projects to be brought under the BUP umbrella, was the project (Air 5398) to acquire a stand-off attack weapon (or weapons) that has seen the purchase of the Rafael/Lockheed Martin AGM-142E Popeye. This weapon was selected in early 1998, and a second tranche was ordered later the same year, allowing both the blast and penetrating (semi-hardened target) warheads to be introduced. It was originally intended that the datalink pod for the weapon would be carried on the rear fuselage station once reserved for the GBU-15 pod, but this has proved to be impossible, and the pod will now have to occupy an underwing pylon. It was initially hoped that the weapon would be in service by 2003, but problems with the integration of the weapon with the aircraft (the RAAF is itself responsible for integrating the missile with the airframe) have been identified and the weapon will not now enter service until 'beyond 2004'. There have been rumours circulating that the RAAF was planning to offload its missiles to Greece (apparently based on an earlier story that the USAF plans to remove the weapon from service), but the 2002 Defence Budget confirms that deliveries of the AGM-142E will continue and that, once integration is complete, modification of a prototype will commence at Amberley.

Scherger, on the Cape York peninsula, is the last of the RAAF's Bare Bases to be finished, complete with flow-through OLA (Ordnance Loading Apron) shelters. Constructed to withstand tropical cyclones, the shelters are camouflaged and protected by earthworks and the surrounding gum trees.

Project Air 5398 was subsequently broadened to involve a whole family of weapons, including an anti-radiation capability (the mooted trial of ALARM mentioned earlier was to have been part of the development phase of this project) and is now known as the 'Follow-On Stand Off Weapon' project, or Air 5418. According to Lockheed Martin, its AGM-158 Joint Air-to-Surface (JASSM) weapon was selected by the RAAF, and further press reports suggested that the Taurus Systems GmbH KEPD-350 would be held in consideration if JASSM did not meet expectations. A certain degree of confusion and controversy has surrounded this project to date, and it has been suggested that the RAAF are having a rethink due to the overlap between current (AGM-84), purchased (AGM-142 and the RAN's Penguin Mk 2 Mod.7) and planned (JASSM) weapons. During an interview with senior RAAF officers for the preparation of this story, the writer was told that Air 5418 is not limited to JASSM, and that other weapons such as Taurus KEPD-350 or MBDA Storm Shadow were still under consideration.

One weapons upgrade that originally formed part of the BUP to have fallen by the wayside is the project to replace the AIM-9M Sidewinder as the self-defence weapon on the aircraft. The MBDA AIM-132 ASRAAM is being acquired for the Hornet force, and it had been anticipated that this weapon would have flowed over to the F-111 fleet. This is now under review, and a decision is yet to be taken, so the AIM-9 will continue to be deployed for at least the foreseeable future.

Block Upgrade Programme

C1	**completed 09/99** AIR.5225/2 AUP
G1	**complete** AIR.5225/3 (A&B) F-111C/G Commonality
	complete (C) F-111G Acquisition
C2	**mid-1999 – early 2001** AIR.5391/2 Interim EWSP Capability CMDS ALE-40 (8 aircraft) AIR.5391/6 Interim EWSP Capability RWR ALR-62(V)6/7 (upgrade of original Dalmo-Victor ALR-62) (18 aircraft)
C2A	AIR.5416 ALR-2002A FSED (BAE Systems)
C3	**mid-2000 – 07/02** (except for A8-132) EWCS Installation (EW MUXBUS), further CMDS upgrade, EW pod installation AIR.5391/6 Upgrade ALE-40 to ALE-47 with Terma Elektronic AS Electronic Warfare Management Unit (AN/ALQ-213[V]) MIS554 Voice and Data Recorder (VADR) Cockpit Data Recorder
C3A	**07/01 – mid-2004** AIR.5391/6 Interim EWSP Capability, Jammer Pod with DFRM (Elta Electronics EL/L-8222) (ex 5391/7) DFCS Upgrade
C4	**mid-2003 – mid-2005** AIR.5398/1A/2 AGM Capability (AGM-142E) JP.2036/2 Advanced Narrow Band Digital Voice Terminal
G2	**2003 – 2006 (planned, now suspended)** AIR.5404/2 F-111G Precision Weapons Modification (similar to C4)
C5/G3	**mid-2005 – 2007 (G3 now suspended)** AIR.5408/2 GPS Enhancement AIR.5409/1 Bomb Improvement Programme (GBU-31/32/35 JDAM and/or JDAM-ER) AIR.5418/1 Follow-on Stand Off Weapon Capability
C6/G4	**mid 2006 – 2012 (G4 will now not proceed)** AIR.5395/3 Air Combat Training System (pod integration) AIR.5404/2 Strike Capability Enhancement AIR.5416 EWSP Enhancement for F-111 (Project Echidna) AIR.5421 Tactical Reconnaissance Strike Support Capability (Reconnaissance System Upgrade)

The future

The introduction of the MiG-29 and Su-27 into the region has not changed F-111 operational doctrine, but has made its execution more difficult, in that crews will now have to work harder to avoid being shot down. Other than this, the RAAF still expects operations out to 2020 to remain viable, and (for example) still considers the Pave Tack pod an effective tool.

Though not yet part of the Block Upgrade Programme, a reconnaissance package upgrade is being sought to replace the wet film system of the 'Recce Jets'. Project Air 5421 (Tactical Reconnaissance and Strike Support) is considering a range of options, including the deployment of a podded system that will be capable of being carried by all the fleet. The aforementioned trials of the DB-110 pod, though not directly part of this project, will no doubt provide much data to enable a system to be identified and selected. Naturally, if such a system can be deployed, the four RF-111Cs would then be de-modified and provide a boost to the Pave Tack-capable F-111C force. The project has now been defined and a decision is expected to be made in the 2004/2005 timeframe.

So what will replace the 'Pig'? Project Air 6000 was formed to seek a (preferably) common replacement for both the F/A-18A/B Hornet (from 2012 to 2015), and the F-111 (2015-2020), and types such as the Rafale, Typhoon, F/A-18E/F, F-22A and F-35A were under consideration. While it was acknowledged that the F-15E Strike Eagle is the closest in capability to the F-111 (and arguably more effective, due to the greater choice of weapons available), types that are in production today, such as the Eagle, were considered to be rendered obsolete before 2020. In late June 2002 the Minister of Defence, Robert Hill, announced that Australia would, barring problems with the programme, procure up to 100 Lockheed Martin F-35s, to enter service from around 2012. At the same time, Australia announced it was joining the SDD (Systems Development and Demonstration) team as a Level 3 partner, for an outlay of US$150 million.

To restructure the Air Force into the configuration required for the introduction of the Air 6000 type, the organisation of the RAAF has recently changed, and this has seen the demise of No. 82 Wing's parent unit, the Strike & Reconnaissance Group at Amberley. In February 2002 the Air Combat Group formed at Williamtown, bringing all fast-jet operations under a single command and, as a result, has also seen the disbandment of the Tactical Fighter Group (F/A-18). Nos 1 and 6 Squadrons continue to operate as they always have on a daily basis, but depending on whether their tasking is Precision Strike or Operational Training, they may report to either HQ 82 (Precision Strike) or 78 (Operational Conversion and Training) Wings. Obviously this will affect No. 6 Squadron more, because of its instructional role, but No. 82 Wing will now gain control of Hornet Precision Strike operations. This new arrangement is designed to be sufficiently flexible for both peacetime and wartime operations, and replaces the Contingency Wing 'shadow' organisation.

Aside from the weapons under consideration for the future, the RAAF provided an F-111G to the USAF during 1999/2000 for trials with the supersonic GPS-guided bomb (or SSB/Small Smart Bomb, now SDB/Small Diameter Bomb). Because the F-111G has a weapons bay of similar size to that of the F-22 (and no Pave Tack), it was chosen in order to fast-track the F-22 weapons bay clearance programme. There are no publicly-announced plans for the RAAF to acquire this weapon but, should it form such plans in the future, it will be no doubt be beneficial that a lot of the testing has already been carried out.

In 2000 F-111Cs from No. 1 Squadron deployed to Hickam AFB, Hawaii, to participate in the RIMPAC 2000 exercise. During this exercise the AGM-84 Harpoon was fired. Here RAAF crew load a missile.

In 2002 – while the F-111C fleet is being fitted with wings from retired F-111Fs – there is some doubt as to whether the 'Pig' can be expected to achieve its planned OSD (out-of-service date) of 2020. The fleet suffered a further setback in late June 2002, when a hasty landing by F-111C A8-112 at Darwin, following a fuselage fuel cell rupture, led to another grounding.

The acquisition of an F-111-compatible AAR capability is also on the horizon, with the commencement of Project Air 5402 to acquire up to five NGTT (New Generation Tanker/ Transport) aircraft such as the Airbus A330 or Boeing 767 from 2006. Funding for this project was made available in the 2002 Defence Budget, and such a capability will confer even longer legs to the Amberley strike force.

Wing cracks again

With the AUP successfully completed, the BUP well under way and a stated commitment until 2020, the future looked bright for the 'Pig'. This outlook suffered a setback in late 2001, however, with the failure of an F-111C (test specimen) wing during fatigue testing at the Aeronautical and Maritime Research Laboratories in Melbourne. The wing was being tested as part of the Sole Operator Programme, and was focussed on a previously identified stress concentration area that had been repaired using a boron patch. The new failure occurred close to the area of concern, with multiple failures emanating from tapered-fastener holes in the aft spar, and was not anticipated. This resulted in the grounding of the entire fleet, until the problem could be understood and quantified, and the RAAF's suspicion is that the original build quality is to blame.

Because the RAAF has always been the sole operator of the long-span aircraft in the tactical role, it has suffered more fatigue-related problems than the USAF ever did. When General Dynamics performed the original fatigue studies for the long-span wings, for example, they were centred around the FB-111A and its (relatively) benign operational profile.

As a temporary measure, an interim wing safe-life limit was established, and caused grave concern when it was realised that all of the F-111Gs and most of the Cs were beyond this limit. The fuselage has a safe life of 10,000 hours, but the present wing safe-life limit is 5,900 hours (F-111C) and 3,950 hours (F-111G). The highest-time aircraft in the fleet, as of May 2002, is F-111G A8-274 with 7,189 hours 'on the clock'. The remaining F-111Cs were returned to service (with fairly severe limitations) in February 2002 and frenetic activity to restore the fleet to an operational status was then undertaken.

Wings from ex-USAF F-111Fs at AMARC are in the process of being removed and shipped both by air and sea to Amberley for installation on the F-111C fleet – priority being given to these models, as they are the operational arm of the strike force. A team of RAAF technicians arrived at AMARC in early April 2002, and the first wing-set arrived at Amberley barely two weeks later. Four sets of wings had previously been removed from aircraft at AMARC and stored in Australia; one set later being fitted to an F-111C, which had re-entered service some time before these latest problems surfaced. The other three sets were delivered to the Weapons Systems Business Unit by the end of April, and have now entered 'production'. At the time of writing in May 2002, a further four sets had been shipped to Australia, and the remainder are planned to arrive at fortnightly intervals until the end of November.

F-111F wings were chosen because of their known history and (relatively) recent build date, but the RAAF now faces the dilemma of whether to fit the extended wing tips of the original aircraft. There are plusses and minuses for both cases (less fatigue concerns and better handling qualities of the short-span wing vs. the increased range of an aircraft with the long-span wing for example) and a decision will be made at the end of July 2002, in favour of whichever option restores the entire fleet to operational status in the shortest time.

Rectification of the current wing problem comes with no guarantee that future fatigue concerns will not cause the premature retirement of the aircraft, but the more recent transition from low-level to medium- and high-level operations (the result of the employment of precision-guided weapons) should go some way toward easing the burden.

The current aim of the re-winging project is to achieve what has been referred to as an 'elegant transition' to the F-35, almost certain to replace both the 'Pig' and the Hornet. Should the aircraft show cause to be withdrawn prematurely, however, the RAAF will face serious problems. Leasing of aircraft such as the F-15E or F/A-/18E/F has been suggested in such a case. However, there is no aircraft either flying now, or on the drawing board, that can directly match all the capabilities of the F-111.

Nigel Pittaway

A rolling upgrade (BUP) aims to keep the F-111C force viable until near-certain replacement by the F-35 Joint Strike Fighter. Many of the important elements of BUP will not now be applied to the F-111G fleet, which will remain largely restricted to training duties.

FAU *Fuerza Aérea Uruguaya*

Photographed by Marnix Sap/MIAS

In 2003 Uruguay celebrates both the 90th anniversary of military aviation in the country (establishment of the Escuela de Aviación Militar on 17 March 1913) and the 50th anniversary of the air force as an independent service (27 November 1953). In recent years the FAU has emerged from a period of stagnation with a gradual modernisation of its inventory.

***Above:** The three Bandeirantes are among the hardest-working of the FAU's transport fleet, being mainly used on the TAMU civilian network. TAMU fulfils a vital task, operating as a scheduled airline to serve remote destinations within Uruguay.*

***Left:** The FAU received two C-130Bs (591 and 592) in 1992 to greatly enhance the air force's transport capability. A third aircraft (593) was delivered in 1994, but was only used as a spares source. The two airworthy aircraft are worked hard from their base at Carrasco, and are heavily involved in resupply flights to the Uruguayan scientific research station in Antarctica.*

Above: Uruguay's transport fleet is heavily reliant on the CASA 212 (illustrated) and the Bandeirante. Both have secondary tasks, two CASA 212s being able to accept a Comint package, as evidenced by the extra aerials on this aircraft. In the past the C-212s, like the rest of the transport fleet, were also assigned civilian registrations to facilitate flights out of Uruguay, but this practice has been discontinued.

Left: The A-37B Dragonfly's reliability, agility and accuracy made it arguably the best close air support/light attack aircraft of its time, and it is still considered ideal to operate in Uruguay's defence from forward bases throughout the country. For 20 years the Dragonfly partnered the Lockheed AT/T-33A in Grupo/Escuadrón Aéreo No. 2, before the elderly 'T-birds' were withdrawn in June 1996.

Right: Escuadrón Aéreo No. 5 operates two AS 365N2 Dauphins. They can be used for SAR duties, hence the winch, and for medical evacuation. Their speed and comfort is also prized for presidential and ministerial transport.

Below: On at least two occasions the FAU has borrowed Pucarás from the Fuerza Aérea Argentina's Grupo 3 de Ataque. FAU226 was in service during 2002, although it was scheduled to return to Argentina later in the year.

The two-aircraft option from Uruguay's initial Pucará purchase was not exercised, although much later, in 1996, the FAU tried to acquire three more. Again, funds were not forthcoming. The type is useful for attack and surveillance and, like the A-37B, was involved in Uruguay's successful anti-drugs efforts. The Escuadrón Aéreo No. 1 fleet numbers these three aircraft.

Escuadrón Aéreo No. 1 (Ataque)

Grupo de Aviación No. 1 (Ataque) was officially established on 5 August 1955 and initially tasked with tactical reconnaissance, but in fact existed only on paper until Grupo No. 2 received its F-80s in 1958. At this point Grupo No. 1 took over eight North American F-51D Mustangs and two North American T-6 Texans. In 1969 its mission was changed to 'Instruction and Training', and in 1978 the unit received its first T-34 Mentors, which it operated until 1981 when 19 of these trainers were transferred to the CIEVA. The remaining Texans were kept airworthy until the last was phased out in 1989.

On 12 November 1980, a U$10.8 million contract was signed with Argentina to purchase six new-build FMA IA-58A Pucarás (220-225), plus two options. On 15 May 1981, the first two Pucarás arrived in Uruguay. By October 1981 all were delivered to their new home at Durazno. Once again the mission was changed and the unit reverted to being an attack squadron. One of the Pucarás (225) was lost on 22 July 1993, but was later replaced by 227. Illustrating the technical capabilities of the FAU, 227 is a composite aircraft built up from a surplus FAA airframe and two grounded Uruguayan examples.

Escuadrón Aéreo No. 2 (Caza)

Grupo de Aviación No. 2 (Caza) was created on 8 May 1951. Its 25 F-51D Mustangs were initially based at the Aeropuerto Internacional de Carrasco (Brigada Aérea I) until they were transferred in 1958 to Grupo de Aviación No. 1. On 23 October 1956 the unit received its first Lockheed T-33 'T-Birds' which, from May 1958, were joined by some 14 Lockheed F-80Cs.

On 31 October 1976 the first batch of new Cessna A-37B Dragonflies (270-277) arrived at Carrasco, but on 5 March 1980 all aircraft relocated to the Aeródromo Militar 'Tte.2° Mario W. Parallada', situated near the city of Durazno (Brigada Aérea II). In October 1987 two (270 and 276) were lost in a mid-air collision, and were replaced in February 1988 by two attrition replacements (278 and 279) from US Air National Guard stocks. The final batch to arrive in Uruguay – delivered in March 1989 – comprised six ex-ANG machines (280-285).

Twelve aircraft remain on the books, but only four are currently in service. Soon the FAU has to decide whether to overhaul and update the remaining fleet or select a suitable alternative. On 6 April 2002 FAU279 returned from ENAER after a 3-year overhaul in Chile.

Right: ***A batch of five CASA 212-200s (530-534) was received in late 1981/early 1982, including one (534) for SAR missions with a radar nose. This aircraft, together with 530, was returned to CASA, leaving three aircraft to continue on military transport and surveillance tasks.***

Escuadrón Aéreo No. 3 (Transporte)
This squadron was created on 24 April 1994 by integrating the operations and assets of Grupos de Aviación No. 3 (Bombardeo), No. 4 (Transporte) and No. 6 (Transporte). The unit operates from Carrasco IAP and has one EMB-120 Brasilia (FAU code C-120), three EMB-110 (C-95) Bandeirantes, three CASA 212-200 (C-212) Aviocars and two C-130B Hercules (*above*). The latter are undergoing overhaul by ENAER in Chile, and are being fitted with artificial ADF (automatic direction-finder) and HSI (horizontal situation indicator). As well as military transport tasks, the unit also carries out photo-survey and mapping tasks using one of the Bandeirantes, and two of the C-212s can be equipped for Comint missions. The Brasilia and Bandeirantes are also routinely used on TAMU (Transporte Aéreo Militar Uruguayo) scheduled civilian flights serving remote destinations.

Above: Five EMB-110C Bandeirantes (serialled 580-584) were delivered to Grupo de Aviación No. 6 from January 1976. 584 was lost in June 1977, and a replacement ordered for delivery in August 1978. The new aircraft, 585, was an EMB-110B1 survey platform, which remains in use with Escuadrón No. 3 alongside regular transports 580 and 583.

Right: The transport unit's latest acquisition is a single EMB-120ER Brasilia, purchased second-hand in the US and arriving on 15 July 1999 to replace the last Fokker F27 (C-27) in Escuadrón No. 3 service. It features an EFIS cockpit.

On 14 July 1997 Escuadrón Aéreo No. 5 took delivery of six Westland Wessex HC.Mk 2s which had served with the RAF's No. 28 Squadron at Sek Kong prior to the UK's handover of Hong Kong to China. The helicopters arrived at Montevideo by sea, but suffered from poor serviceability, and consequently saw little service. In April 2002, during a visit to Uruguay, the UK Chief of the Air Staff expressed his favourable disposition to the transfer of three further HC.Mk 2s to Uruguay when No. 84 Squadron at Akrotiri retires the type in 2003.

Escuadrón Aéreo No. 5 (Helicópteros)

This unit was initially created as a provisional search and rescue group in 1955, with two Bell H-13Gs, two Ryan L-17B Navions and two Piper PA-18s operating from Aeródromo Militar Cap. (Av.) Juan Manuel Boiso Lanza (currently HQ FAU). In 1960 the unit was formally commissioned and renamed Grupo de Aviación No. 5 (Búsqueda y Rescate), and in December 1963 moved to its current location at Base Aérea No. 1, Carrasco International Airport. On 25 August 1976, six UH-1B 'Cañoneros' were delivered from the USA, of which the last flew until 22 November 1990. By 1980 two civil Bell 212s had strengthened the force and, every year, one Twin Huey is airlifted by C-130 to Antarctica in support of the Uruguayan scientists based there during the summer period. For this role both 212s received a high-viz colour scheme (*below*). When not deployed south, the helicopters are extensively used for SAR duties. Ex-RAF Wessexes were delivered in 1997, and in November 1997 the FAU received the first of two new AS 365N2 Dauphins (*left*) for SAR and VIP transport duties, the second arriving in January 1998. On 29 June 1998, Escuadrón No. 5's complement was expanded by five UH-1H Hueys (*above*) delivered by C-5 Galaxy. Three of them had been overhauled prior to their delivery and were put directly into service, joining the three survivors from an earlier batch. The Huey is a much appreciated workhorse and is used for tactical transport, police duties, special day and night operations, and also for SAR, medevac and fire-fighting duties.

Today, Escuadrón Aéreo No. 5 can rely on some 30 pilots and 50 technicians operating some 10 helicopters to provide search and rescue services across the vast national territory. One helicopter and crew are on a permanent 5-minute SAR alert at Carrasco IAP, with two additional crews on a 30-minute alert. A single Huey is on detachment at the 'El Jagüel' airport on a 5-minute alert, close to the popular beach resort of Punta del Este.

Escuadrón Aéreo No. 7
On 5 December 1998 Uruguay took delivery of 10 newly-built U-206H Stationairs at Cessna's Independence plant in Kansas. The FAU had awarded this U$3 million contract in November, over offers from Raytheon, Piper and a Brazilian manufacturer. The Stationair can seat up to six and features the new 300-hp (224-kW) Textron Lycoming IO-580-A1A engine, a key factor in Uruguay's decision to accept the Cessna bid. On 22 December, all 10 aircraft arrived at Montevideo-Carrasco IAP after a 14-day ferry flight. The newly established Escuadrón Aéreo No. 7 at Mellila (Brigada Aérea III) received six of the Stationairs (*above*), operating them alongside the four Cessna T-41D Mescaleros the unit had adopted from the EMA (*right*). The U-206s can be fitted with an optional ventral cargo pannier.

Escuadrilla de Enlace
Created within Brigada Aérea II at Durazno, the Escuadrilla de Enlace began operations with four Cessna U-17As (185C) and single examples of the Cessna 206, Cessna 210 Centurion and Piper L-21 Super Cub. The unit's varied tasks include medical evacuation, transport of organs, VIP and staff transport, border and coastal surveillance, and general liaison duties.

On 31 December 1999 the unit received four new Cessna 206Hs (U-206), which replaced the last of the U-17As. The L-21 is still flown by the unit, retained in immaculate condition (*left*). It is the oldest aircraft in the FAU inventory.

Escuadrón de Vuelo Avanzado (EVA)
Formerly known as the Centro de Instruccion y Entrenamiento de Vuelo Avanzado (CIEVA), the advanced training school at Durazno took delivery of six Pilatus PC-7Us (AT-92) in 1992, at which time the unit's Mentors were transferred to the EMA. The advanced school was renamed Escuadrón de Vuelo Avanzado, and provides graduates from EMA with a comprehensive advanced syllabus including IFR and night flying instruction, plus tactical training. For the latter the PC-7s can be fitted with rocket pods and guns under the wings for exercises at the nearby La Carolina range, located some 40 km (25 miles) north of the base.

Escuela Militar Aeronáutica (EMA)

Winner of an international tender, the FAU selected the Italian Aermacchi SF-260EU (*above and right*, FAU type code T-260) to replace the ageing Beechcraft T-34 Mentor (*bottom*) in the basic training role. The 13 SF-260s were delivered by boat as knockdown kits, and were locally assembled by FAU technicians. The last four aircraft to be assembled were delivered on 6 July 2000 to serve with the Escuela Militar Aeronáutica (EMA, Military Aviation School) at Pando.

In their fourth and last year at the school, cadets begin their Curso de Vuelo Basico (Basic Flying Course), which comprises some 20 flights before they can go solo. In total, students receive between 80 and 96 hours of practical training before graduation and, in the course of 2002, some 16 students hope to receive their wings. It is foreseen that this number will double in 2003. The FAU is very proud to have an increasing number of female pilots successfully completing the EMA course and two have already found their way to the FAU combat units flying the A-37 and IA-58.

As well as the SF-260s, the EMA also operates two Beechcraft B-58 Barons (type code UB-58) delivered in August 1998 for multi-engine training and staff liaison missions (*below*). The survivors of the T-41D Mescaleros previously in EMA service have been transferred to the newly established Escuadrón No. 7 at Melilla. All of the current training course (including pilot screening) is now performed by the SF-260s.

Greece

The Hellenic Air Force (HAF) was established early in the second decade of the 20th century during the Balkan Wars. Later on, the Hellenic Air Force, which became a separate branch of the Greek Armed Forces, established a reputation for bravery and determination during World War II. Today, in the dawn of the new millennium, the Elliniki Polemiki Aeroporia continues an ambitious modernisation programme which started almost 20 years ago, and remains one of NATO's most important members.

Konstantinos Dimitropoulos

Elliniki Polemiki Aeroporia (Hellenic Air Force)

In the summer of 1974 an event occurred that rocked the North Atlantic Treaty Organisation to its foundations; Greece withdrew militarily from the Alliance. Pulling the plug on the landlines traversing its territory, and thus blacking out a vital section of the NATO integrated communications system, this reduced the effectiveness of the NATO Air Defence Ground Environment (NADGE) network by halting the lateral cross-feed of early warning information. Greece had dramatically demonstrated its supreme importance to the continued viability of the Western Alliance. Greece had been the linchpin of Allied Forces Southern Europe.

Tasked with providing a protective curtain along a 1,700-mile (2740-km) arc, from the Resia Pass in northern Italy, across the northern borders of Greece and Turkey, to Mount Ararat in eastern Turkey, AFSouth was plunged into disarray. Denied the early warning links, surveillance stations, strike bases and other installations sited in Greek territory, and with a substantial proportion of its assigned forces peremptorily withdrawn, it no longer presented a coherent – even credible – defensive barrier.

Cyprus conflict

This most traumatic event in the then quarter-century history of the Alliance, which Greece – together with Turkey – had joined in 1952, had been precipitated by Turkish use of military force on the predominantly Greek-populated island of Cyprus. Sworn enemies for centuries, the two nations collaborated within NATO remarkably well, despite some souring of relations in the 1960s as a result of the Enosis movement on Cyprus, which finally resulted in the Turkish military action of July 1974.

Aggrieved by the lack of support received from fellow Alliance members, which rejected Greek representations calling for NATO intervention, Greece promptly withdrew its forces from the integrated military organisation. Various *ad hoc* measures were hurriedly taken to bridge the gap punched in NATO's southern flank by the Greek withdrawal. The landline links were reinstated and the ACE-High communications system restored to full operational capability, but it was a nightmare situation and patently obvious to the Atlantic Alliance that its most southerly command would lose credibility without Greek forces.

Meanwhile, despite a somewhat precarious economic situation, Greece initiated a major programme of upgrading and modernising its armed forces in general, and the air arm in particular. Although US personnel were re-admitted to the military base facilities that they had formerly occupied in Greece, albeit under modified ground rules, little progress in improving Greco-Turkish relations was made until the 1977 US pledge to "actively and unequivocally oppose" a military solution to the dispute over the Aegean. Whereupon the respective national prime ministers, Konstantinos Karamanlis of Greece and Bulent Ecevit of Turkey, agreed to codify the principles for good relations between their countries.

Many were the bones of contention between Greece and Turkey, the principal being control of the Aegean Sea, which the latter accused the former of attempting to turn into a "Greek Lake". Matters worsened with the unilateral extension by Greece to 10 miles (16 km) of the airspace over the Aegean islands. Impasse again appeared to have been reached when, on 20 October 1978, Mr. Ecevit declared that Turkey would block Greek NATO reintegration unless Greece agreed to share with Turkey operational jurisdiction of the Aegean, Mr. Karamanlis responding that Greek relations with NATO could not be subject to conditions imposed by Turkey. Fortunately for the Alliance, a more pragmatic attitude on the part of the military junta that soon thereafter assumed power in Turkey, coupled with Greek acceptance of the desirability of reintegrating prior to joining the European Economic Community, led to agreement that negotiations concerning settlement of the long-standing and complex Aegean jurisdiction dispute be postponed until after Greece resumed its military role within NATO.

This followed acceptance of an interim formula proposed by General Bernard Rogers, the Supreme Allied Commander Europe, under which Greek and Turkish officers in their respective Air Headquarters at Larissa and Izmir would exchange "full and unlimited" information between units under their respective commands and covering a zone extending 30 miles (48 km) each side of the national boundaries in the eastern Aegean. The NATO Defence Planning Committee gave its full approval to the return of Greece to the military wing of the Alliance, and in October 1980 this policy was endorsed by a parliamentary vote of confidence in Athens. Thus, six years after their withdrawal, Greek forces began the process of reintegrating with those of other NATO members and, to Western relief, it seemed that the defences of the Alliance's southern flank would at last be fully reinstated.

However, Greek reintegration was not initiated without reservations. While most Greek military leaders believed NATO to be as important to Greece as Greece was to NATO, the positions of Greek and Turk concerning the Aegean still appeared irreconcilable, and the Greek government warned the USA that failure to reach agreement on the subject of US bases in Greek territory would result in complete Greek withdrawal from the 'Alliance'. The Aegean problem remained unreconciled and negotiations over the US bases broke off.

Haggling over the price tag that Greece placed on the US bases negotiations continued through most of the first half of 1980. Greek demands included a substantial shopping list of battle equipment to be supplied in a rolling five-year programme: provision of sophisticated electronic warfare equipment and Grumman E-2 Hawkeye early warning aircraft; the transfer of US technology to aid the expansion of the Greek armaments industry; a one-time grant of some $300-400m for advanced military equipment in addition to normal military aid; and reassurances of support in the unresolved Greco-Turkish dispute over the Aegean. The US Defense Department had been reluctant to commit itself to the availability of some of the equipment that the Greek armed forces desired to obtain and, in June 1981, negotiations were suspended pending the November elections, which were won by the Democratic Socialists.

Taking a firm stance

Headed by Andreas Papandreou, a gifted leader who reinstated good relations with the US but also kept an austere stance on Turkey, the Greek government seemed determined to end Turkey's challenges over the Aegean, despite the latter's continuous attempts to raise tension, on an almost daily basis. Another very serious stand-off occurred in March 1987, when once again the Turks deployed seismographic vessels in the area, a move that nearly led to war between the two NATO allies. In April 1990 a more US-friendly conservative government was elected, signing an eight-year extension to the existing base agreement, paving the way for further US aid. The Socialists regained power almost three years later to stay, with Andreas Papandreou until 1996, and his successor Kostas Simitis until today.

There remained a constant stream of minor irritatations between the two countries, especially after 1 June 1995 when the UN's Convention on the Law of the Sea allowed Greece to extend its territorial waters from 6 to 12 miles (10 to 19 km). The Turkish Parliament declared that Turkey's vital interests in the Aegean made the extension of Greek territorial waters unacceptable, and that "the country's interests would be defended by any means required". Then, in January 1996, a new stand-

Above: While the original Greek F-4Es are concentrated at Andravida, the ex-US ANG F-4Es fly from Larissa with 337 Mira. A few retain their original Hill Gray II scheme (illustrated), but are adopting the new Ghost scheme. Ex-US machines lack Dias RWR fairings.

Right: Tactical reconnaissance capability is concentrated in 348 Mira at Larissa, which flies the HAF's RF-4E fleet. The original three-tone green and brown scheme (illustrated) is being applied to the ex-Luftwaffe aircraft which form the bulk of the inventory.

Above: The Greek Phantom fleet is in the process of receiving the F-16-inspired 'Ghost' scheme, displayed by a Sidewinder-armed 338 Mira aircraft on patrol over the Aegean. This squadron is the first to get upgraded PI 2000 Phantoms.

While the Peace Icarus 2000 Phantoms can fire the AIM-120, unmodified aircraft retain the AIM-7 Sparrow as the primary BVR weapon. This F-4E is assigned to 337 Mira. HAF units are now referred to by this designation, the role-defining suffixes having been dropped.

Above: With the name **Princess of Andravida** *on the fin, the first F-4E Peace Icarus 2000 conversion undertook trials at Manching after modification by EADS.*

off occurred over an outcrop of rocks in the Aegean, known as Imia. These small islets have been Greek since 1947, but Turkey seems to 'ignore' the international treaties. These, and such disputes as the concerning oil exploration rights along the Aegean shelf, serve to keep Greco-Turkish tension on the boil.

Three 339 Mira F-4Es display the 'Aegean Interception' colour scheme. Most of Greece's Phantoms are fitted with the Dias (Zeus) radar warning receiver kit, with antennas on the intake sides, under the radome and at the tip of the fin.

Evolution of the Hellenic Air Force

In 1911 the Hellenic Government assigned the establishment of an Air Service to French experts. As a result, six officers were posted to France in order to be trained as pilots, while the first military aircraft were ordered from the French firm Maurice & Henri Farman. On 8 February 1912 aviator Emmanuel Argyropoulos performed the first flight in Greece with a 50-hp (37-kW) Nieuport. The same day, Argyropoulos carried out a second flight with the then Prime Minister, Eleftherios Venizelos, on board. On 13 May 1912 1st Lieutenant Dimitrios Kamperos flew the first military Henri Farman aircraft, and later participated in army manoeuvres. In June of the same year – while the establishment of a Naval Air Service was being examined – Kamperos, having converted the first Henri Farman *Daedalus* into a seaplane, flew at an average speed of 110 km/h (68 mph), thus achieving a new world record. In late September 1912, the 'Aviators Company', consisting of four flying officers, was established at Larissa. The first combat air unit was born.

Activities of the Army Air Service during the Balkan Wars pioneered the use of aircraft as a war asset. On 5 October 1912 the first combat air reconnaissance mission took place on the Thessalia front. Other missions followed with drops of improvised bombs on the Bizani strongholds, as well as provisions and newspapers to the besieged inhabitants of Ioannina. On 24 January 1913 a naval co-operation mission took place above the Dardanelles, a world first of its kind. On 4 April 1913,

Emmanuel Argyropoulos was killed in his Bleriot XI, becoming the first casualty in the history of Hellenic aviation. His death simultaneously marked the end of air operations in the Balkan wars of 1912-1913.

Greece entered World War I on the Allied side and new perspectives arose for the organisation of the Military and Naval Air Services. Immediately after the completion of training at the flying schools of Sedes and Mudros, the first Hellenic combat units were created. No. 531 Fighting Squadron, along with No. 532 and No. 533 Reconnaissance-Bombing Squadrons, flew hundreds of missions above enemy lines, consisting mainly of patrols, reconnaissance and intelligence-gathering, as well as attacks against enemy trenches. No. 2 Squadron of the Naval Air Corps performed night raids along the border, anti-submarine warfare and attacks against targets of vital importance, as well as interceptor missions. Among the most significant were the night raids on the Gallipoli-Constantinople peninsula in June 1917, heavy bombing of the enemies' positions, the raid against the German cruisers *Goeben* and *Breslau*, constant raids during the battle of Skra Di Legen, as well as the bombing of Smyrna.

Fighting in Asia Minor

In May 1919 the Hellenic Armed Forces landed at Smyrna, with the consent of the Allies. Despite the inadequacy of assets and ageing equipment, the HAS participated actively in the operations with four squadrons of the Army Air Service and one from the navy. From 1919 until the end of 1922, it undertook a variety of reconnaissance, strafing and bombing missions, effectively covering a front which reached 435 miles (700 km) in length in 1921. Exceptional among the air activities were the bombing missions against Kioutachia and Eski Sechir. Support by the Hellenic Air Service did not abate during either the Hellenic Army's advance or its retreat that followed.

The Asia Minor Campaign was followed by a long period of peace, during which both the Hellenic Army and Naval Air Services were reorganised and upgraded. New types of aircraft from Britain and France started to arrive from 1925, including Gloster Mars Nighthawks, Fairey IIIFs, Morane-Saulniers, Hawker Horsleys, Avro 621s and 626s, and Breguet Bre 19s. In co-operation with Blackburn, the Aircraft Factory was set up at Faliron, where T3A Velos seaplanes, Atlases and Avro 504s were manufactured.

In 1930 the Air Ministry was founded and the Hellenic Air Force was established as an independent branch of the Hellenic Armed Forces. In co-operation with the linguist Manolis Triantafyllides, a new RAF-style ranks structure – still in use today – was established. In 1931 the Air Force Academy was founded at Tatoi and became the 'seedbed' for new officers. At the same time, the establishment of the Weather Bureau constituted an achievement of vital importance for flight safety. In 1934 the Hellenic Air Force General Staff was established and new aircraft types – namely the PZL P24, Dornier Do 22, Avro Anson, Potez Po 63, Henschel Hs 126, Fairey Battle and Bristol Blenheim – were delivered. At the same time significant infrastructure work was carried out, to modernise facilities and air bases all over the country.

Upon the declaration of war between Greece and Italy, 77 front-line Hellenic aircraft were available to confront 463. Despite the overwhelming superiority of the Regia Aeronautica, the Hellenic Air Force managed to resist, due to the bravery and self-sacrifice of its crews. Worth noting is the mission of 2 November 1940, during which a Breguet Bre 19 performing a low-altitude flight along the Albanian frontier, detected the Italian Mountaineers Division "Julia", at Pindos Mountain, on its way to occupy Metsovon. This event radically changed the developments on the front in favour of the Greeks. On the same day, Flying Officer Marinos Mitralexis used the propeller of his PZL to bring down a three-engined Italian bomber, having exhausted the supply of ammunition for his machine-guns.

German intervention in 1941 forced a number of Greek aircraft to seek shelter in the Middle East. With five Avro Ansons, one Dornier Do 22 and three Avro 626s the Hellenic force began to reorganise under the Ministry of Air Force based in Cairo, Egypt. Three Hellenic squadrons were formed, namely the 13th Light Bombing Squadron with Avro Anson, Bristol Blenheim and Martin A-30 Baltimores, and Nos 335 and 336 Squadrons (RAF) with Hurricanes and Spitfires, which were integrated with the allied air forces. Aircraft maintenance and repair facilities, as well as an Air Force General Hospital, were established in Cairo. Flying and technical personnel undertook high-level training at the Greek training centres established in Southern Rhodesia and South Africa.

Greek squadrons flew almost every type of mission in the Middle East, such as anti-submarine warfare, close air support, reconnaissance, bombing attacks and interceptions of enemy aircraft. In summer 1943 they also participated in the attacks against German military targets in Crete, the most significant German operating base in the Middle East. From May until November 1944 the activities of the Greek squadrons continued in Italy, where bombing, reconnaissance and patrols were undertaken.

In 1948 the American Military Aid Mission replaced the British Mission in Greece and organised the delivery of the first of a flood of US-built aircraft, commencing with Curtiss Helldiver dive-bombers, which proved invaluable in the operations against Communist forces in the 1946-1950 civil war. In November 1950 Greece, wishing to assist the United Nations, sent the 13th Transport Aircraft Flight to South Korea. It consisted of seven Douglas C-47 Dakota aircraft, 67 officers and soldiers. Until May 1955 the Greek Flight carried out a total of 2,916 missions under exceptionally harsh conditions, comprising air evacuations, transport of personnel, POWs and most kinds of material, drops of supplies and ammunition, replenishment of allied bases and intelligence-gathering. Twelve officers and NCOs, as well as two aircraft, did not return to Greece.

Rebuilding in the Jet Age

Greek fighting power, paradoxically, underwent substantially more capability enhancement during Greece's years of estrangement from NATO than it did during the preceding decade when it was fully integrated with the Alliance. While major upgrading was enjoyed by all Greek armed services, it was particularly dramatic in the case of the Elliniki Polemiki Aeroporia/EPA (Hellenic Air Force/HAF). When Greece withdrew from NATO, the HAF acutely felt its lack of combat aircraft possessing adequate range to intervene in Cyprus. The HAF was suffering the after-effects of five years in which US military and economic aid had been largely withheld as a result of the overthrow in April 1967 of the Greek government by a military junta. As consequences of this denial, much of the HAF's equipment was suffering from advanced senility and there were spares shortages and maintenance problems, a combination that was inevitably resulting in poor readiness levels by NATO standards.

Backbone of the HAF's combat element was the 28th Tactical Air Force, which had previously been assigned to NATO's 6th ATAF. It was principally equipped with the aged F-84F Thunderstreak, which had by then seen almost a score of years in Greek service, and the more youthful but, in Greek eyes, "delightful but comparatively inefficacious" Northrop F-5A, of which large numbers were provided in the late 1960s and 1970s, along with over 100 F/TF-104Gs, mainly from NATO stocks.

Starfighters were finally withdrawn from service in March 1993 when the last two F-104 units (335 and 336 Mires at Araxos AB) converted to ex-USN A-7E/TA-7C Corsairs. Freedom Fighters went on to serve until 29 March 2001, when the last F-5 unit (343 Mira at Mikra AB) disbanded after 35 years of continuous operations with Northrop's lightweight. Some measure of all-weather air defence was afforded by a small number of well-worn ex-USAF 'hand-me-down' F-102A Delta Daggers, which were both difficult to maintain and unsuited to Greek operating conditions, and advanced training was provided by elderly T-33As which were running out of hours.

Also, the HAF possessed no heavylift capability, its logistic support force being confined to C-47s utilised since the immediate post-war years, supplemented by only marginally more capable ex-Luftwaffe N 2501D Noratlases. By the early 1970s, force modernisation had assumed the utmost urgency if the HAF was to maintain any credibility as an effective fighting arm. What was, in view of Greece's precarious economy, an ambitious re-equipment plan was formulated, coupled with plans to provide an industrial base affording some self-sufficiency in a wide variety of aeronautical skills relating to aircraft maintenance.

Re-equipment priority

Insofar as the ruling military junta in Greece was concerned, the primary urgency was associated with the need to establish effective defences against neighbouring Turkey in case the fluctuating tension between the two countries should erupt into hostilities. On the other hand, the US government had become increasingly aware of Greece's strategic importance in the light of the massive Soviet incursion into the Mediterranean basin, and was anxious to conclude an agreement under which homeport facilities would be granted to the US Sixth Fleet at Piraeus. Some liberalisation measures were taken by the Greek government to cater for American sensibilities and on 5 March 1972, the US government, tempering moral principles concerning the military regime in Athens with expediency, announced that it would resume military and economic aid to Greece forthwith.

Above: 341 Mira at Nea Anchialos flies Block 50 F-16s. SEAD with AGM-88 HARMs is its speciality, although it also takes its turn in the air defence mission, for which this aircraft carries AIM-9s and AIM-120s.

Air defence is the main role for the two Mires flying the AMRAAM-capable F-16C Block 30 – 330 at Nea Anchialos (above) and 346 at Larissa (below). Note the derelict F-5s in the background of the 330 Mira aircraft.

Hellenic F-16 Block 50s are used for a variety of tasks, including air defence, SEAD and attack. 347 Mira at Nea Anchialos is the lead attack unit, its F-16s routinely carrying LANTIRN pods for night/bad weather precision attack duties with laser-guided weapons. This example is heading towards the range with a live GBU-10.

Above: 345 Mira has five TA-7Hs for type conversion at Souda, this example carrying Vietnam-era M117 bombs.

Below: 336 Mira flies the ex-US Navy A-7E, which originally flew in this scheme.

Within the month, it was revealed that the HAF was to be supplied with 36 F-4E Phantoms at a price of $150m which, partly financed by US military sales credits and including spares and ground support equipment, were to provide the Greek air arm with the long-range fighter capability that it had sought for so long. In the event, 18 of the F-4Es were to have reached Greece by the time Greek forces were pulled out of NATO, but initial operational capability was still to be achieved, and thus the HAF was to unable to commit the Phantoms against Turkish aircraft operating over Cyprus, an action which would undoubtedly have proved extremely embarrassing to the US government.

Valued at some $500 million, the principal contracts on which the major re-equipment programme was to be based were, in fact, placed only a month before Greek withdrawal from the NATO military alliance. The first of these, signed on 13 June 1974, called for delivery to the HAF of 40 Mirage F1CG interceptors as part of a $295 million arms agreement with France. Such was the urgency attached to acquisition of these fighters owing to fears that tension with Turkey would result in military confrontation that the first 11 were diverted from an Armée de l'Air contract. HAF pilot training in France commenced barely seven months later, on 29 January 1975, with the first two Mirages being ferried to Tanagra by HAF pilots on the following 5 August, to begin the replacement of the F-102s.

On the day following the signing of the contract with Dassault-Breguet, the Greek government endorsed US letters of offer covering the supply of 65 Vought A/TA-7H Corsair strike aircraft, at a cost of $259.2 million, and 40 Rockwell T-2E Buckeye advanced trainers with secondary strike capability. In the following year, contracts were issued for eight RF-4E Phantoms at a cost of $91 million and 12 C-130H Hercules transports valued in excess of $71 million. In November, the establishment of a national aerospace industry became effective.

Known as the Hellenic Aerospace Industry (HAI) at Tanagra, 50 miles (80 km) north of

Athens, it was intended, first and foremost, to afford substantial self-sufficiency in the overhaul and repair of all HAF airframes, engines, instrumentation, and airborne and ground-based electronic equipment. Its tasks were also envisaged as embracing civilian work and some sub-contract manufacturing. The complex involved one of the largest single construction programmes ever undertaken in Greece, and Hellenic Aerospace Industry was to be officially opened on 18 December 1979.

Thus, during the second half of the 1970s, the HAF, which consistently absorbed a very substantial proportion of the annual defence budget, underwent a dramatic transformation. This process continued well into the 1980s with the addition of 40 Mirage 2000EG/BGs (Talos programme of July 1985, worth $1,380 million) and a similar number of F-16C/D Block 30 Fighting Falcons (Peace Xenia I programme of January 1987, worth $940 million for 34 single-seat CGs and six DGs, all with GE F110-GE-100 engines), the first of which were delivered during 1988 and 1989, respectively. In the 1990s more aircraft arrived in the form of ex-US ANG F-4Es (27 in 1992), ex-Luftwaffe RF-4Es (27 in 1993) and ex-USN A-7E Corsair IIs (54 single-seaters and eight TA-7Cs in 1993-94).

A follow-on order for another 40 F-16C/D Block 50 Fighting Falcons (Peace Xenia II, also for 34 single-seat and six twin-seat aircraft but with more powerful GE F110-129 engines), worth $1,366 million, was signed in 1992, with the first aircraft arriving in June 1997. In the same year the German company DASA was awarded a $335.9 million contract, initially named Peace Icarus 2000, to update 39 F-4Es to a standard similar to the Luftwaffe's F-4E ICE, with APG-65 radar and AIM-120 capability, while HAI was to carry out a SLEP (Structural Life Extension Programme) for 70 Phantoms.

Advanced F-16s for Greece

More advanced F-16s were ordered in March 2000, when Greece signed a letter of agreement for 60 F-16C/Ds Block 52+ (40 F-16Cs and 20 F-16Ds) under Peace Xenia III. Deliveries of these aircraft will begin in autumn 2003 and be completed in spring 2004. The Block 52+ aircraft will – surprisingly – have Pratt & Whitney F100-PW-229A engines, developing 29,100 lb (129.5 kN) thrust in afterburner (all earlier Greek F-16s are GE-powered). They will also feature the upgraded AN/APG-68(V)X radar with improved detection range, two 4-in (102-mm) colour MFDs and a terrain-mapping facility. They will also have improved range performance, achieved by the use of conformal fuel tanks (CFTs) mounted on the upper fuselage, providing a total of 500 Imp gal (2271 litres) of extra fuel. When used in combination with a pair of underwing fuel tanks, the CFTs confer an extra 50 per cent range over the standard Block 50. The 20 twin-seat F-16Ds ordered are expected to come configured for precision strike and defence suppression missions, equipped with a box-shaped dorsal spine housing additional avionics.

Currently on order are 15 new Mirage 2000-5s, which will boast advanced RDY radar, MICA air-to-air missiles (100 EM-guided and 100 IR-guided), plus 56 SCALP EG conventional cruise missiles. Another 10 M2000-5s will be produced by HAI conversion of existing M2000EGM airframes and all 'Dash 5s' will be equipped with the latest ICMS Mk 3 self-protection suite. There is also an option for the purchase of three more and the conversion of another five EGM/BGMs, plus 50 MICAs. When completed, the new aircraft will form a 25-aircraft squadron (Mira), with the rest of the original delivery EGM/BGMs forming another similar-sized unit, both with enhanced multi-role capabilities. The first step towards this enhancement took place in January 1997, with the order of 39 Exocet Block 2 missiles for anti-ship operations.

Another recent acquisition concerns four EMBRAER EMB-145 aircraft in AEW&C (Airborne Early Warning & Control) configuration, with Ericsson ERIEYE dorsal side-looking radar. Saab has already provided two suitably converted 340H Argus aircraft from Swedish Air Force stocks to act as trainers for the Greek crews, the aircraft operating with the newly established 380 Mira from 112 PM, Elefsina. Also new are 45 T-6A Texan II trainers, which are gradually replacing the obsolete T-41 and T-37 aircraft, in the basic and primary phases of pilot training. Deliveries of the new aircraft will soon be completed and the retirement of the two Cessna trainers is expected in 2002.

Taktiki Aeroporiki Dynami

In 2002 headed by Lt Gen Panagiotis Papanikolaou as Chief of Air Staff, the HAF possesses a personnel strength of 26,800 officers and men, with some 30,000 reservists available to be called upon in times of emergency, and in excess of 500 aircraft, of which almost half are combat types, with others possessing secondary combat roles. The HAF comprises three Commands: the Tactical Air Force Command (Taktiki Aeroporiki Dynami), the Air Training Command (Diikisi Aeroporikis Ekpaidefseos) and the Air Support Command (Diikisi Aeroporikis Ipostirixis).

The largest of these Commands, the Tactical Air Force Command (TAFC) was founded 50 years ago in 1952, along with headquarters at Larissa, as the 28th TAF of NATO's 6th ATAF. It operates from six main bases, each of which possesses its own combat wing (Pteriga Mahis) and – usually – a single Agusta-Bell AB 205A on readiness for SAR duties. The premier combat wing, the 110 Pteriga Mahis operating from Larissa, is one of the two mainly Phantom-equipped Pteriges, the original batch of 38 F-4Es (the initial contract having been supplemented with two aircraft as attrition replacements) delivered in 1974-76 having been followed by a further 18 aircraft in 1978-79.

A multi-role unit, the 110 PM comprises three squadrons (Mires) of which one, 337 'Fantasma' (Phantom) Mira is proud of its long tradition dating back to World War II, when it flew Spitfires in the Middle East. Today it is fully equipped with Phantoms and possesses a strength of over 18 F-4Es for the air defence task, with close air support as a secondary role. All come from ex-US ANG stocks (these Phantoms were originally operated by 338 Mira at Andravida but have since been exchanged), and have been painted in a most effective Ghost camouflage, which replaced their original Hill Gray II camouflage. 337 Mira also has a permanent deployment at 130 Sminarhia Mahis (combat group) at Limnos (shared with F-16s). The second component squadron of 110 PM, 346 'Iason' (Jason) Mira, operates a similar number of AMRAAM-capable F-16C/D Block 30 Fighting Falcons in the same roles.

110 PM's remaining squadron provides the tactical reconnaissance element of the HAF, the 348 Mira Taktikis Anagnoriseos (Tactical Reconnaissance Squadron), which flies the RF-4E Phantom on day and night all-weather tactical reconnaissance tasks. Its complement includes four survivors from the original US-supplied batch and the ex-Luftwaffe aircraft delivered in 1992-94 (27 aircraft of which about seven were used for spares), which are gradually being painted in the standard SEA camouflage in place of their original 'Lizard' colours. Some have been locally modified to carry AIM-9P/L missiles as a self-defence measure, while one was used to test the Thales (ex-Thomson-CSF) ASTAC Elint pod, which is likely to be procured by the HAF (four pods).

A VIP-configured C-47 Dakota, detached from Thessaloniki-Sedes, catered for the transport needs of Air Command's VIPs for a number of years, but has since been withdrawn from use and replaced by a similarly configured C-130H from 112 PM, Elefsina.

At Nea Anchialos, near Volos on the Gulf of Pagasai, the 111th Pteriga Mahis fulfils both interception and close air support tasks with its three resident squadrons, all equipped with F-16s. 330 Mira 'Keravnos' (Lightning) was established on 9 January 1989 and received – along with 346 Mira – the first Greek F-16s (Block 30). The 341 Mira 'Velos' (Arrow) was a former F-5A/B user which was disbanded in August 1993 and re-established on 30 March 1998 as an F-16C/D Block 50 squadron. Although it is a dedicated SEAD (Suppression of Enemy Air Defences) unit, a role for which HARM missiles are used, it also flies interception missions with AIM-120 AMRAAM and AIM-9L missiles.

The last 111 Pteriga resident is 347 Mira 'Perseas', which flies Block 50 F-16s equipped with LANTIRN pods for ground attack missions. Like other F-16 Mires, 347 Mira also flies interception missions, operating when necessary from FOLs (Forward Operating Locations) at 130 and 133 Sminarhies Mahis at Limnos and Kasteli, respectively. For bombing missions conventional 250-, 500- and 1,000-lb bombs are used, as well as Paveway I/II LGBs.

Training for the F-16

Also stationed at 111 PM is the SMET (Scholio Metekpaidefseos ston Tipo, type conversion unit) which, like others of its kind, is only activated when needed, 'borrowing' two-seaters and personnel from the relevant squadrons. The F-16 course includes about 100 hours of theoretical lessons as well as 78-80 flight hours for new pilots. Experienced pilots converting from other types are subject to a 40-hour course. Maintenance is undertaken at three levels on the F-16s: squadron personnel carry out the first level; the base maintenance squadron fulfils the second level, while HAI is responsible for the third.

Most northerly of the HAF bases is the 113 Pteriga Mahis at Mikra, Thessaloniki. This ceased operations on 29 March 2001, when the F-5 was retired. 343 'Asteri' (Star) Mira was the last to operate the legendary Northrop lightweight, after the disbandment of 341 (now an F-16C/D Block 50 operator) and 349 'Kronos' (Saturn) Mires in 1993 and 1997, respectively.

The original camouflage pattern applied to the A-7Es was amended with a reduced-height upper surface pattern (above, on a 335 Mira aircraft), mirroring the camouflage applied to the original A-7Hs (below, 345 Mira). The Corsair II is expected to retire shortly as the Block 52+ F-16s are delivered.

331 Mira's main role is air defence, for which its Mirage 2000EGMs carry Magic II (illustrated) and Super 530D missiles. A secondary role is anti-ship attack, for which the AM.39 Exocet is used. The fleet is being significantly enhanced by the delivery of 15 Mirage 2000-5s and the upgrade of 10 existing machines to this standard.

332 Mira parents the Mirage 2000 SMET (OCU) in addition to its front-line roles, and operates the four 2000BGMs . Both Mirage units deploy routinely to their FOL at Skiros, at the heart of the Aegean.

Now up for sale, the Mirage F1CG fleet is concentrated in one unit, 342 Mira at Tanagra. Most of the 24 aircraft remaining in the fleet carry the names of Greek islands, the example at right being **Samothraki**. ***Since the disbandment of the Crete-based F1 squadron in 2000, its place has been taken by increased detachments.***

Greece was a major operator of the Freedom Fighter, with the first 52 MAP aircraft entering service in June 1965. They were accompanied by 16 RF-5As and nine F-5Bs. More arrived in 1975 when 10 F-5As and two F-5Bs were acquired from Iranian surplus stocks, while three years later (November 1983) Jordan supplied 13 F-5As and six F-5Bs. Nine more F-5As were transferred from Norway in May/August 1986 and an additional batch of 16 F-5As and four F-5Bs arrived in 1989, again from Jordan. A further 11 Dutch NF-5As and a single NF-5B completed the fleet (over 150 aircraft) in 1991. Tasked with ground attack, reconnaissance and day fighter roles, the Freedom Fighters proved very efficient and had an excellent safety record until the very end. The fate of 113 PM is currently unknown, since no details about its future use have been disclosed yet.

The air superiority mission is assigned to 114 Pteriga Mahis at Tanagra. Situated 50 miles (80 km) northwest of Athens, it is one of the largest bases in Greece and is located on the edge of Schimatari town, adjacent to the Hellenic Aerospace Industry (HAI), with which it shares a main runway. Home of all Mirages, 114 PM hosts 331, 332 and 342 Mires, the first two equipped with the Mirage 2000EGM/BGM and the latter with the surviving 24 Mirage F1CGs (from a total of 40 originally delivered in 1975-77). 334 Mira 'Talos', the second HAF F1 unit, operated from Iraklion, on the island of Crete until its disbandment in July 2000. Its surviving aircraft were thus transferred to 342 Mira 'Sparti'. The type is expected to withdraw from active duty by the end of 2002 and the HAF is already seeking possible buyers, since the remaining aircraft are well maintained and can operate for several more years. Initial plans to upgrade these machines to undertake anti-shipping missions were abandoned, after the Mirage 2000s of 331 Mira were certified for the Exocet anti-ship missile, of which 39 were purchased in Block 2 form during 1997.

The first four 2000s (two single-seat EGs and two two-seat BGs) were delivered on 27 April 1988 to 331 Mira 'Aigeas', which had been established a few days earlier, on 18 April.

332 Mira 'Geraki' (Hawk) followed a few months later, receiving its first aircraft on 11 August 1989. The 40 aircraft (36 EGMs and four BGMs) were evenly distributed between the two squadrons and armed with MATRA Magic 2 and Super 530D missiles. They were tasked primarily with the defence of Athens, a role they retain today. 331 Mira is also tasked with the anti-shipping role, while 332 Mira is responsible for converting new pilots to type.

Most prospective Mirage 2000 pilots come from other fighter types such as the Mirage F1, F-4 or the F-104, and have little problems converting. Training includes theoretical lessons and eight to 10 simulator hours, plus 65-70 hours with the Sminos Metekpaidefsis ston Tipo (SMET). The course lasts approximately six to eight months before graduates fly solo in the single-seaters.

332 Mira also hosts the F1 SMET (OCU). There are no twin-stick F1s so, since 1978, all conversion has been undertaken at Tanagra with the aid of an LMT simulator. New Mirage pilots were selected from those possessing at least 150-200 hours on combat aircraft, and normally received 10 hours in the simulator, then two to three hours taxiing down the runway in order to gain familiarity with the aircraft (in the past five or six hours were flown in the T-33A, until the 'T-bird' was retired). The first solo flights were always in pairs with an experienced wingman and consisted of typical General Handling (GH) exercises. With impending retirement, no new pilots are being trained on the F1. However, in an attempt to boost the morale of 342 personnel and to give the aircraft a more 'modern' look, two F1CGs have been painted in a very effective Ghost camouflage (*à la* F-16).

Thorough training also awaits the crews of 331 Mira tasked with flying anti-ship missions. Each pilot of this elite team has over 500 flying hours in the Mirage 2000. Early disappointment with the poor performance of the M2000's RDM-3 radar led to a temporary halt half-way through the programme, but today the problems have been overcome, although not to the total satisfaction of the Greek side.

Despite this important glitch, the French fighter has matured into a vital weapon in the defence of the Aegean and is praised by its crews for its excellent handling, the effective self-protection suite and the excellent nav-aid systems. These features are 'tested' on an almost daily basis, when they are called to intercept Turkish fighters violating Greek airspace. Both Mirage 2000 squadrons deploy regularly to 135 Sminarhia Mahis (SM) at Skiros, which acts as their FOL (Forward Operating Location). It is certain that the arrival of the far more capable 2000-5 will dramatically boost the capabilities of 114 Pteriga Mahis, which will play a very important role in the defence of Athens and the Aegean in the immediate future.

Hellenic Corsairs

115 Pteriga Mahis at Souda Bay, near Chania, Crete, was the initial operating unit of the A-7H Corsair, comprising 345 Mira Vomvardismou 'Lailapa' (squall), which re-equipped from F-84F Thunderstreaks during 1975-77. Its sister A-7H squadron, 340 Mira Vomvardismou 'Alepou' (fox) disestablished on 30 November 2001 and its aircraft were transferred to 345 Mira. Both Mires had also received the Corsairs of 347 Mira back in 1992, after its disbandment on 31 July of the same year, thus becoming the principal operating units of all A/TA-7Hs.

A total of 59 single-seaters and six twin-seaters was received in the 1970s, of which 41 A-7Hs and four TA-7Hs remain today. The Corsairs have provision for AIM-9P/L Sidewinder AAMs to provide both a self-defence capability and to fulfil a secondary air defence mission, although their main roles are Interdiction Strike (IDS), Close Air Support (CAS) and Tactical Air Support of Maritime Operations (TASMO).

Souda Bay, from which 115 Pteriga Mahis operates, was also the home of the last Hellenic Lockheed T-33As. They were equipped with Dornier tow-targets for use over the nearby NAMFI (NATO Missile Training Centre) range, until 19 January 2000, when the type retired from active duty, after almost 50 years of continuous operations. Today only 345 Mira operates from 115 PM, but it is also expected to de-activate in autumn 2002, pending arrival of the new F-16C/Ds (Block 52+). The first of these are expected in February 2003 and a number of former A-7 pilots and technicians has already been chosen for conversion – their training taking place at 111 PM and in the US.

Also on Crete, at Iraklion, is 126 Sminarhia Mahis (SM), which from 1989 until July 2000 was the home of the Mirage F1-equipped 334 'Talos' Mira. The squadron completed 66,968 flight hours until its disestablishment (its aircraft transferred to the sister 342 Mira at 114 PM, Tanagra), of which more than 3,000 were intercepting Turkish fighters over the southeastern Aegean. The 133 SM at Kasteli is the third HAF base on Crete and it is used as a FOL by the F-16s of 111 Pteriga Mahis.

Corsair concentration

All surviving A/TA-7H Corsairs will eventually move to 116 Pteriga Mahis at Araxos, on the Gulf of Patrai. This was the principal operating unit of the legendary F/TF/RF-104G Starfighter, which flew in Greek colours in large numbers with an excellent safety record, from 1964 until March 1993. Its replacement came in the form of ex-USN A-7E Corsair IIs, of which 54 single-seat and eight TA-7C trainers were delivered in 1993-94. These aircraft are operated by 335 'Tigris' (tiger) and 336 'Olympos' (Olympus) Mires, and differ slightly from the H version in having an aerial refuelling capability and Stencel SIIIS ejection seats. All A/TA-7Hs are equipped with McDD ESCAPAC ejection seats and have their air refuelling probes removed, although they retain their navy-type folding-wing mechanisms). They are also optimised for buddy air refuelling (six D-704 pods) and night/adverse weather attack missions with the aid of an undisclosed number of AN/AAR-45 FLIR pods.

The remaining HAF tactical combat unit, the 117 Pteriga Mahis, is also based on Peloponnisos, at Andravida, and is the principal Phantom element, consisting of 338 'Aris' (Ares) and 339 'Aias' ('Ajax') Mires, each with about 18 F-4Es and having a primary task of close air support and secondary air defence duties. 338 Mira was equipped with the 28 ex-US ANG Phantoms delivered in 1991, but has since traded its aircraft with those of 337 Mira, which were judged more suitable for the Peace Icarus 2000 programme.

PI 2000 involves the upgrade of 39 F-4Es to a standard similar to the Luftwaffe's F-4E ICE. The heart of the upgrade is the Hughes (Raytheon) AN/APG-65Y radar, with enhanced power output and improved modulation correlation, user interface and lower maintenance needs compared to the standard -65 radar. Avionics upgrades include a new GEC CPU-143A digital air data computer and a 1553B databus, connected to a new Elbit Systems fire-control computer, Honeywell H-764G LINS/GPS and an improved Hazeltine AN/APX-113(V) IFF system combined with new Litton ALR-68(V)-2 RWRs. In the cockpit new multi-function colour displays (MFDs), El-Op Systems HUD and state-of-the-art HOTAS (Hands On Throttle And Stick) technology bring the old F-4 into the 21st century.

Some software problems encountered early in the integration stage delayed the programme for almost a year, but have now been overcome and the prototype Peace Icarus aircraft (01523), named *Princess of Andravida* is undergoing evaluation at 117 PM. The upgraded aircraft will be fully compatible with the AIM-120 AMRAAM, IRIS-T and AFDS and are expected to serve until 2015 at least. In parallel, HAI is carrying out a SLEP (Structural Life Extension Programme) for 70 Phantoms, and the upgrade of the rest (36 out of a total of 39) aircraft with support from DASA.

The 134 Sminarhia Mahis (SM) at Thira (Santorini) is used as a FOL for both 117 Pteriga Mahis F-4E Mires. Also part of 117 PM is the KEAT (Kentro Aeroporikis Taktikis – Centre for Advanced Tactics) set up in 1983. Its mission is to organise advanced weapons and tactics training courses, which are usually carried out each summer through SOT (Scholio Oplon-Taktikis – Weapons and Tactics School), and involve all major types of combat aircraft and helicopters, as well as the P-3Bs.

Completing the air defence force are three SAM Mires Katefthinomenon Vlimaton (guided missile units): the 21st at Athens, the 22nd at Skiros and the 23rd at Thessaloniki. Each is equipped with two Raytheon Patriot PAC-3 batteries, which replaced the 36 obsolete Nike-Hercules SAMs. Three more Patriot PAC-2 batteries, which had been leased for the training of the Greek crews, remain at 23rd MKV. They will eventually be distributed among the three Patriot Mires in the near future.

Also on strength are two Russian S-300PMU-1 SAM batteries, which were originally procured for Cyprus, but ended up in Crete, after strong opposition from Turkey. Its capabilities allow the multiple engagements of 12 missiles against six different targets, at heights ranging from 32,810-88,580 ft (10,000-27,000 m), with excellent ECCM resistance. For SHORAD (SHOrt Range Air Defence) the French Thales (ex-Thomson-CSF) Crotale NG (New Generation) was recently selected, with nine vehicles on order. These will complement the 40 Contraves Skyguard systems, which are better known as the Velos (arrow) in Greece.

Diikisi Aeroporikis Ekpaidefseos

From its headquarters at Tatoi-Dekelia, the Diikisi Aeroporikis Ekpaidefseos (Air Training Command) is responsible for all flying and ground training. Greece has always trained its own aircrews, unlike some other NATO countries. This has been standard practice since

Glossary of EPA/HAF unit designations

AKEP	Aeroporiko Kentro Epihiriseon Periohis	Sector Air Operations Centre
ASEPE	Aerometaferomeno Sistima Egairis Proedopiisis & Elenchou	Airborne Early Warning & Control
ATA	Arhigio Taktikis Aeroporias	Tactical Air Force Headquarters
DAE	Diikisi Aeroporikis Ekpaidefsis	Air Training Command
DAI	Diikisi Aeroporikou Ylikou	Air Material Command
EKAE	Ethniko Kentro Aeroporikon Epihiriseon	National Air Operations Centre
KEAT	Kentro Ekpaidefseos Aeroporikis Taktikis	Air Tactics Training Centre
MAE	Mira Arhikis Ekpaidefsis	Primary Training Sqn
MAEDY	Mira Exipiretisis Dimosion Ypiresion	Civil Service Assistance Sqn
MAI	Mira Anahaitisis Imeras	Daylight Interception Sqn
MDV	Mira Dioxis/Vomvardismou	Ground Attack/Pursuit Sqn
MED	Mira Erevnas/Diasosis	Search & Rescue Sqn
MEE	Mira Epihirisiakis Ekpaidefseos	Operational Training Sqn
MK	Mira Krousis	Strike Sqn
MNAS	Mira Aeroporikis Naftikis Sinergasias	Naval Air Co-operation Sqn
MPK	Mira Pantos Kairou	All Weather Sqn
MPE	Mira Prohorimenis Ekpaidefsis	Advanced Training Sqn
MPAK	Mira Anahaitisis Pantos Kerou	All Weather Sqn
MTM	Mira Taktikon Metaforon	Tactical Transport Sqn
MVE	Mira Vasikis Ekpaidefsis	Basic Training Sqn
MV	Mira Vomvardismou	Bombing Sqn
PAE	Pteriga EkpaidefseosAeros	Air Training Wing
PM	Pteriga Mahis	Combat Wing
SEA/E	Scholio Epihiriseon Aeros/Epifanias	Air/Surface Operations School
SEE	Sminos Epihirisiakis Ekpaidefsis	Operational Training Flight
SI	Scholi Ikaron	Air Force Academy
SIP	Scholio Ilektronikou Polemou	Electronic Warfare School
SM	Sminarhia Mahis	Combat Group
Sminos	Sminos	Flight
SMET	Sminos Metekpaidefsis ston Tipo	Type Conversion Flight
SOT	Scholio Oplon-Taktikis	Tactical Weapons School

Deliveries of the 45 Raytheon T-6A Texan IIs (above) is under way to 361 Mira at Kalamata to begin the replacement of the Cessna T-37B/C (below) in the basic training role. The T-37 has been in use for nearly 40 years, the fleet having been acquired from the US and Jordan.

Above: The Cessna T-41D remains in service with 360 Mira at Dekelia on flight screening duties, but is in the process of being replaced by the T-6A.

Right: T-2E Buckeyes continue to provide advanced and weapons training with 362 and 363 Mires. Little-known is a secondary emergency war role.

1931, when the relevant Ethniki Aeroporiki Academia (National Air Force Academy), better known as Scholi Ikaron (Icarus School), was formed by the amalgamation of the separate army and navy flying schools. At the Academy, which in recent years has also trained – among others – several students from the air forces of Cameroon, Chad, Gabon, Jordan, Senegal, Tunisia and Zaïre, students spend four years and they gain their first hours of military flying in one of the four Grob 103 Twin Astir gliders, operated since 1983 by 360 MAE (Mira Archikis Ekpaidefseos – Primary Training Squadron). Those who fail the flight-training course or decide not to fly can pursue other courses, such as engineering, administrating, logistics, etc.

Successful candidates will then proceed to the screening phase, flying the Cessna T-41D Mescalero, of which about 18 remain in service out of a total of 20 purchased in 1970. These are gradually being replaced by the far more capable turboprop T-6A Texan II, of which 45 are on order. About half of the latter will also replace the entire T-37 fleet, which has been used for primary training since 1964.

Graduates are commissioned as second lieutenant before posting to 120 PEA (Pteriga Ekpaidefseos Aeros – Air Training Wing) at Kalamata, in the south of Peloponnisos. At Kalamata, about 80 hours basic training, spread over a six-month course, are flown on the Cessna T-37B/Cs of the 361 Mira Vasikis Ekpaidefseos 'Mystras' (Basic Training Squadron). About 30 'Tweety Birds' remain in service, all procured in the 1960s and 1970s from ex-USAF and Royal Jordanian AF stocks.

Advanced training

For advanced training the HAF relies on the T-2E Buckeye, of which 40 were delivered in 1976-77, making Greece the only European operator of the type. Survivors are spread between 362 Mira Prohorimenis Ekpaidefseos (Advanced Training Squadron) and 363 Mira Epihirisiakis Ekpaidefseos (Operational Training Squadron). Though officially 362 Mira 'Nestor' is responsible for advanced training and 363 Mira 'Danaos' for weapons training, both can conduct either role. The trainees amass about 130 T-2 hours, acquiring such skills as low-level navigation, formation flying and, of course, air-to-air and air-to-ground tactics. Both of these components of 120 PEA have a secondary operational role in the event of hostilities, their Buckeyes being tasked for close air support and flown by instructors.

120 Pteriga Mahis flies an average of 2,000 training sorties monthly, making it the busiest air base in Greece. After the completion of the 18-month training course at Kalamata, successful students progress to become fighter pilots, though some will fly with support or agricultural types, depending on the air force's needs at the time.

ATC's ground training organisation consists of the 123 Technical Training Wing at Dekelia, at which an average of 400 airmen and 200 NCOs undergo instruction each year; the 124 Basic Training Wing at Tripolis, which has six annual intakes of 1,500 men; the 125 Administrative Training Group at Thessaloniki-Sedes for all ranks; the 128 Telecommunications and Electronic Warfare Training Group at Kavouri, where 800 conscripts and 200 NCOs receive training each year; an English Language School; and a Women's Training Centre.

Diikisi Aeroporikis Ipostirixis

With headquarters at Elefsina, the Diikisi Aeroporikis Ipostirixis (Air Support Command) includes within its jurisdiction both the air transport component of the HAF and the miniscule ASW force. Virtually the entire transport force is gathered within the five-squadron 112 Pteriga Mahis, together with the ASW element. The latter, 353 'Triaina' (trident) Mira Naftikis Aeroporikis Synergasias (Naval Air Co-Operation Squadron), fulfils ASW, SAR, fishery and EEZ (Economic Exclusion Zone) patrol missions under naval control with the six examples of the ex-US Navy P-3B Orions delivered in 1996-97. They replaced the obsolete HU-16B amphibians, 13 of which were transferred to Greece from Norway in 1969 and went on to serve until 17 December 1997, when the last operational Albatross flew for the last time.

HAF Orions are lightweight TACNAVMOD aircraft, and underwent an overhaul at Waco, Texas, before delivery. The first upgraded aircraft (BuNo. 153441) was officially handed over to the HAF on 22 May 1996. The Orions fly with navy systems operators and air force pilots and are painted in the standard US Navy Tactical Paint Scheme (TPS), with low visibility grey colours and insignia. Although dramatically more capable than the Albatross it replaced, the P-3B is in need of an avionics upgrade, since its systems are obsolete by today's standards. It can carry eight Mk 46 Mod 5 torpedoes and/or up to five Mk 55 mines. The US Navy also supplied four P-3As, which serve for ground instruction training and as a source of spares. The modernisation of the Orion fleet is now a priority matter.

355 'Ifaistos' Mira Taktikon Metaforon (Tactical Transport Squadron) is tasked with fire-fighting missions, an extremely demanding task, especially during the very hot Greek summers. To carry out this mission the highly experienced crews fly a mixed fleet of Canadair (now Bombardier) CL-215 and CL-415 fire-fighters. The first two CL-215s (1015 and 1039) arrived in 1973, resplendent in a striking yellow/red finish. Nine more followed and four more were added to the fleet from ex-Yugoslavian stocks in 1995 (15 in total).

Of these four have been written-off in fatal accidents in 1977, 1984, 1993 and 2000, while the rest continue to fly alongside the dramatically improved CL-415s, of which 10 were ordered in 1998, with the first landing at Elefsina AB in record time – less than a year later, on 30 January 1999. Powered by a pair of Pratt & Whitney Canada PW120 turboprop engines each delivering 2,830 shp (2111 kW), in place of the World War II-vintage P&W R-2800s of their predecessors, the new aircraft can carry more than 2,750 Imp gal (12500 litres) of water to fight forest fires.

From 1 May until 30 October more than 10 aircraft are kept on a 15-minute readiness, with pilots manning two shifts per day from first light to last, whether it be at Elefsina or the various detachments at Samoa, Thessaloniki, Volos, Rhodes, etc. Service of these hard-working aircraft is performed at base level, with HAI (Hellenic Aerospace Industry) at Tanagra being responsible for the major overhauls, which are usually performed during the winter.

355 Mira has also a secondary transport role, which is performed by a handful of ex-Luftwaffe Dornier Do 28Ds. Only three out of a total of 15 supplied in 1985 remain operational, one of them sporting a smart white/blue finish for VIP missions.

Also part of 355 Mira, but operating from Sedes, Thessaloniki, are the last four HAF C-47 Dakotas. Over 85 of these legendary transport aircraft were delivered to Greece after World War II from the USA and Great Britain. The four survivors are painted in SEA-style camouflage and are part of 355/1 'Atlas' STM (Sminos Taktikon Metaforon – Tactical Transport Flight), operating in transport, mapping, SAR and forest observation roles. They are expected to retire soon, probably with the arrival of the C-27J.

The heavy-lift support mission is provided by the immortal C-130 Hercules of 356 Mira Taktikon Metaforon 'Iraklis' (Hercules). Twelve C-130H (H=Hellenic) aircraft were acquired in 1975-77. Ten survive today, since two were written-off in fatal accidents in 1991 and 1997. One (741) is optimised for ELINT/SIGINT missions and two more are equipped with the MAFFS (Mobile Airborne Fire-Fighting System). During the summer they contribute to the CL-215/415s' fire-fighting tasks.

Between August and November 1992 the C-130 fleet was augmented with five ex-USAF/ANG/AFRes C-130Bs. These are gradually receiving the standard SEA-type camouflage of the H version, in place of their original European One colours. The 'Polemiki Aeroporia – Hellenic Air Force' logos, written in white capital Greek (port) and Latin (starboard) lettering, now adorns the nose section of the entire C-130B/H fleet. Hercules operate in a variety of transport roles, which include carrying VIPs and personnel, picking up aircraft spares from the USA, paratroop-dropping and humanitarian relief in places such as Somalia, Turkey, Bosnia and, more recently, Afghanistan.

Upgrading the Hercules

An electronics upgrade programme, known as AUP, is pending, as declared by the Ministry of National Defence on 15 June 2001. The estimated $100 million upgrade will be carried out over a three-year period by the Canadian SPAR Aerospace for the entire C-130 fleet and includes new Electronic Flight Information System (EFIS), Flight Management System (FMS), Electronic Traffic Collision Avoidance System (ETCAS), EW systems including a Missile Proximity Warning System (MPWS), new radios and radar, new weather radar, digital autopilot, GPS and Inertial Navigation System, Enhanced Ground Proximity Warning System (GPWS), VOR/ILS Instrument Landing System and IFF systems. It is yet unknown if the C-130 fleet will also be subjected to a SLEP (Structural Life Extension Program) as well, and if and how HAI will participate in the programme. SPAR's choice came as a surprise, since the C-130 AUP was expected to be associated with the new MTA (Medium Transport Aircraft) programme choice, described below.

Also part of 356 Mira is the one remaining NAMC YS-11A. This is used occasionally for calibration and navigational training and, sparingly, as a VIP transport. A total of six of these ex-Olympic Airways aircraft was acquired in 1981, but today all but the one described above are discarded, and lie derelict around the airfield. Owing to their airliner origins, the YS-11s were simply not tough enough for many military tasks and were of little use to the HAF.

356 Mira strength will be increased in the immediate future by the addition of 12 new Medium Range Transport Aircraft (MRTA), and two ex-Italian C-130s. The joint Lockheed Martin/Alenia SpA C-27J, aptly named Spartan, the EADS-CASA CN-235 and C-295, and the Russian Antonov An-32 competed for the HAF MRTA. The Spartan was declared winner on 4 March 2002, and the contract calls for 12 aircraft with an option for three more. The C-27J Spartan is based on the proven twin-engine C-27A/G222 airlifter. It has the same dimensions as the latter, but with substantially improved performance and a digital 1553B-based integrated avionics suite. It incorporates the same propulsion system and advanced technology that is used in the C-130J Hercules, including Rolls-Royce AE2100 engines with six-bladed composite propellers, advanced avionics and a new 'glass' cockpit. The 'Ifaistos' Mira also operated a single Gulfstream I acquired in 1968 for VIP flights, but it was withdrawn from use in 1997 and joined the air force museum at Dekelia-Tatoi.

112 PM's newest resident is the 380 Mira ASEPE (Aerometaferomenou Sistimatos Egairis Proedopiisis & Elenchou – Airborne Early Warning & Control), which was established on 25 June 2001. The relevant contract, worth $600m for the supply of four AEW&C aircraft was signed on 1 July 1999, after a prolonged competition between the C-130J AEW&C, the E-2C Hawkeye 2000 and the EMB-145/Erieye AEW&C, with the latter declared winner. Four EMBRAER EMB-145s with dorsal Ericcson ERIEYE phased-array radars and Thales self-protection systems will be delivered to the HAF, with the first arriving in 2003. To train the Greek crews (most ex-A-7, F-4 and C-130 crews), two SAAB 340H/Erieye aircraft (100003 and 100004, known as S 100B Argus in Swedish service) from Swedish stocks arrived at Elefsina AB on 26 September 2001.

As an offset, a VIP-configured ERJ-135 executive jet was delivered to 356 Mira 'Iraklis', with another one arriving in 2002. These replaced the single Dassault Falcon 900B (donated as part of the 1992 Talos deal for the amendment of the RDM-3 radars of the Greek Mirage 2000s), which is now on sale by the HAF, following the freak accident of 15 September 1999, when Alternate Foreign Minister Yannos Kranidiotis, his 25-year-old son and four other passengers were killed on their way to a meeting in Bucharest, Romania, when the aircraft descended violently from 23,000 ft (7010 m) to 6,000 ft (1828 m). The aircraft was flying with civilian pilots and personnel from Hellenikon airport. The civilian crew has since been replaced by HAF pilots and the new VIP EMBRAER ERJ-135s are expected to remain at 112 PM, Elefsina, although the possibility to move to the new 'Eleftherios Venizelos' airport at Spata seems likely.

Search and rescue

The last resident flying unit of 112 Pteriga Mahis is the 358 Mira Erevnas-Diasosis 'Faethon' (Search and Rescue Squadron), which operates a mixed bag of about 20 helicopters, spearheaded by the quartet of Eurocopter AS 33C1 Super Puma Mk 1s. These are the property of the Ministry of the Merchant Marine, but they are operated by air force crews, and have taken over the bulk of the

Above: Only one of the six NAMC YS-11As transferred to the air force from Olympic Airways remains in use with 356 Mira, the others lying derelict at Elefsina. It is flown sporadically for navigation training and calibration.

The Hercules is the backbone of the transport fleet, assigned to 356 Mira. Among the fleet is a single aircraft (below) outfitted for ELINT duties (note non-standard antennas beneath the fuselage). Two ex-Italian C-130s are expected to join the squadron soon.

Above: Flown by the air force's 353 Mira, but controlled by the navy, Greece's P-3B Orion fleet is marked with a small anchor under the tail to signify the navy crew manning the systems. One aircraft briefly wore a large anchor and 'PN' (Polemiki Naftikon) legend on the nose, but it was swiftly removed by air force officials.

Above: Four C-47s remain in service, flying on various duties from Thessaloniki-Sedes. They are due to retire when C-27Js are delivered.

Right: Two S 100B Argus AEW&C platforms are on loan from Flygvapnet to 380 Mira to train operators in the use of the Erieye radar system prior to delivery of Greece's Erieye-equipped EMB-145s.

355 Mira originally had 15 Do 28Ds in its inventory, but only three remain in use. Two are camouflaged while one wears a VIP scheme.

355 Mira's primary role is fire-fighting, for which a mix of CL-415s (foreground) and CL-215s (right background) are in service.

356 Mira's mixed transport fleet includes two EMBRAER ERJ-135s which are used for VIP and staff transport. Their supply was part of the Erieye package.

demanding SAR role from the old AB 205s. The first two were delivered in December 1999, followed by the remaining pair in March 2000.

HAF's Super Pumas are fitted with a nose-mounted Bendix 1500B radar, a Thales (ex-Thomson-CSF) Chlio S FLIR system, a Spectrolab searchlight, loud-hailers, a 600-lb (272-kg) capacity hoist, jettissonable liferafts and EMF (Emergency Medical Facilities). They are powered by a pair of Turbomeca Makila 1A1 engines (each rated at 1,877 shp/1400 kW) and feature a Sextant Avionique Nadir Mk 2 navigation system coupled to a SFIM 155 autopilot, which can perform fully automatic transitions and hovering. Two are always on duty at Elefsina, while the other pair shares one of seven SAR detachments the squadron supports (Limnos, Chios, Rhodes, Iraklion, Kalamata, Araxos and Nea Anchialos). Future deliveries include six CSAR-configured AS 532A2 Cougar Mk 2s, with stretched cabin, FLIR, weapons pylons and chaff/flare dispensers.

Complementing the Super Pumas on SAR missions are 11 AB 205As, operating from the above-mentioned detachments (and wherever else necessary). Also operated by 358 Mira are four VIP AB 212s and a single AB 206 JetRanger, which is used for VIP flights, training and liaison duties.

Also part of 112 PM, but based at Dekelia-Tatoi, is 359 MAEDY 'Dimitra' (MirA Exipiretisis Dimosion Ypiresion – Civil Service Assistance Squadron). Its main task is fire-fighting during the summer months, but the unit also works for the Ministry of Agriculture, undertaking crop-spraying, seed-planting on large fields and fish-feeding tasks. To achieve its mission, 359 Mira is equipped with five Bell 47Gs, three OH-13H

Sioux for training, 12 G.164 AgCats and 20 PZL M-18 Dromaders. The latter were received in 1982, making the squadron the first within NATO to equip with a Warsaw Pact aircraft. Most of the pilots flying for 359 MAEDY are retired ex-military personnel.

Ten non-flying units fulfil the supply functions of the Air Support Command. They consist of the 202 Maintenance Depot at Elefsina for the overhaul of airframes, engines and electronics; the 201 Supply Depot which holds all spares and supplies, apart from munitions, which are held by the 204 Ammunition Depot; the 129 Supply Wing for logistic support and ground transportation; the HAF Construction Unit; the HAF Procurement Agency; the Air Support Forwarding Depot for handling 'special' supplies and equipment; the Calibration Support Unit; an agency for the translation, publication and distribution of English language documentation; and a Plant Representation Office for liaison with Hellenic Aerospace at Tanagra.

Future plans

Greece's European commitment was a key factor in the selection of Eurofighter as the HAF's new fighter in the dawn of the new millennium, and represented the Typhoon's first sales success. The decision to purchase at least 60 aircraft (with another 30 as an option) in a deal worth an estimated $10.2 billion was announced by the Greek government in February 1999.

Under the initial deal, Greece would have made annual payments from 2001, to ensure deliveries from 2006. The acquisition was confirmed on 8 March 2000 but a few months later, in January 2001, Greece began negotiations with EADS, which headed the Eurofighter sales programme, to postpone payments and re-negotiate the price. On 29 March the Prime Minister, Kostas Simitis, announced the postponement of the deal until after 2004, to allow the funding of various social programmes, including the Olympic games in 2004. Greece is also due to stage a general election that year, the winner of which will have to reaffirm the Typhoon purchase.

Postponement of the deal opened the way to Dassault, which will certainly try to re-offer its new Rafale fighter to the HAF, despite being an EADS subsidiary. The US is also pressing, in order to pursue more F-16 sales (Block 60 or upgraded Block 50 aircraft) as a stop-gap measure until the advent of the F-35 in 2010-12. The latter was recently presented to officials.

To further enhance the capabilities of the upcoming Block 52+ F-16s, the HAF placed an order in August 2001 for 55 (with an option for 10) Joint Helmet-Mounted Cueing System (JHMCS) units. Developed by Boeing and Vision Systems International, the JHMCS combines a magnetic head tracker with a display projected onto the pilot's visor, providing a targeting device that can be used to aim sensors and weapons wherever the pilot is looking. With JHMCS, the pilot can aim the radar, air-to-air missiles, infra-red sensors and air-to-ground weapons by pointing his head at the target and pressing the switch on the flight controls. Additionally, the pilot can view any desired data (airspeed, altitude, target range, etc.) while 'head-up', eliminating the need to look into the cockpit during visual air combat.

For the self-protection of the F-16C/D Block 52+ fleet the HAF selected the proven Litton ASPIS electronic warfare system. A total of 60 units was ordered on 1 March 2002, with the first scheduled for delivery in February 2003. The ASPIS is also standard on all earlier HAF F-16C/D Block 30/50s and was selected after a hard-fought competition with the Israeli Elisra ASPS (a derivative of the SPS-2000/3000 used on the F-15/F-16s). The latter was initially selected but it was rejected twice in 2002 by the air force evaluating committee. It is highly likely that ASPIS will also equip the upgraded Peace Icarus 2000 F-4Es, C-130 Hercules and the CSAR-configured AS 532A2 Cougar Mk 2s on order by the HAF.

Targeting pods

Rafael is to supply 15 (with 10 more as an option) Litening II targeting pods, developed in collaboration with Northrop Grumman, for the Block 30 Fighting Falcons and the PI 2000 F-4Es. Litening II contains a high-resolution, forward-looking infra-red (FLIR) that displays an infra-red image of the target to the pilot; it has a wide field-of-view search capability and a narrow field-of-view acquisition/targeting capability of battlefield-sized targets. The pod contains a CCD TV camera used to obtain target imagery in the visible portion of the electromagnetic spectrum. An on-gimbal inertial navigation sensor has established line-of-sight and automatic boresighting capability. It is also equipped with a laser designator for precise delivery of laser-guided munitions and an automatic target tracker.

These features simplify the functions of target detection and recognition, and permit pilots of single-seat fighters to attack targets with precision-guided weapons in a single pass. The Litening II will make the most out of a number of 'smart' weapons like the Paveway III family of LGBs and the AGM-130 rocket-boosted glide bomb, under procurement by the HAF. Also from Israel (BVR Systems) comes an advanced flight simulator for the training of F-16 pilots at 111 Pteriga Mahis.

Further additions in the air-to-ground arsenal include 100 AFDS (Autonomous Free-Flight Dispenser System) units and 56 MDBA SCALP EG conventional cruise missiles. The former is based on the DWS 39 and has a maximum range of 13 miles (20 km) at high altitude or 6.3 miles (10.2 km) from a low-altitude release. It is an area-denial fire and forget weapon, which can carry up to 16 BLU-108/Bs or STABO two-stage runway breakers, depending on the mission profile. It is destined for use on the A-7, PI 2000 F-4E and possibly the F-16.

MDBA's SCALP EG will be used exclusively by the Mirage 2000-5 and will provide the HAF with a true strategic capability. It is a long-range (over 155 miles/250 km), conventionally armed cruise missile, optimised for use in pre-planned attacks against high-value and heavily defended targets. It can be used day and night and in adverse weather, with a large launch envelope, from low to medium altitudes, and is a fire-and-forget weapon. After launch, the missile descends to very low altitude, cruising at high subsonic speed (Mach 0.8), powered by a TRI 60-30 turbojet. Guidance is provided by inertial, GPS and TERPROM (Terrain Profile Matching) systems, with excellent ECCM capabilities. The passive infra-red imaging homing head is activated during the final approach to the target, and uses its library recognition algorithms to compare the actual scene with the memorised scene, identify the designated target and select the proper impact point for the BROACH warhead.

Regarding the HAF's air-to-air weaponry, a contract was signed on 21 December 2001 for the purchase of 350 (plus 100 as an option) IRIS-T short-range AAMs, to equip the F-16s and the upgraded PI 2000 F-4Es. IRIS-T is managed by a six-nation consortium formed in 1998 by Canada, Germany, Greece, Italy, Norway and Sweden, and headed by Germany's BGT. The missile is designed to operate from all aspects, following pilot line-of-sight or cueing signals from target acquisition systems such as radar, infra-red search and track, or threat warning devices. A unique feature is its ability to deploy in 'lock-on-after-launch' mode (LOAL), common in medium-range, air-to-air and surface-to-air missiles.

Exceptional manoeuvrability enables the pilot to shoot at targets flying behind, or in very tight manoeuvres, where the aircraft cannot acquire the target. The flight control system, integrating rocket thrust vectoring and aerodynamic control surfaces, facilitates extreme directional and attitude changes, even at low velocities and high angles of attack. Recent tests have demonstrated the missile's controls handling loops under severe conditions up to its *g* and angle of attack limits.

IRIS-T is equipped with an advanced seeker, which consists of a 128 x 128-scanning array, with advanced ECCM. The seeker is equipped with a target library that includes all known military aircraft, each viewed from different angles. Each target has eight focus points, the most vulnerable of which is determined by the 'smart' proximity fuse, in order to achieve the most lethal hit. In addition, another 100 AIM-120C5 AMRAAM missiles were ordered from Raytheon, in a $41.6 million deal. Deliveries will be completed in August 2003.

Meanwhile, the HAF is offering for sale its entire force of 24 Mirage F1CG fighters, together with 35 ATAR 09K50 engines, ground support equipment and spare parts. Also for sale are two NAMC YS-11A transport aircraft, together with five Rolls-Royce Dart Mk 542-10K engines and spare parts.

Plans for HAI

A scheduled HAI (Hellenic Aerospace Industry) $300 million privatisation package is part of the Greek government's widespread public sector divestment and it is – albeit unofficially – linked to Greece's imminent order for new fighter aircraft. Central to the outcome of the HAI deal will be the role of the restructured company in tandem with its new business partner. HAI will be closely involved in the offset arrangements linked to the fighter purchase, but the exact nature of its involvement is uncertain at this point.

Of major importance and closely linked with HAI's future is the re-location of 202 KEA (Kratiko Ergostasio Aeroskafon – National Aircraft Industry), which until now operated at Hellenikon airport. The latter's closure, in favour of the new 'Eleftherios Venizelos' airport at Spata, implies the transfer of KEA to a new location. The 202 KEA was established in 1925 and today performs major overhauls to a

Above: 358 Mira is the HAF's main helicopter unit. AB 205s serve on SAR duties, detached around the nation to provide local SAR dets at the main bases.

358 Mira's helicopter fleet includes the AS 332C1 (above), which is operated on overwater SAR duties on behalf of the Ministry of Merchant Marine, and the AB 212 (below) which is used for VIP and staff transport.

Above: 359 Mira – the MAEDY – is an unusual squadron operating a mix of agricultural aircraft on a variety of civilian tasks. Crop-spraying and seed-planting is performed on behalf of the Ministry of Agriculture. This is one of the unit's M-18A Dromaders.

359 Mira also flies Grumman AgCats (above) and Bell 47Gs (right) alongside the Dromaders. As well as agricultural work, the fleet is active on fire-fighting missions during the summer.

number of military aircraft types, including F/RF-4Es, CL-215s, C-47s and AB 205/212s. Together with HAI it is of strategic importance to the HAF, and it is planned to move to 112 PM, Elefsina, in new facilities.

The future of the Hellenic Air Force will undoubtedly be influenced by a number of factors, both political and economic, but few would question the vital strategic importance of Greece to the Western Alliance, nor the value of Greek fighting power on NATO's southern flank. Now a full member of the EU, Greece is becoming a major player in the Balkans and the European continent in general, and it remains to be seen if the winner of the 2004 general elections continues this dynamic policy.

Indicative of the current situation was the 24 May 2002 meeting of the Greek Governmental Committee for Foreign Affairs and Defence (KYSEA), which decided to extend the current five-year procurement plan (EMPAE 2001-2005), worth Dr 3.95 trillion ($10.16 billion), until 2008. Although there will be no programme cuts, most new programmes will not be initiated until after 2005. It is certain that the next decade will prove decisive in the evolution of the Hellenic armed forces, and the HAF in particular, with continuous modernisation and re-organisation being key elements in the evolution of the Elliniki Polemiki Aeroporia at the dawn of the 21st century.

Elliniki Polemiki Aeroporia

Taktiki Aeroporiki Dynami – Tactical Air Force

110 PM, Larissa	
337 Mira 'Fantasma'	F-4E Phantom II
346 Mira 'Iason'	F-16C/D Block 30
348 MTA 'Matia'	RF-4E Phantom II
ATA Flight	C-130H, C-47B
111 PM, Nea Anchialos	
330 Mira 'Keravnos'	F-16C/D Block 30
341 Mira 'Velos'	F-16C/D Block 50
347 Mira 'Perseas'	F-16C/D Block 50
Det 358 MED	AB 205
114 PM, Tanagra	
331 Mira 'Aigeas'	Mirage 2000EGM/BGM
332 Mira 'Geraki'	Mirage 2000EGM/BGM
342 Mira 'Sparti'	Mirage F1CG
Det 358 MED	AB 205
115 PM, Souda	
345 MV 'Lailapa'	A-7H, TA-7H
Det 358 MED	AB 205
116 PM, Araxos	
335 MV 'Tigris'	A-7E/H, TA-7C
336 MV 'Olympos'	A-7E/H, TA-7C
Det 358 MED	AB 205
117 PM, Andravida	
338 Mira 'Aris'	F-4E, F-4E PI 2000
339 Mira 'Aias'	F-4E, F-4E PI 2000
SOT (Scholio Oplon-Taktikis)	F/RF-4E, F-16C/D, F.1, A-7E/H, M2000, T-2E, UH-1H, AH-64A, P-3B

Diikisi Aeroporikis Ipostirixis – Air Support Command

112 PM, Elefsina	
353 MNAS 'Triaina'	P-3B Orion
355 MTM 'Ifaistos'	CL-215, CL-415, Do 28
356 MTM 'Iraklis'	C-130B/H, YS-11A, ERJ-135
356 MTM/1 Sminos 'Atlas'	C-47 *(based at Thessaloniki)*
358 MED 'Faethon'	AS 332C1, AB 205, AB 212
380 Mira ASEPE 'Dias'	Saab 340H Argus *(on loan)*

Diikisi Aeroporikis Ekpaidefseos – Air Training Command

120 PEA, Kalamata	
361 MVE 'Mistras'	T-37B/C, T-6A Texan II
362 MPE 'Nestor'	T-2E
363 MEE 'Danaos'	T-2E
Det 358 MED	AB 205
Dekelia	
360 MAE	T-41D, T-6A Texan II, Grob 103
359 MAEDY	AgCat, M-18A Dromader, Bell 47

Diikisi Elikopteron Naftikou (Navy Helicopter Command)

Tasked with the defence of Greek territorial waters and the hundreds of islands in the Aigaio Pelagos (Aegean archipelago), the Hellenic Navy regained an independent air arm in March 1976, following on from the July 1975 establishment of the Diikisi Elikopteron Naftikou (Navy Helicopter Command). Dictated by the latest developments in naval warfare, it became clear that the aerial assets of the navy would best be controlled and operated by a separate command.

First helicopters for the service were four Aérospatiale SA 319B Alouette IIIs (serialled PN-01 to PN-04), which arrived at Athens-Hellenikon airport and then moved to Amfiali heliport (near Piraeus, Athens) in May 1977, to form the Alouette Sminos (Alouette Flight). The SA 319B was primarily tasked with ASW and was suitably equipped with a nose-mounted Omera ORB 31 search radar and a towed MAD (Magnetic Anomaly Detector) system, fitted to one of the pylons. The 600-shp (447.6-kW) Turboméca Astazou engine was barely adequate for overwater operations in the Aegean, but at least provided enough power to allow carriage of two Mk 44 torpedoes for anti-ship missions. PN-01 and PN-02 were written-off in accidents in 1991 and 1994, respectively, leaving only PN-03 and PN-04 operational today, now used only for training since the arrival of the S-70B-6.

Sikorsky's dedicated S-70 naval variant is called the Aegean Hawk and is a LAMPS (Light Airborne Multi-Purpose System) Mk III, based on the US Navy's SH-60B/F version, from which it differs in a number of systems, the most important being the ASW suite. The US-versions use a MAD, while the S-70B-6 is equipped with a Bendix AN/ASQ-18(V)3 active dipping sonar. The Eaton AN/APS-143(V)3 surveillance radar is compatible with Kongsberg Penguin Mk 2 Mod 7 anti-ship missiles. Aegean Hawks can also carry Mk 46 Mod 5 torpedoes and are capable of Anti-Submarine Warfare (ASW), Anti-Surface Unit Warfare (ASUW), Search and Rescue (SAR) and Medical Evacuation (MEDEVAC) missions. The helicopter has a crew of three, comprising the pilot, the Airborne Tactical Officer (ATO) and the sensor operator. Sea deployments are undertaken from MEKO 200-class frigates.

Signed on August 1992. the initial contract for the first five S-70B-6s was worth $161.5 million. Aegean Hawks PN-51, PN-52 and PN-53 were the first to arrive at Amfiali by sea on 18 March 1995. They were reassembled (rotors and sub-systems) and flown to Kotroni Heliport, near Marathon, which replaced Amfiali Heliport as the primary naval air arm base in 1985, and formed S-70 Sminos (S-70 Flight). The remaining two followed in February 1996. At the same time, an additional order for two more was placed, with the first arriving in September 1997. On 12 June 2000 the Greek government signed a contract to purchase two more Sikorsky S-70B-6 Aegean Hawks, with an option to buy yet two further aircraft. The contract also calls for FLIR upgrades to the eight S-70B-6s already in service.

Before the arrival of the Aegean Hawks, the Agusta-Bell AB 212 was the primary ASW platform in Hellenic Navy service. Twelve of these aircraft were delivered between July 1979 and 21 June 1984, the batch including two AB 212 EWs, suitably equipped for electronic warfare. These aircraft formed the AB 212 Sminos (AB 212 Flight). Serialled PN-21 to PN-30, the Greek AB 212 ASWs are fitted with AN/APN-208(V)2 Doppler search radar and a Bendix AN/AQS-13B dipping sonar. Ten remain operational today, since the loss of PN-27 in 1982 and PN-21 on 31 January 1996. Known affectionately as the 'Merlin' among navy crews, PN-21 was part of the naval force deployed to Imia Island, the most recent subject of serious tension between Greece and Turkey. Following a bad weather night recon mission over the island, the aircraft crashed in the Aegean, a few kilometres away from the Greek fleet, killing all three on board. Bullet-like holes discovered in the tailboom after the recovery of the wreckage proved to be the result of rivets which, due to friction, had snapped through the thin Duralumin skin.

DEN training

Pending the arrival of the sophisticated S-70B-6, the DEN training programme was modified accordingly. Previously all DEN pilots had been trained on helicopters at the army's Scholi Aeroporias Stratou (SAS) at Stefanovikio. In 1992, however, the SEN (Scholi Elikopteron Naftikou – Naval Helicopter School) was established, as part of DEN's heliport at Kotroni, Marathonas, to qualify pilots for all three types flown by the navy. Alouette III pilots are subjected to two months ground school and about 90 flight hours, while AB 212 crews undertake three months on theoretical courses and about 160 flight hours.

Training for the S-70B-6 is more thorough. To assure a smooth transition, the navy prefers to assign experienced crews to the advanced Aegean Hawk fleet, although novice pilots are not ruled out. Candidates begin with two months of ground school and about 50 flight hours on the Alouette III, going solo after the 15th hour. If successful, they proceed to fly the AB 212 for another 60 hours, combined with three months of ground school. Then they are qualified as operational pilots in the S-70B-6 fleet. Also trained by SEN are the systems operators for the P-3B Orions, which operate from 112 PM, Elefsina, with air force pilots. This has been standard practice since 1969, when the HU-16 Albatross was introduced into service. The navy hopes to both fly and operate the Orions in the near future, although this seems unlikely under the current operational doctrine.

The Greek Coast Guard is also undergoing a major re-organisation, which includes the expansion of its airborne assets. A first step was made with the recent addition of four Reims F406 FLIR-equipped aircraft, which operate alongside the two older SOCATA TB-20s and a pair of Cessna 172RGs from Tatoi-Dekelia.

Diikisi Elikopteron Naftikou

DEN, Kotroni, Marathonas	
Sminos Alouette III	Alouette III
Sminos AB 212	AB 212 ASW, AB 212 EW
Sminos Aegean Hawk	S-70B-6 Aegean Hawk
SEN	all types

Elliniki Aeroporia Stratou (Hellenic Army Aviation)

The origins of the EAS go back to 1950, when the first Aerial Observation Unit (M.EPAR – Monada Enaerias Paratirisis) was established at Megara with eight Piper L-18B Super Cubs, to assist the aiming of Greek artillery by identifying targets. The first Greek Army officer completed EPAR training in the United Kingdom in 1949 and 11 years later (1960) the M.EPAR transformed into the Elliniki Aeroporia Stratou (Hellenic Army aviation), which was organised on a US Army aviation model, but with limited capabilities. Initial rotary-wing equipment comprised 88 Bell OH-13H/S Sioux, delivered during 1961-69, and a few Agusta-Bell AB 204s. The latter were soon replaced by later and more capable 'Hueys', with over 50 of the UH-1H version and over 70 of the similarly configured Agusta-Bell AB 205 entering service. Fifteen AB 206A/B Jetrangers were also received, arriving in 1976. 'Hueys' were initially painted in a three-tone camouflage pattern, consisting of brown/grey/green upper surfaces and light grey lower surfaces, but since 1992 the fleet has been repainted in the standard low IR-signature Olive Drab. Additional 'Hueys' arrived in July 1993, in the form of 30 refurbished ex-US Army UH-1Hs, while a 1996 deal for the purchase of 100 more, from surplus US-stocks, was abandoned due to fatigue-related concerns. The 30 ex-US Army UH-1Hs differ from the original batch aircraft in having wire cutters installed, and three-digit tail codes beginning with '8' instead of '6', as is the case with the older machines. Most EAS 'Hueys' feature NVG-compatible cockpits and can be equipped with M60D machine-guns in the doors for self-protection. Today about 100 remain in EAS service, while a pair of medevac aircraft was recently donated to FYROM.

Contrary to some recent reports, which referred to the EAS Jetranger fleet as being withdrawn from use, about 10 of the 15 AB 206A/B JetRangers delivered in 1976 remain operational in the liaison and observation roles. A few are also used by the SAS (Army Aviation School) as operational conversion trainers. A single, reconfigured Agusta-Bell AB 212 undertakes VIP flights, painted in a HAF-style blue/white finish, along with two UH-1Hs and two AB 205s. Both of the latter retain their 'military' Olive Drab colour but have their cabins converted for their new tasks, with five and six-seat arrangements, respectively.

Fixed-wing EAS assets comprise about a dozen surviving Cessna U-17A/B Skywagons, of the 50 delivered in 1965 for liaison and observation duties, and a single Raytheon (Beech) C-12C Huron. The Skywagons are the favourite fixed-wing trainers as well, while the

Above and above right: The Agusta-Bell AB 212ASW has served the Hellenic Navy well since its introduction in 1979, flying principally on shipborne ASW/surveillance duties. Eight remain in service. Two AB 212EW aircraft are outfitted for electronic warfare, retaining their roof-mounted radomes but with a plethora of antennas added, especially around the nose.

Below: Taking over from the AB 212 as the main shipborne helicopter is the S-70B-6 Aegean Hawk, essentially an SH-60B/F hybrid with dunking sonar instead of MAD. Weapons capability is provided by Penguin anti-ship missiles and Mk 46 torpedoes. Shore-based at the main DEN base at Kotroni, the Aegean Hawk fleet also deploys aboard the navy's MEKO 200-class frigates, providing ASW, ASUW and SAR coverage.

Above and below: Only two of the four SA 319B Alouette IIIs remain in DEN service, now relegated to training duties at Kotroni. One aircraft (above) is fitted with a towed MAD system, while the other (below) has Omera ORB 31 search radar fitted in the nose.

Above: Considerably enhancing the Coast Guard's patrol capability was the delivery of four Reims-Cessna F406 Caravan IIs, equipped with FLIR turret under the nose, search radar under the belly and SLAR under the rear fuselage for pollution surveillance.

Above: The Hellenic Coast Guard operates from Dekelia with a small but growing fleet. Its main tasks are coastal patrol and monitoring shipping. The four lightplanes in use comprise two Cessna 172RGs and two SOCATA TB-20 Trinidads (illustrated).

Right: For many years Greece's army aviation remained a small force providing limited mobility to the army. With the arrival of the Chinook and Apache, the EAS is now a major force in southern Europe, with considerable firepower and heavylift capability at its disposal. More attack helicopters are required, for which the Eurocopter Tiger and Longbow Apache are front-runners.

Huron, assigned to the 4th TEAS, only undertakes VIP duties. Entering EAS service in 1977, the C-12C is a military variant of the Beech 200, with a luxurious cabin arrangement for eight VIPs. In the summer of 1998 the aircraft underwent a major upgrade, with the replacement of the Pratt & Whitney PT6A-38 engines (750-shp/559.5-kW each) with the PT6A-41 version offering an additional 100 hp (74.6 kW), and the reconditioning of the interior and the exterior with a smart blue-white scheme. Equipped with a Rockwell-Collins WXR-270 weather radar, the aircraft undertakes international flights, operating mainly in the Balkan area, although it has flown to Rome and Paris.

To complement the strength of 4th TEAS two more C-12R/APs are expected as part of a $11.4m contract. The new aircraft will feature aerial photography capability, and an easy-to-install mission equipment package. A single Aero Commander 500 and a pair of Aero Commander 680FLs, which operated mainly in the agricultural photo-survey role, with secondary liaison and light transport roles, were officially withdrawn from service in 1999 after almost 30 years of service, although they had practically been inactive since 1997.

EAS organisation

The Elliniki Aeroporia Stratou (Hellenic Army Aviation) has undergone – like the HAF – a major modernisation in the last decade, the most significant being the introduction of the AH-64A+ Apache into service. This implied a major reorganisation of the entire Army Command and the creation of the Directorate of Army Aviation, as well as the establishment of a Rapid Reaction Force and a separate Airborne Brigade. The latter, the 1st TAXAS (TAXiarhia Aeroporias Stratou) was established on 2 January 1998 to command and operate all of the Army's air assets.

Today the Hellenic Army is divided into five front-line Tagmata (battalions) and the always busy SAS (Scholi Aeroporias Stratou – Army Aviation School). All Tagmata are under 1st TAXAS command, and each TEAS (Tagma Elikopteron Aeroporias Stratou – Army Aviation Helicopter Battalion) includes two helicopter Lohous (companies) and a mixed helicopter/fixed-wing Loho. The 1st TEEP (Tagma Epithetikon Elikopteron – Attack Helicopter Battalion), which operates all of the Greek AH-64s from Stefanovikio, is a little different, consisting of two Apache Lohous and a UH-1H Loho, with no fixed-wing aircraft under its command.

Established in 1971 at Stefanovikio, near Volos, the 1st TEAS (Tagma Elikopteron Aeroporias Stratou – Army Aviation Helicopter Battalion) acted as the core for the further development of the Aeroporia Stratou. The 2nd TEAS based at Pahi, Megara, followed in 1977. Next was the 3rd TEAS, which established in 1982 at Alexandria Imathias, near Veria, with the last – and newest – 4th TEAS forming in December 1997, also at Pahi, Megara.

EAS training

EAS training is the responsibility of the SAS (Scholi Aeroporias Stratou – Army Aviation School), which is housed in its own facilities at Stefanovikio, alongside the 1st TEAS and 1st TEEP. For basic rotary-wing training the SAS operates 25 surviving Nardi-Hughes NH-300Cs, of the 30 originally delivered in 1969, along with a number of UH-1Hs, AB 205As, AB 206Bs and U-17A/Bs. Since 1998, SAS has trained graduates of the army schools assigned exclusively to Army Aviation: previously candidates were selected from various departments.

Initial training lasts 40 weeks and includes 360 hours of theoretical lessons. The flight syllabus commences with 35 hours on the NH-300C, with the student going solo after the 15th hour. If successful, they continue to the basic training phase and fly the 'Huey'. Unsuccessful candidates continue their army career in the infantry. The UH-1H/AB 205 stage begins with instruction in the simulator and includes 40 hours of basic flying, followed by another 33 hours of IFR (Instrument Flight Rules) flying and 22 hours of tactical training. Fixed-wing pilots learn to fly the U-17A/B in a 150-flying hour course.

SAS also runs a post-graduate class for helicopter flight instructors (with 26 flight hours on both the NH-300C and UH-1H/AB 205A) who wish to convert from rotary-wing to fixed-wing aircraft (56 flight hours on the U-17A/B), or to the AB 206 Jetranger (10 flight hours on the type). Until 1992 the army was also responsible for training naval helicopter pilots, when the navy established its own SEN.

Apache and Chinook

In 1991 the Ellinikos Stratos (Hellenic Army) became the first European export customer for the Boeing (McDonnell Douglas) AH-64 Apache. An LOA (Letter of Acceptance) was signed on 24 December between the Hellenic Ministry of Defence and the US Army, for the purchase of 12 Apaches, with an option for eight (which could be increased by another four). Deliveries began by sea in June 1995 and included six aircraft, with the remaining 14 arriving in December that year, as part of a programme costing $400 million. The 20 AH-64A Apaches were delivered to the 1st TEEP (Tagma Epithetikon Elikopteron – Attack Helicopter Regiment), which had been established a few days earlier (1 June 1995) at Stefanovikio. A follow-on contract for 24 aircraft was expected to be signed during 1996-97, but this was stalled when the Hellenic Army decided to wait for the advanced AH-64D Longbow version. Instead, four ex-US Army AH-64s were procured in 2000, to complete the planned 24-aircraft inventory of the 1st TEEP.

Introduction of the Apache into Greek service had a major impact on EAS operational doctrine, since until then the Army operated mainly in the troop-carrying role, with no attack capabilities. Despite the lack of experience in such roles and the logistic support and maintenance concerns, the Greek pilots adapted very well to the new weapon and its tactical use. Initially, six army pilots were sent to Fort Rucker, Alabama, on a three-month training course. Of these, four went on to graduate as instructors and another two as test pilots. These Apache pilots were responsible for training others at Stefanovikio, despite the lack of a simulator. This necessitates sending all EAS Apache pilots to the US Army in Germany, where virtual training in emergency procedures takes place. Today all 24 Apaches are operated by three companies under 1st TEEP, alongside five UH-1Hs, which support their operations.

Greece's AH-64A fleet was delivered with full-colour national insignia, in contrast to the low-visibility markings that adorn the Apaches of the other users around the world. To counter this, each Greek AH-64A is equipped with a can of spray paint of the appropriate Olive Drab colour. The crews are ordered to use it in the event of hostilities, to cover the fuselage and tail blue/white insignia.

To carry out their primary anti-tank mission, the Greek Apaches are armed with various versions of the Hellfire II missile (439 of the AGM-114F version, 88 of the AGM-114K1 version and 160 of the AGM-114KBF version). The anti-ship AGM-114L is being considered for the anticipated AH-64D Longbow. Also in use are 2.75-in (70-mm) Hydra rockets, and their associated 19-round M261 pods. Maintenance work on the Apache (and all other EAS aircraft), is conducted by 307 Tehniko Sigrotima Ypostiriksis Aeroporikou Ylikou (TSYAY – Aviation Technical Support Unit), also based at Stefanovikio.

For heavy-lift missions and paradropping, the EAS relies on the Boeing CH-47D Chinook. Four CH-47Cs were purchased in 1981 from Agusta's Elicotteri Meridionali division. They were augmented by five C models transferred from the Polemiki Aeroporia in 1988. Seven more CH-47Ds were ordered on 20 October 1999, under a Foreign Military Sale (FMS) programme in excess of $200 million, thus expanding the EAS Chinook fleet to 16.

In 1995 Boeing upgraded the nine -Cs to CH-47DG standard as part of a programme costing more than $151 million. The first two CH-47Cs were sent to Boeing in February 1993, and were returned as CH-47DGs in December of the same year: deliveries of the rest were completed in 1995. The upgrade consisted of digital FADEC (Full Authority Digital Engine Control) and uprated Textron Lycoming T55-L-712 engines, NVG-capable 'glass' cockpit, radar, INS/GPS, composite rotor blades, bulged 'saddle' tanks with enlarged fuel capacity, three cargo hooks instead of one, and new avionics. The latter includes an effective self-protection suite with Sanders AN/ALQ-156(V)2 missile warning systems, APR-39(V)2 RWRs and ALE-47 chaff/flare dispensers. M60D and M24 machine-guns can also be carried, providing some protection and suppressive fire for assault troops.

Besides their military tasks, the EAS CH-47Ds provide civil relief services, including fire-fighting. All are painted in low-IR Olive Drab colour today, in place of the 'Asia Minor' camouflage that adorned the early CH-47Cs. Army pilots are required to have flown at least 100 hours in the left-hand seat of the 'Huey' (and 600 hours in total) before they proceed to the left-hand seat of a CH-47D. Operational conversion lasts about 16 weeks and takes place with the 4th TEAS at Pahi, Megara, where all EAS Chinooks are based.

Future plans

With the 'Huey' celebrating its fourth decade in Greek service, the rapidly changing NATO operational doctrine and the threat posed by Turkey, the Elliniki Aeroporia Stratou must continue its modernisation well into the next decade. Of utmost priority is the replacement of the 'Huey' fleet with a new type of utility helicopter, while the establishment of a second (and then a third) TEEP, to effectively cover the

Above: Whether built by Agusta or Bell, the AB 205/ UH-1H is the most numerous type in EAS service, flying with five assault regiments. The 'Huey' is overdue for replacement, with several types currently under evaluation.

Left: Today the AH-64As serve with the 1st TEEP, with UH-1Hs operating in support. The creation of at least one further attack helicopter regiment is planned.

Above: This EAS AB 206, which wore these 'Police' titles for a while in the 1990s, is unusual in having tall undercarriage skids with flotation equipment.

Right: As well as routine military tasks, the 4th TEAS CH-47D fleet is also made available to civilian agencies. The fleet sports enlarged 'saddle' tanks.

A few AB 206 JetRangers (above) remain in EAS service for liaison and training tasks, as do a number of Cessna U-17As (right). The EAS is one of the last army air arms to retain a fixed-wing observation capability.

Thrace and the Aegean regions, as well as subsequent Rapid Reaction Force commitments, is obvious. Current plans call for the procurement of at least 66 new utility helicopters, to be chosen from the Eurocopter AS 532 Cougar Mk 2, Sikorsky UH-60 Black Hawk and Mil/Kazan Mi-17, which underwent thorough evaluation at Stefanovikio in 2000.

Interest has also been expressed in four to six helicopters configured for dedicated Combat Search and Rescue (CSAR) missions, following the HAF's decision to acquire six Eurocopter AS 532A2 Cougar Mk 2s. In addition, four to six dedicated medevac helicopters are sought, a role currently carried out by a few suitably configured 'Hueys'. The planned establishment of a 2nd TEEP will bring more attack helicopters into EAS service. For some time during 1997 Denel's Rooivalk attack helicopter was considered to equip the 2nd TEEP, following a formal collaboration agreement between the Greek and the South African governments, but this seems highly unlikely today. Also evaluated were the Bell AH-1W Super Cobra, the Agusta A109 and the Eurocopter Tiger. Together with the AH-64D Longbow Apache, the Tiger was deemed favourite, but no acquisition plans have been disclosed yet.

Army anti-aircraft protection is provided by a number of Improved Hawk batteries, plus 12 OSA-AK (known in NATO as the SA-8 'Gecko'), 25 TOR-M1 (NATO code name SA-15 'Gauntlet') systems and over 500 ZU-23 23-mm guns. The latter, along with the OSA-AKs, were obtained from former East German stocks, while the TOR-M1s were bought new in 1999, following an army request, to fulfil the service's SHORAD requirement. Also on order are 54 German Atlas Short-Range Air-Defence (ASRAD) systems.

Elliniki Aeroporia Stratou

SAS, Stefanovikio	
1 Lohos Elikopteron	NH-300C
2 Lohos Elikopteron	UH-1H
3 Lohos Aeroskafon/Elik.	U-17A/B
4 Lohos Aeroskafon/Elik.	Aero Commander 680FL (wfu)
1st TEAS, Stefanovikio	
1 Lohos Elikopteron	AB 205, UH-1H
2 Lohos Elikopteron	AB 205, UH-1H
3 Lohos Aeroskafon/Elik.	U-17A/B
1st TEEP, Stefanovikio	AH-64A Apache, UH-1H
2nd TEAS, Megara	
1 Lohos Elikopteron	AB 205, UH-1H
2 Lohos Elikopteron	AB 205, UH-1H
3 Lohos Aeroskafon/Elik.	AB 206A, U-17A/B
3rd TEAS, Imathia	
1 Lohos Aeroskafon/Elik.	UH-1H, AB 205, AB 206A, U-17A/B
4th TEAS, Megara	
1 Lohos Elikopteron	CH-47D
2 Lohos Elikopteron	CH-47D
3 Lohos Aeroskafon/Elik.	AB 212, C-12C, C-12R/AP

Tupolev Tu-95/142 Russia's Mighty 'Bear'

Above: Blessed with very long range and economic operation, the Tu-95 was a natural choice for a maritime patrol aircraft. Successful experience with the Tu-95RTs 'Bear-D' in the oceanic surveillance and targeting role led to the development of the Tu-142 'Bear-F', dedicated to hunting submarines. A series of new versions has attempted to keep the Tu-142 abreast of Western sub developments, and central to the AVMF's war plans.

Main picture: With four of the most powerful turboprops ever put into production driving 32 giant propeller blades, the Tu-95 remains an awe-inspiring spectacle a half-century after the type's first flight. Yet with a bellyful of missiles – plus others in clusters under its trademark swept wings – the Tu-95MS 'Bear-H' is also still a formidable weapon. In the 21st century it is belatedly following its B-52 counterpart into the conventional missile arena.

Comparisons between the 'Bear' and the B-52 are inevitable: both were conceived as their respective nations' principal vehicles of strategic nuclear deterrence in the pre-missile era and, despite the development of more survivable aircraft for the penetration mission, both were to regain a major strategic role as the natural platforms for the air-launching of cruise missiles. Whereas the B-52 retained and expanded its conventional bombing capabilities, the 'Bear' branched out into the reconnaissance arena, in turn spawning an important family of anti-submarine warfare platforms. In both missile platform and ASW roles, the Tu-95/142 remains an essential fixture in today's Russian air forces.

At first glance the Tu-142 looked very similar to the Tu-95RTs upon which it was based, especially if fitted with the undernose fairing for sensor tests. However, there were some notable changes, the most obvious being the 12-wheel main undercarriage. More importantly, the wing was of revised section and slightly enlarged, with a new flap arrangement. The earlier bag tanks were abandoned in favour of integral tanks in the wing box. The second Tu-142 introduced an extended forward fuselage, a modification required to house the large mission crew and its equipment.

Production of the Tu-95RTs 'Bear-D' reached 52, plus one converted Tu-95M, and it became the most visible version on 'snooping' missions around NATO nations and their navies. The variant was the first with a dedicated maritime mission, and naturally provided the basis for development of the Tu-142 'Bear-F'. With its primary mission of over-the-horizon targeting no longer relevant, the Tu-95RTs was retired from service in the mid-1990s. Here an example is escorted by a Tornado F.Mk 3 from the RAF's No. 5 Squadron.

Tu-142 - sub-hunting 'Bear'

In the early 1960s the Soviet navy (Voyenno-morskoy flot – VMF) perceived a need for a long-range anti-submarine aircraft, in order to counter the threat posed by the West's growing fleet of nuclear strike-capable submarines. Initial requirements stood at 160 aircraft to equip two navy fleets, although this was later revised to 114, and finally to 100.

On 28 February 1963 the Soviet government released a requirement covering the improvement of ASW forces and of the means of detecting and destroying high-speed submarines. One of the items on the list was the creation of an anti-submarine aircraft based on the Tu-95 bomber. Known as the Tu-95PLO (*protivolodochnaya oborona*, anti-submarine warfare), the aircraft was to be presented for joint tests in 1964. A prototype of the Berkut-95 system was prepared for tests in the Il-38 'May', and this was expected to be installed in the Tu-95PLO. The Berkut system relied on a mix of human and automated control. Decision-making in navigational and tactical matters was largely left to the crew, in accordance with the conditions, while the very primitive computer software could be used when deemed useful.

In September/October 1963 the outline plan for what had been redesignated Tu-142 was submitted. In answer to a VVS stipulation, the Tu-142 should be able to operate from unpaved strips able to support 6 kg/cm² (85.34 lb/sq in). In November of the same year it was revealed to a model committee that the ASW variant would be based on the Tu-95RTs 'Bear-D' maritime reconnaissance and target-illuminating version.

To meet the unpaved field requirements, the designers developed a complex and heavy landing gear. Twelve aircraft from the first series had each of their main legs equipped with 12 low-pressure wheels mounted in pairs on three axles. The nose leg had two wheels with a turn angle of 45° and the tail section had a two-wheel retractable bumper. Retraction and extension of the main undercarriage and tail support was electro-mechanical, while the forward leg was hydraulically activated. Emergency release of the landing gear was provided. In order to accommodate a main gear of this kind, it was necessary to increase the size of the aft portion of the inboard engine nacelles, which housed the gear. This in turn led to a reduction in flap area.

Progress on Tu-142 development was delayed more than once. On 17 January 1968, under pressure from the Soviet navy and air forces, the Military-Industrial Commission (VPK) noted that "the prototype, constructed in 1967, and first production aircraft are still not ready for factory tests. The Berkut equipment is in a stage of setup and adjustment..." "The second production example is being modified with a lengthened cockpit, and the static tests of the airframe has not begun." The VPK decided that the Tu-142 had to be presented for production tests in 1968 (prototype in March, first production aircraft in May and the second production aircraft, with the lengthened forward section, in July). The Tu-142 fitted with the Berkut PPS (*poiskovo-pritselnaya systema*, search and targeting system) was to be presented for joint state tests in the second quarter of 1968.

First flight

The first Tu-142 aircraft were under construction at plant No. 18 at Kuibyshev, where main production of the type was planned. The first flight of the prototype (No. 4200), took place on 18 June 1968 from Kuibyshev's airfield, flown by a crew led by test pilot I. K. Vedernikov. The second aircraft from the factory (No. 4201), with a cockpit lengthened by 1.7 m (5 ft 7 in), took to the air on 3 September. The third aircraft (No. 4203), with the lengthened cockpit and a complete set of equipment, flew on 31 October. These aircraft carried out most of the tests required by the programme. Subsequently aircraft No. 4200 (and later No. 4243, the first Tu-142MK 'Bear-F Mod 3' which had been seen at Zhukovskiy for years) were used as flying laboratories and testbeds for the engines intended for the Tu-22M3 'Backfire-C' bomber and Tu-144 'Charger' supersonic airliner.

By September of 1970 factory tests were completed. These revealed that the aircraft did not meet its set specifications, particularly in range. At a 182-tonne (179.1-ton) take-off weight the range was 9860 km (6,127 miles). Early aircraft had 5.8-m (19-ft) diameter AV-60P series 01 propellers. In order to improve range, they were replaced by 5.6-m (18-ft 4-in) diameter AV-60K series 01 propellers, which were more efficient in cruising mode.

Tu-142 No. 4231 was chosen to go through a new series of tests in order to prove that the aircraft could meet the requirements. It was stripped of various items such as the Gagara infra-red monitor, active countermeasures and radio-intelligence equipment. This reduced weight by 3685 kg (8,124 lb) and allowed the Tu-142 to meet the required numbers. General Designer Andrey Nikolayevich Tupolev appealed to the minister of defence with a letter, in which he expressed a disagreement with VVS requirements regarding the necessity to operate the heavy aircraft from unpaved runways. In his reply dated 6 October 1970, the minister recognised Tupolev's point. As a result, the triple-axle landing gear was replaced by a lighter, double-axle unit, taken from the Tu-114 'Cleat' airliner. Accordingly, the size of the main gear fairings and flaps were restored to those of the Tu-95RTs.

By government resolution dated 14 December 1972, the Tu-142 fitted with the Berkut-95 PPS was accepted into service with the Soviet navy. Basic performance characteristics were:

- 12300-km (7,643-mile) range at 182-tonne (179.1-ton) take-off weight with a 5500-kg (12,125-lb) combat load and 5 per cent reserves
- 4-hour 30-minute time on station at 4000 km (2,485 miles) from base
- 830 km/h (516-mph) maximum speed
- 11000-m (36,090-ft) ceiling
- 2350-m (7,710-ft) take-off run at 182 tonnes
- 1200-m (3,937-ft) landing run at 106 tonnes

By early 1973 18 Tu-142s had been completed at Kuibyshev (two in 1968, five each year in 1969, 1970 and 1971, and one in 1972). This number included 12 aircraft with the larger main undercarriage.

In the early 1970s, the Ministry of Aviation Industry (Ministerstvo Aviatsionnoy Promyshlennosti – MAP) decided to transfer Tu-142 manufacture to plant No. 86 at Taganrog. According to some documents the Taganrog plant could build six/seven Tu-142Ms per year in 1976-1980, and in subsequent years up to 15. A new runway had to be constructed at the plant's airfield. The manufacturing documentation was transferred to Taganrog along with a sample aircraft (No. 4242). Due to the time required to upgrade the airfield in 1975-1976, the plant rolled out 10 aircraft (Nos 4232 through 4241). All of them were made according to the sample aircraft, and featured the forward cabin lengthened by 2 m (6 ft 7 in), and main landing gear with four-wheel bogies.

ASW equipment

When an aircraft was equipped with the PPS, it became known as a 'complex', or weapon system. The ingredients of a complex included the aircraft, the PPS, ground resources and control facilities. The offensive loads were not officially part of the complex.

***Above:** A Tu-142MK sweeps over* Kuznetsov *in a display of Russian naval power. The MK serves in slightly greater numbers than the later Tu-142MZ, and remains an important asset, despite having less capable ASW systems although it was upgraded in the early 1990s. Much of the 'Bear-F' fleet is mothballed, and flying is severely restricted compared with Cold War times.*

***Far left:** For years a routine feature of any 'Bear' crew member's life was the constant attention from NATO aircraft when operating over international waters. US Navy carrier battle groups naturally drew the attentions of maritime patrol 'Bears', which in turn led to regular intercepts by US Navy aircraft. As well as fighters, supporting tankers regularly got in close for a look at the giant patrollers, as demonstrated by this KA-6D from* Forrestal*'s VA-85 'Black Falcons'.*

Known to NATO has 'Bear-F' (sometimes with the retrospective Mod 0 suffix), initial Tu-142 production (12 aircraft) featured massive fairings for the enlarged 12-wheel rough-field undercarriage. This is one of two aircraft which were used for various tests, including the fitment of various sensors in an undernose fairing.

For the Tu-142 the anti-submarine equipment initially comprised the Berkut-95 PPS and equipment for the carriage and release of submarine detection/destruction equipment. Onboard search equipment included the 15-GHz pitch-roll-stabilised radar (RLS); radar scope (SPIU); a TsVM-263 computer with communications coupling unit; geographical coordinates panel; and other devices.

The SPIU used 40 channels (38 used, two in reserve) to receive and distribute data sent by sonobuoys to the onboard computer. An ARK-UB compass unit was used in conjunction with the radar scope for homing the aircraft to the buoy. During flight in automatic mode, the TsVM-263 processed flight data and communicates heading corrections to the autopilot. When in a search mode, the computer – using data received from sonobuoys – calculated the position of the submarine and direction in which it was moving. Based on these results, the TsVM automatically adjusted the flight profile and tracked the target. On the basis of available data, the computer estimated the probability of a successful attack, and sent out signals for opening of the bomb bay and release of weapons.

Sonobuoys were the only source of the information about underwater conditions. There were three types of sonobuoys: RGB-1, RGB-2 and RGB-3. Each of them contained a hydroacoustic channel, data transmitter, beacon, parachute box and power supplies.

In standard form the Tu-142 lacked the undernose radome. The result was the 'cleanest' of the maritime 'Bears', its lines spoilt only by the Berkut-95 radar and the bulge for the enlarged nosewheels. This is a very early aircraft, built with the much extended nacelles for the rough-field landing gear. Operations with the new aircraft got under way in 1970, and the type soon came into contact with NATO interceptors over the North Sea and north Atlantic.

■ **RGB-1 – passive, non-directional:** the acoustic system descended to 35 or 70 m (115 or 230 ft) and received noise in the 6.25-7.75 kHz range. Basic modes of operation were a 'duty mode' with an automatic start-up, which activated the data transmitter if received signals exceeded a preset level, and a 'marker' mode, in which the buoy's beacon was constantly on, the acoustic channel being left in 'duty' mode. In 'duty' mode the buoy worked for up to three hours. During tests in the Black Sea, the range of detection of a diesel-powered submarine moving at 11-15 km/h (6-8 kt) was 1500-2100 m (1,640-2,297 yards). Range of communication between the aircraft and buoys was at least 40 km (25 miles). Each buoy was 1.2 m (3 ft 11 in) long, 150 mm (5.9 in) in diameter and weighed 14.2 kg (31.3 lb). A complete set contained 24 buoys.

■ **RGB-2 – passive, directional:** these were intended to measure and transmit magnetic bearing data on 'noisy' contacts. After entering water the RGB-2's transmitter worked continuously. The acoustic system descended to 20 m (66 ft) and rotated at 8 rpm, for an endurance of 45 minutes. Magnetic bearing definition was accurate to 3°. The RGB-2 measured 1.85 m (6 ft 1 in) in length, 230 mm (9 in) in diameter and 45 kg (99 lb) in weight. A complete set contained 10 buoys.

■ **RGB-3 – active:** after entering the water the RGB-3 worked in a 'noise-bearing finder' mode, but was switched to an 'echo-bearing finder' mode by command from the aircraft. Reflected acoustic signals from a submarine were transformed into electrical fluctuations received by the aircraft. In 'radiating' mode the RGB-3 worked for five minutes. Range of submarine detection was not less than 1500-2000 m (1,640-2,187 yards). Length of the RGB-3 was 2.84 m (9 ft 4 in), diameter was 392 mm (15.5 in) and weight was 185 kg (408 lb). A complete set contained four buoys.

Although never used in service, in search-only configuration the aircraft could carry a maximum of 440 RGB-1 sonobuoys. The optimal loadout – search and destroy – comprised 176 RGB-1s, 13 RGB-2s, four RGB-3s, three torpedoes (or APR-1 Condor anti-submarine missiles), 14 PLAB-250-120s (*protivolodochnaya avia bomba*, anti-submarine bomb – the first number means that the weapon has the dimensions of the FAB-250 high-explosive bomb, the second indicates the weight) and 14 marker bombs. These represented the weapons in use by Soviet naval aviation at the time of the Tu-142's development.

Underwater weapons

Torpedoes were considered to be the main weapon against enemy subs. The acoustic torpedo AT-1 became operational in 1962. The range of its automatic homing system was estimated at 300 m (328 yards), speed was 52 km/h (28 kt), depth of operation was up to 200 m (656 ft), weight was 560 kg (1,234 lb) and range was 4000 m (4,374 yards). With these specifications the torpedo did not satisfy the requirements for the Tu-142. Nevertheless, initially it was included in some mission loading configurations of the aircraft.

A more advanced weapon, the AT-2, entered service in 1973. After descending to its initial depth it started a programmed search, entering a left-hand cylindrical spiral 140 m (460 ft) in diameter, travelling at 43 km/h (23 kt). The programmed trajectory began at a depth of 40-45 m (130-148 ft). A change of spiral pitch on the first section of the trajectory occurred due to the automatic change in the trim of a torpedo from 11° to level. The homing system worked in an active/passive mode in cycles. For 35 per cent of the time it worked in active mode but for the remainder was passive.

This method of search provided observation across a range of depths. If the level of noise emitted by a potential target exceeded the reflected signal, the active-passive cycle was broken and the homing was controlled by the passive channel. In the case where the target was identified through a reflected signal, all further actions were based on commands from the active channel. If the target was lost, the equipment was switched back to the active/passive mode. Depth of operation for the AT-2 was 20-400 m (66-1,312 ft), diameter was 533 mm (21 in), weight was 1000 kg (2,205 lb) and speed was up to 74 km/h (40 kt).

Subsequently, the AT-2 was modernised to become the AT-2M. The pneumatic systems were replaced by hydraulic ones, the explosive warhead was increased in weight by 20 kg (44 lb) and the silver-zinc accumulators were replaced by new batteries. Production ended in 1978 after 975 torpedoes had been produced.

On 29 June 1971, by order of the defence minister, the APR-1 Condor anti-submarine missile became operational. The Condor's trajectory was established before launch: search in a straight line at a speed of 20 kt (37 km/h). On detection of the target the thrust of the engine was increased for six to seven seconds, boosting speed to 60 kt (111 km/h). The missile then performed a 900-m (984-yard) dash while homing on the target. The APR-1 was effective against targets submerged by up to 400 m (1,312 ft). By the time production ended in 1977, 263 missiles had been manufactured.

Taking into account the shortcomings of Berkut PPS, revealed in operations of the Il-38, the Tu-142 was equipped with the ANP-3V-1K automatic navigational device. It had two systems of coordinates: rectangular and polar. These two systems were essential for en route navigation and for manoeuvring in the target area. Among other things, the system allowed 'Bear-F' to fly a circular path of constant radius. Developers of the Berkut had divided the tasks to be handled by the system into three stages: flight in a given area, action in the tactical area and return to base.

Flight in a given area was carried out on one of several connecting great circles (lines of shortest distance between two points on a sphere). Navigational calculations were handled by the TsVM computer. When the aircraft deviated off course the computer issued a turn command, which was then passed to the autopilot, which then corrected the heading. At a distance of 90-130 km (56-81 miles) to the next waypoint on a route, the new data was entered.

On arrival in the search area, RGB-1 buoys were deployed. Buoys were sown either as a barrier or a field (to uniformly cover the area of search in calculated intervals and distances). In order to set a barrier, the TsVM was given the coordinates of its beginning and end, as well as desired interval between buoys. After that, the computer issued control commands for the aircraft, and calculated the points for opening the bay doors and dropping the buoys. Setting of a field could be also made in automatic mode.

After all the buoys had been deployed, the TsVM was tasked to supervise the buoys by flying the aircraft along and across the barrier. The TsVM issued appropriate commands to the aircraft to make the next turn. The navigator and the SPIU operator watched the display of the first channel to catch the moment a buoy indicated a response (the moment was also entered into the TsVM's memory) and to classify the contact. If the contact appeared authentic, the aircraft was directed to the active buoy under the directions of the ARK-UB radio compass.

After the decoder of the buoy's transmitter beacon was activated, the position of the buoy in relation to the aircraft was shown on the radar display. The crew then had an opportunity to choose one of several actions: to drop a 'confirming' RGB-2 to check the data, sow a barrier of buoys perpendicular to the probable heading of the target, or to drop an RGB-3 active buoy for accurate detection. In the case when the sub's heading was unknown, the crew commanded a 'Scope', in which RGB-1s were dropped encircling the area of detection, followed by the 'Flight around target' command which allowed subsequent supervision of the circular buoy pattern.

With an RGB-2 buoy in the water directional data could be received, from the SPIU to the TsVM. By overlapping the bearing marks with a mark of the target, the coordinates of the submarine and its speed could be calculated. If necessary, a 'confirming' RGB-3 was released. After processing of all data, the choice of weapons and the task of their preparation was done by the radar operator. According to test data, the time spent by a crew from initial detection of the submarine to weapon release ranged from 19 to 27 minutes.

Radar-tracking searches for surface targets were not complicated. Search of the area was performed using alternating port and starboard tacks. Surface attacks were made with the help of the radar, without taking into account the speed of the target's movement. The major shortcoming of the Berkut-95 was its inability to perform the task of tracking in an automatic mode.

Tu-142 in service

In May 1970, the first two Tu-142s were shipped from the manufacturer to the Northern Fleet's air branch, where the 76th Independent Long-Range Anti-submarine Warfare Air Regiment (Otdelniy Protivolodochniy Aviatsioniy Polk Dalnego Deystviya – OPLAP DD) had been undergoing

A revised flight deck layout with a raised roof was the main external feature of the 'Bear-F Mod 2'. The changes gave the pilots a much improved look forward. Confusingly, the Tupolev OKB and Taganrog plant referred to these aircraft as the Tu-142M. However, in service they retained the Tu-142 designation, as they were still fitted with the original Berkut-95 system. The fairings on the end of the tailplanes initially contained antennas for the Arfa and Kristall systems, but in later Tu-142s served the Lira long-range navigation system.

The 'Mod 2' became the standard for the first-generation Berkut-equipped aircraft. Delivery of newer Tu-142Ms to the Kipelovo unit in the early 1980s allowed the transfer of Tu-142s to the Pacific Fleet's 310th OPLAP DD, which had been established in 1978. As well as operations from its home base at Kamennyi Ruchei (near Vladivostok), the Pacific unit flew sorties from Cam Ranh Bay and Da Nang in Vietnam from 1979. From 1982 to the early 1990s this became a permanent detachment, with four aircraft assigned.

Berkut-equipped Tu-142s remained in service into the 1990s, although the fleet dwindled from 1988.

Below right: The stalky undercarriage and ample ground clearance under the Tu-95/142's fuselage made it a natural for use in the engine testbed role. Initially, the redundant second Tu-95 prototype was converted to Tu-95LL standard (illustrated), being used to test engines for the Tu-22, Tu-144 and Tu-22M. It was lost in an accident in 1973, and was replaced by the first Tu-142 prototype which, after conversion to Tu-142LL standard, went on to test engines for the Tu-22M3, Tu-144 and Tu-160. It was retired in the mid-1980s due to fatigue concerns.

Below and below far right: To replace the first Tu-142LL, a second testbed was converted from the prototype Tu-142MK. ASW equipment was removed and replaced by test equipment, while the refuelling probe and MAD boom were removed. In 1990 it set several time-to-height records for its weight class but has since fallen into disrepair at Zhukovskiy.

activation since June 1969 at Kipelovo naval air station. Special attention was paid to selecting aircrews. The positions of crew commanders were filled with pilots qualified as 'pilot 1st class' and experienced in flying heavy aircraft. Often, they were former air detachment commanders.

November 1971 saw the second squadron formed, with the regiment having received nine aircraft by then, and another one being delivered in 1972. Two others were operated by the Combat Training Centre, so by 1973 the AVMF had 12 Series I aircraft (Nos 4211 through 4222). When the third squadron formed, the regimental activation was complete.

Conversion training was easy enough since the fleets' air branches had seasoned instructors trained on the Tu-95, while difficulties – both objective and subjective – would normally crop up afterwards during the learning of the basics of combat. The conversion trainees were assisted by navigators and engineers from an Il-38 regiment garrisoned at the same naval air station, whose Il-38s were outfitted with Berkut systems.

Service Berkut-95 software programmes differed somewhat from those of the prototype. In particular, a conclusion was made that the kill probability against submarines in training torpedo releases could be enhanced by making the attack run perpendicular to the vessel's course. During their training, Tu-142 aircrews would use practice bombs against a land bombing range, with all data being entered into the surveillance and targeting system as if they were live ordnance. This allowed complicated anti-submarine torpedo employment to be practised without using the precious weapons.

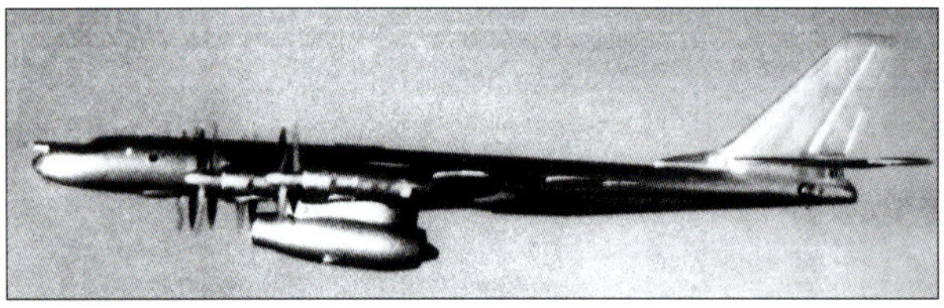

From time to time this led to amusing incidents. For instance, it occurred to a navigator that his aircraft was approaching the target from the wrong direction, so he 'corrected' the heading accordingly. The computer automatically acted on his initiative – and released the bombload on the barracks of the bombing range support team, completely shattering the armoury. Thankfully there were no casualties, as the buildings were empty. There were some problems with training aircrews since the number of buoys supplied for Naval Aviation's combat training was limited to 30-40 annually. The bulk of the buoys supplied would be used to resupply combat buoy kits.

Tracking submarines

Under the combat training schedule, special attention was paid to practising submarine-tracking techniques. In an hour-long tracking session around 35-65 RGB-1 buoys were expended, although seasoned aircrews would expend 30-40 buoys in an hour, achieved by a number of techniques. Thus, sometimes the surveillance and targeting systems were not used. Instead, a more speedy approach was at times resorted to. With a short time delay, buoys would be deployed as a triangle or a rectangle, dubbed '*semi-tavr*' and '*tavr*', to pinpoint the contact. The flight profile was maintained via the ANP-3V-1K. Following that, the ANP-3V-1K would be used to set linear or arch surveillance barriers. Radars proved to be ineffective in setting up surveillance barriers due to the wrong imagery scale on the radar scope.

On training missions and during actual tracking of unidentified submarines, tension would greatly increase between crew members, and the careful selection of the crew played an important role in their effectiveness. Everybody's nerves would be set on edge in expectation of

a contact on the next sonobuoy barrier. If a foreign submarine was being tracked, incompetent commanders would get in the way with their endless requests for minute reports of where buoys were deployed, and recommendations where they should be deployed. A key factor was the crew's ability to navigate accurately en route, especially when operating at the full combat radius.

The first sortie to the North Sea in search of foreign submarines was undertaken on 27 July 1970 by a pair of Tu-142s flown by regimental commanding officer V. I. Dubinskiy and squadron commander Lieutenant-Colonel V. A. Shimanskiy. To maximise the precision of their approach to the designated area, the crews approached the shoreline a few times and updated their positions through the use of known radar-contrast reference points. As experience was gained, the intensity of flight operations gradually grew. Tu-142 aircrews got excellent practical experience while tracking an unidentified submarine spotted by a Be-12 amphibian on 8 August 1974 in the Barents Sea. Occasional brief search missions in limited areas of the North Sea and the northern section of the Atlantic resulted in the detection of unidentified submarines, and subsequent tracking.

Northern Fleet Tu-142s would also fly in small tactical groups to the Pacific Fleet, via the so called Arctic Route, to take part in the Fleet's planned operations to monitor the submarine situation in the area. Also, they would operate on their own, flying a few sorties to the northern part of the Philippine Sea in 1977.

To augment the Pacific Fleet's anti-submarine warfare (ASW) capability, a decision was taken to reassign several Tu-142 aircraft from the Northern Fleet. In 1976 saw a Tu-142 squadron activated at Khorol airfield. The air and ground crews completed their ground school in August 1977 and commenced their flight training. Following the completion of the conversion programme, the 310th OPLAP DD deployed in 1978 to its permanent base – Kamennyi Ruchei NAS. By then, the regiment had received 14 Tu-142 aircraft.

Pacific Fleet aircrews gradually extended beyond the Seas of Okhotsk and Japan, and in January 1979 launched routine operations from the air bases at Da Nang and Cam Ranh in Vietnam, flights to which were accomplished across the Korean Straits or around Japan. In the former case, a sortie's duration totalled nine hours, while in the latter it took another 90 minutes. The 'Bears' would typically carry 268 RGB-1 and 10 RGB-10 sonobuoys. Owing to Da Nang's limited field length, take-off weight was normally limited to 165 tonnes (162.4 tons), including 70000 kg (154,321 lb) of fuel for 10 hours aloft. Afterwards, four Tu-142s were assigned to the 169th Air Regiment (Composite), deployed on a temporary basis to Vietnam by agreement.

From the first combat sortie to 1993, Tu-142 aircrews detected 56 foreign submarines and tracked them for some time. After that time, the detection rate grew somewhat. Tu-142s began to be retired in 1988.

Tu-142P

A little-known version was the Tu-142P, equipped with the Atlantida searching and targeting system (PPS). Today it is almost forgotten as there was only one aircraft

A characteristic cloud of rubber smoke signifies the arrival of a 76th OPLAP DD Tu-142MK back at its Kipelovo base. The 'Bear-F Mod 3' entered service with the Northern Fleet in 1979, and the 76th OPLAP DD completed its conversion to the updated version in 1981.

The Kongelige Norske Luftforsvaret (Royal Norwegian air force) was in NATO's front line in terms of intercepting the 76th OPLAP DD's Tu-142s as they headed out on missions from Kipelovo. Here a Tu-142MK 'Bear-F Mod 3' is watched closely by an F-16A from Bodø-based 331 Skvadron. The 'Foxtrots' plied much of their trade in the north Atlantic, the scene of intense submarine activity during the Cold War. In the 1990s such missions, which had once been routine, rapidly became a rarity.

A Tu-142MK rests on a snowy Kipelovo. As with the first-generation of 'Foxtrots', there was some confusion surrounding its designation. To the design bureau the aircraft was the Tu-142MK (K = Korshun) but, initially, the navy referred to it as the Tu-142M, having not used this designation for the revised Berkut aircraft. Subsequently, the AVMF did adopt the MK suffix, to differentiate between the Korshun-equipped aircraft and the later Tu-142MZ with Zarechye system.

(No. 4362). Its avionics incorporated an ECCM (electronic counter-countermeasures) system, infra-red jammer and the Atlantida anti-submarine system, which was essentially the Berkut upgraded to employ new weapon systems such as the APR-2 and UMGT-1 torpedoes.

The aircraft underwent tests in 1976 and was not mass-produced. It featured several external differences by comparison with the Tu-142: its horizontal stabiliser did not mount the Arfa tip fairing, the aircraft was fitted with fairings for SPS-151, SPS-152 and SPS-153 radar jammers (mounted in place of the navigator's front canopy, between frames 13-14 on both sides, and in the rear). Aerials were moved from the forward canopy to the nose. A radio equipment fairing was located on the port side. The wing leading-edge flap featured recesses for attaching aerials, which were enclosed by dielectric panels, requiring the anti-icing system near the recesses to be switched off. One of the aerials of the Siren system was also situated on the tip of the tailfin.

Additional equipment increased the aircraft's weight by 1300 kg (2,866 lb). The fuel load was also increased. The single point refuelling system could admit 83640 kg (184,391 lb), while an additional 4000 kg (8,818 lb) could be filled by gravity. The aircraft's powerplant comprised four NK-12MV engines from the 04-65 series, with AV-60K propellers. The aircraft's cruising range was 11800 km (7,332 miles) at a take-off weight of 182 tonnes (179.1 tons).

Tu-142M – 'systems from scratch'

By the early 1960s, foreign-made submarine search systems had become more reliant on low-frequency buoys and hydroacoustic explosive sound generators (ESG). Similar research was underway in the Soviet Union as well. By the mid-1960s, the conceptual design of the Udar system (a Soviet response to the US Navy's Julie system) had been completed, as well as a number of other efforts that laid the groundwork for the Korshun ASW surveillance and targeting system. A government resolution of 14 January 1969 ordered the derivation of the Korshun-equipped Tu-142M from the baseline Tu-142. The first prototype was to commence its joint trials in the third quarter of 1972, and the second in the fourth.

There were delays, of course. In May 1974, the Military Industrial Commission noted: "The Taganrog Mechanical Plant has failed to complete the two Korshun-equipped Tu-142M aircraft on schedule [then the fourth quarter of 1973], there have been slipped schedules on the technical records – which were completed a year behind schedule, the Ministries of Aircraft and Radioelectronic Industries have failed to supply VNPK-154s to the Taganrog plant, etc." It was not until 4 November 1975 that test pilot I. K. Vedernikov performed the first flight. The first-stage tests involved three aircraft (Nos 4243, 4244 and 4264) from first-flight date to 23 October 1977. Those trials resulted in recommendations by the navy and air force commanders-in-chief, and the aviation industry minister to launch construction of the Tu-142M. At the time, only seven of the 31 performance requirements had been met. The bulk of the second-stage trials were completed by November 1978.

In a governmental resolution, dated 19 November 1980, the Tu-142M long-range ASW aircraft was officially fielded with the Naval Aviation with the following performance characteristics:

- service range of 12000 km (7,457 miles) at a 185-tonne (182-ton) take-off weight and five per cent reserves

Flying the Tu-142

There were no significant differences between flying the Tu-95RTs and Tu-142, which made the conversion smooth, but not without some peculiarities.

Taxiing to the holding point is done with all engines running, but only the controls of the two inboard engines are used. Sharp turns are out of the question due to the considerable lateral load on the outer wheels and the landing gear's shock strut. At a speed of 5-10 km/h (3-6 mph) the turn radius is at least 15 m (50 ft), growing to at least 50 m (165 ft) as the speed increases. Having taxied to the standing start, the flaps are extended to the take-off position of 27°, with the stabiliser going into the 2° position. As soon as the flaps begin extending, the nosewheel is automatically restricted to a turning angle of 8°.

With a 130-tonne (128-ton) take-off weight, the take-off is performed with the engines in the normal power level (NPL) mode. If the runway is less than 2500 m (8,200 ft) in length, or the aircraft weighs over 130 tonnes, the full-thrust take-off mode is resorted to.

During the first half of the run, direction is maintained by nose wheel steering and brakes, while the rudder is used when the aircraft picks up speed to 130-150 km/h (81-93 mph). In a 130-tonne take-off, when the speed is 10-15 km/h (6-9 mph) lower than the lift-off speed, the nose wheel lifts off at an angle of 5-6°, but at a heavier weight it rises to 6-8°. At some airfields with low-quality runway surfacing (for instance, Khorol naval air station in the Pacific Fleet region) crews often report rather hard jolts when crossing the joints of the concrete slabs of the runway.

When taking off with the maximum take-off weight, the pilot has to pull the yoke back all the way with a load of 3-4 kg (6.6-8.8 lb) – the aircraft then pulls up its nose at a speed of 290-300 km/h (180-186 mph), with the load on the yoke decreasing. With the normal take-off weight at 130 tonnes, lift-off speed is 270-275 km/h (168-171 mph). Heavier take-offs require a lift-off speed of 300-310 km/h (186-193 mph). Critical decision speed during the run is 250 km/h (155 mph): if an engine fails and the runway is longer than 3500 m (11,483 ft), then the run should be continued. If the runway is shorter, the take-off should be aborted.

Taking off can be difficult in a crosswind owing to the large empennage and fuselage, which have a tendency to weathercock. Furthermore, the anhedral of the wing (which facilitates rolling manoeuvres), places an extra load on the undercarriage on the side the wind is coming from. This turns the aircraft towards the wind, too. Therefore, direction during the take-off run is maintained by using brake pedals and pre-emptive acceleration of the outer engine on the side the wind comes from. The aircraft also tends to turn during the lift-off, therefore the nose wheel is lifted off at high speed to minimise this effect.

At an altitude of 10-15 m (33-50 ft) the wheels are braked and landing gear is retracted to prevent vibration. During the retraction process the bogies somersault and drag grows considerably, which results in a reduction in forward and vertical speed. With the gear retracted, the centre of gravity shifts 2-3 per cent rearwards and a pitching moment appears, to compensate for which the stabiliser is moved from the take-off position into the flight one. The engines are switched to the normal thrust mode. Customarily, the landing gear remains extended for circuits. Flaps are retracted at an altitude of at least 100 m (328 ft) in two or three stages, with the retraction of the flaps on both wings being carefully

The 240th GvOSAP (Guards Independent Mixed Aviation Regiment) at Ostrov is the AVMF's training unit for shore-based patrol and attack aircraft, with several types on charge. Among the fleet are a handful of 'Bear-Fs', including this Tu-142MK seen at the base's open day in July 2001.

- 4-hour 10-minute loiter time 4000 km (2,485 miles) from base
- 855-km/h (531-mph) maximum speed
- 735-km/h (457-mph) cruising speed
- 2530-m (8,300-ft) take-off run
- 1830-m (6,004-ft) landing roll at 130 tonnes (127.9 tons)

The aircraft had a wing span of 50 m (164 ft), was 53.1 m (174 ft 2 in) long and 13.6 m (44 ft 7 in) high, while its maximum take-off weight accounted for 185000 kg (407,848 lb), with fuel capacity equalling 86000 kg (189,594 lb) and payload totalling 4395 kg (9,689 lb). Maximum payload was 9000 kg (19,841 lb).

The 'complex' comprised the aircraft with its organic avionics (navigation aids, hydrologic reconnaissance equipment, Strela-142 communications suite, Sayany-M ECM system), the 2Kn-K Korshun-K ASW surveillance and targeting system (the 2Kn-K was mounted on aircraft No. 4293, while aircraft No. 4244's surveillance and targeting system was called Korshun, and those of aircraft Nos 4243, 4291 and 4292 was dubbed Kn-K), the MMS-106 Ladoga magnetometer and ground test equipment. Despite a higher degree of automation, the crew complement grew to total 11 personnel.

As compared to the 1971-vintage stretched-cockpit Tu-142, the Tu-142M's primary performance and dimensions remained relatively unchanged, with some minor alterations. Take-off weight rose and the forward cockpit was redesigned to house nine personnel, rather than the earlier eight. They were accommodated in seats with their backs inclined at 45°, while the pilots' seats were fitted with electric drives to move them fore and aft by 160-220 mm (6.3-8.7 in). Footboards and kneeboards were provided to reduce fatigue. The altered inclination of the instrument panel enhanced its visibility for the pilots. Two navigator/tactical action officers remained seated facing the tail, which proved to be inconvenient and increased fatigue. At the pilots' station the cockpit was widened by 180 mm (7 in) and an enlarged hatch provided some improvement in emergency escape. The view angle over the nose was increased by 1.5°, important for landing.

Cruising at medium altitude, a Tu-142M/MK displays this version's most distinctive feature: the slightly upturned MMS-106 Ladoga MAD mounted at the top of the fin. Two Tu-142s had been used to test this equipment which, as fitted to the Tu-142M prototype (4243), was initially aligned with the top of the fin, rather than angled upwards. The Ladoga proved troublesome at first, requiring several systems to be switched off before it could be used, and was of limited tactical value.

In the fuselage mid-section the surveillance and targeting system was fitted with a new fairing and the main cargo bay was drastically redesigned. The second cargo bay was merged with the technical compartment, and its lower part modified to serve as a bay for small-size bombs used as ESGs. The upper section of the bay accommodated components of the 2Kn-K, Kvadrat-2 ELINT system, communications suite, and other systems.

Revised controls

Controls differed only slightly from those of the Tu-142, for instance, the yoke travel was reduced and it was provided with a loading device generating an extra load on the yoke at high angle of attack, when fed relevant signals by the AUASP-142 sideslip and *g*-load automatic device. The Ka-142 was retained to enhance stability and controllability. To ensure sufficient power for the autopilot's actuators to operate the control surfaces, the elevators were equipped with extra flutter tabs and the trailing edges of the rudder and ailerons were modified. Control of flaps, stabilisers and primary systems was retained from the baseline variant.

A new NPK-142M flight navigation system was installed, comprising a main and back-up sub-systems. The main sub-system centred around an astro-inertial navigation system with an automatic correction device, the L14MA computer and TsVM-10TS-42 single-address 125,000-operations

synchronised. Resulting pitch changes are compensated by retrimming. To save fuel, the aircraft flies to its assigned zone in a continuous climb. Having reached 7100 m (23,294 ft), the aircraft continues its climb while maintaining Mach 0.63 – about 700 km/h (435 mph).

Some helicopters with a tail rotor are prone to what is called 'dynamic drift'. A similar phenomenon is featured by the Tu-142 as well, with the aircraft tending to deviate from its course to the right. This is due to the fact that the front propeller rotates to the right, while the rear one rotates in the opposite direction. The rear propeller both straightens the front propeller's vortex out and deflects it to the left. The offset vortex hits the vertical tail and turns the aircraft to the right. The trend can be corrected via the rudder's trim tab or ailerons.

Return from the patrol zone is performed at an altitude of 11000 m (36,090 ft). It is recommended that high-altitude flights in turbulent conditions be carried out at Mach 0.7, with the rudder and elevators to be operated smoothly. When entering a high temperature area (for example, when crossing the Atlantic), the aircraft can lose speed unless the engines' rpm is increased or altitude reduced. With the aircraft flying on three engines, its fuel consumption per kilometre rises by 3 per cent.

Unless there are restrictions, the commencement point and mode of descent are of special importance. In a powered descent from 11000 m (36,090 ft) to 500 m (1,640 ft), with the throttle controls moved to the trip catch (18°), the Tu-142 will cover 290 km (180 miles) in 26 minutes, burning 1550 kg (3,417 lb) of Avgas in the process. However, if the engines are throttled back completely for the descent, the aircraft will burn 400 kg (882 lb) of fuel to cover 150 km (93 miles) in 11 minutes. If this is followed by no further descent, the aircraft will cover another 230 km (143 miles) with 1150 kg (2,535 lb) of fuel used up, thus adding a distance of 90 km (56 miles) for the same amount of fuel.

Landing gear is deployed at 400 km/h (248 mph) or below, with the stabiliser being set at 2°. Between the downwind and base legs of the circuit (when descending after a circuit, the flaps deploy at 20°) the speed is set at 350-370 km/h (217-230 mph). After turning from base to final, the flaps are deflected to 27°, the speed drops down to 330-340 km/h (205-211 mph) and the speed of descent becomes 2-3 m (6.5-10 ft) per second with the outboard engines being set at the trip catch. Before flying past the outer marker, the flaps are extended to 35° while the stabiliser is set at 4°.

Owing to a low lateral stability reserve, directional errors made during the landing approach can be remedied by 3-5° turns without rolling. The aircraft features low lateral stability, which makes the landing approach with a crosswind much easier to perform, and the drift can be rectified both by altering the course and by rolling. The latter does not cause the oscillations from one wing to the other ('Dutch roll'), for which the Tu-16 was notorious. The outer marker is overflown at an altitude of 200-220 m (656-722 ft) at a speed of 300-310 km/h (186-193 mph), which then decays to 270-280 km/h (168 mph-174 mph), with the vertical speed increasing to 3-4 m (10-13 ft) per second. The middle marker is overflown at an altitude of 70-80 m (230-262 ft).

Flare-out begins at an altitude of 10-15 m (33-50 ft), with the inboard engines being set at trip catches. At a typical weight of 106 tonnes (104.3 tons) touchdown occurs typically at 237 km/h (147 mph) at an angle of 8°. Following the nose wheel touching, the inboard engine controls are throttled back completely and the propellers are taken off the catch. The outboard engine controls are throttled back somewhere between 180-200 km/h (112-124 mph) to further decelerate the aircraft.

During the landing approach minor course corrections can be effected by rudder, without altering the wings-level attitude. The aircraft is prone to weather-cocking once on the runway, especially during the take-off run. At over 100 tons typical landing weight, touchdown is usually 'firm'.

This is the radar operator's station in the original Tu-142, showing the scope for the Berkut radar.

Above: Facing aft at the rear of the Tu-142MZ's forward compartment are the work-stations of the Korshun radar operator (starboard) and tactical navigator (port). The latter has a POTO tactical display screen. The Zarechye sonar system operator sits further forward.

Most removable boxes in the Tu-142 are painted with the aircraft's Bort number. These SPIU (samolyotnoye priyomno-indikatornoye ustroistvo, aircraft receiving-indicating unit) scopes are from Tu-142MK Bort 22.

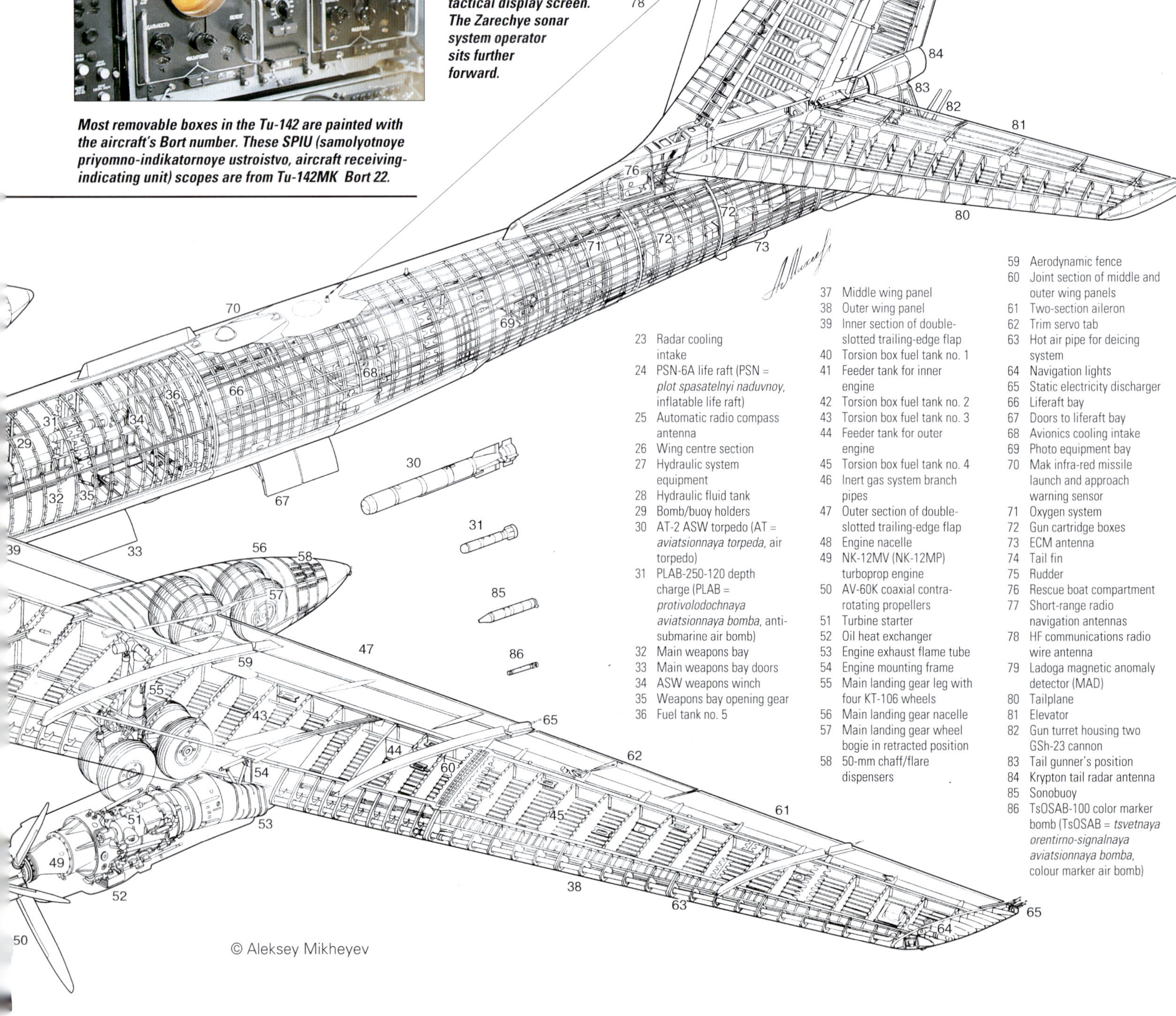

23 Radar cooling intake
24 PSN-6A life raft (PSN = *plot spasatelnyi naduvnoy*, inflatable life raft)
25 Automatic radio compass antenna
26 Wing centre section
27 Hydraulic system equipment
28 Hydraulic fluid tank
29 Bomb/buoy holders
30 AT-2 ASW torpedo (AT = *aviatsionnaya torpeda*, air torpedo)
31 PLAB-250-120 depth charge (PLAB = *protivolodochnaya aviatsionnaya bomba*, anti-submarine air bomb)
32 Main weapons bay
33 Main weapons bay doors
34 ASW weapons winch
35 Weapons bay opening gear
36 Fuel tank no. 5
37 Middle wing panel
38 Outer wing panel
39 Inner section of double-slotted trailing-edge flap
40 Torsion box fuel tank no. 1
41 Feeder tank for inner engine
42 Torsion box fuel tank no. 2
43 Torsion box fuel tank no. 3
44 Feeder tank for outer engine
45 Torsion box fuel tank no. 4
46 Inert gas system branch pipes
47 Outer section of double-slotted trailing-edge flap
48 Engine nacelle
49 NK-12MV (NK-12MP) turboprop engine
50 AV-60K coaxial contra-rotating propellers
51 Turbine starter
52 Oil heat exchanger
53 Engine exhaust flame tube
54 Engine mounting frame
55 Main landing gear leg with four KT-106 wheels
56 Main landing gear nacelle
57 Main landing gear wheel bogie in retracted position
58 50-mm chaff/flare dispensers
59 Aerodynamic fence
60 Joint section of middle and outer wing panels
61 Two-section aileron
62 Trim servo tab
63 Hot air pipe for deicing system
64 Navigation lights
65 Static electricity discharger
66 Liferaft bay
67 Doors to liferaft bay
68 Avionics cooling intake
69 Photo equipment bay
70 Mak infra-red missile launch and approach warning sensor
71 Oxygen system
72 Gun cartridge boxes
73 ECM antenna
74 Tail fin
75 Rudder
76 Rescue boat compartment
77 Short-range radio navigation antennas
78 HF communications radio wire antenna
79 Ladoga magnetic anomaly detector (MAD)
80 Tailplane
81 Elevator
82 Gun turret housing two GSh-23 cannon
83 Tail gunner's position
84 Krypton tail radar antenna
85 Sonobuoy
86 TsOSAB-100 color marker bomb (TsOSAB = *tsvetnaya orentirno-signalnaya aviatsionnaya bomba*, colour marker air bomb)

© Aleksey Mikheyev

Inside the Tu-142

Right: The Tu-142's cockpit is relatively sparse, with basic flight instruments provided and a forest of warning lights above the console. The pilot has a tactical display screen to the right of the main instruments for flight direction. In the centre is a crawlway through to the navigator's position, above which is the control for the autopilot. This is mounted on a pull-out arm and can be swung to left or right for operation by either pilot. With access needed to the nose, there is no central console. Each pilot has throttle controls at his side, aft of the large trim wheel. This cockpit has been 'zapped' by the Swedish air force's SK 60-equipped 'Team 60' display team.

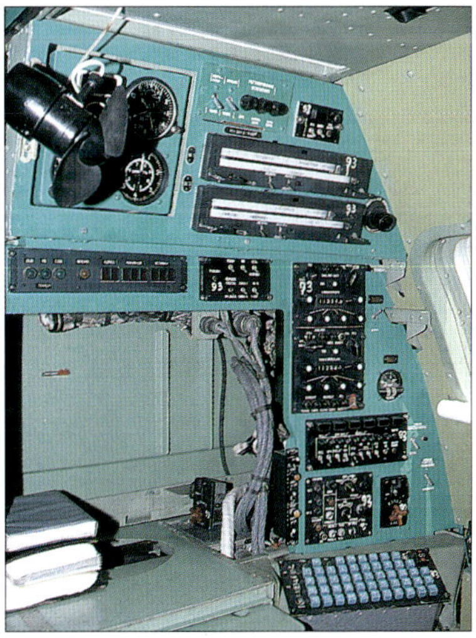

Far left: The flight engineer sits behind the pilot on the port side with engine control the main task. As well as comprehensive instruments, throttles are provided.

Left: Opposite the engineer is the radio operator. The empty space in the console is normally occupied by the sensitive cryptographic terminal.

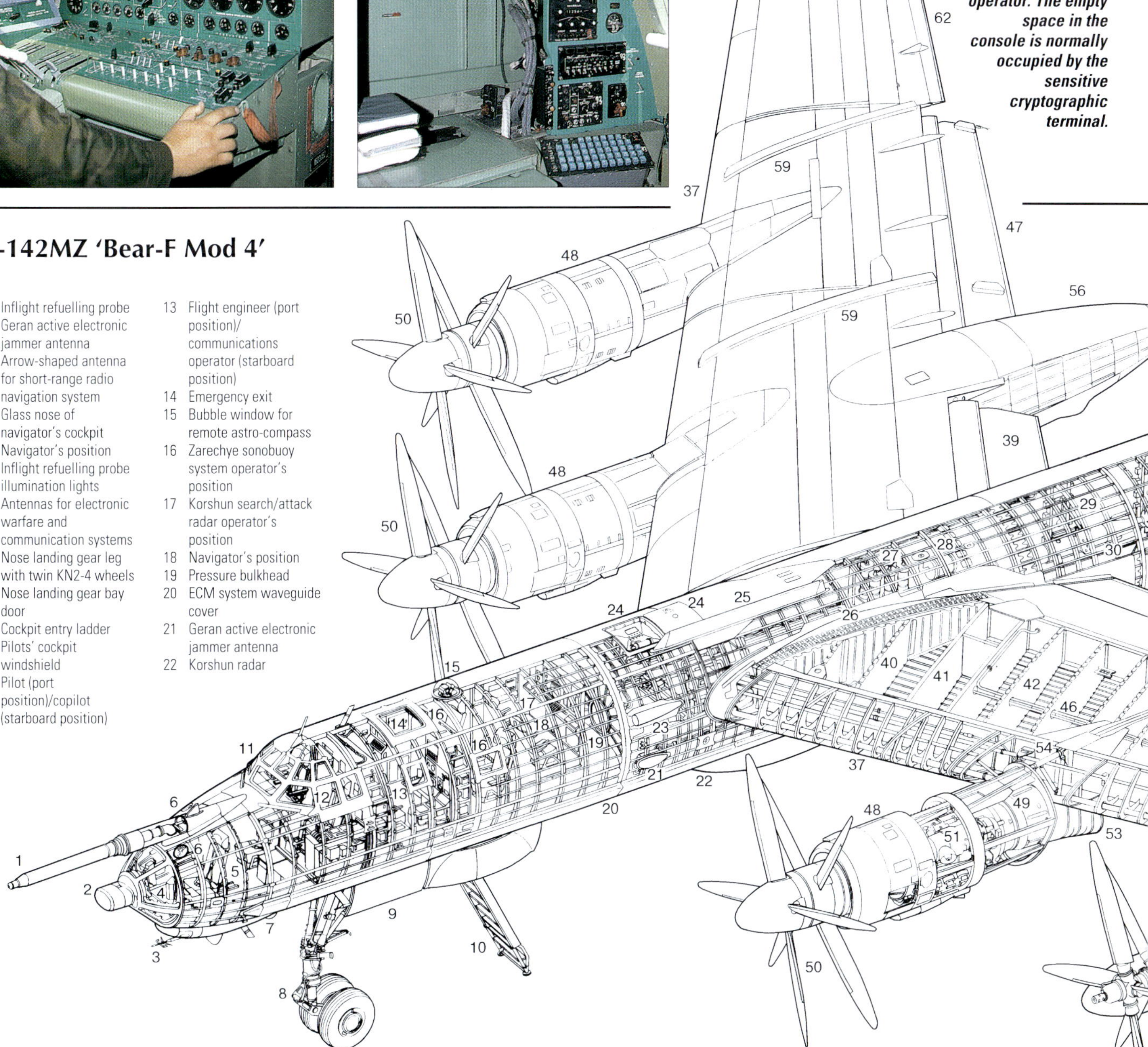

Tu-142MZ 'Bear-F Mod 4'

1. Inflight refuelling probe
2. Geran active electronic jammer antenna
3. Arrow-shaped antenna for short-range radio navigation system
4. Glass nose of navigator's cockpit
5. Navigator's position
6. Inflight refuelling probe illumination lights
7. Antennas for electronic warfare and communication systems
8. Nose landing gear leg with twin KN2-4 wheels
9. Nose landing gear bay door
10. Cockpit entry ladder
11. Pilots' cockpit windshield
12. Pilot (port position)/copilot (starboard position)
13. Flight engineer (port position)/ communications operator (starboard position)
14. Emergency exit
15. Bubble window for remote astro-compass
16. Zarechye sonobuoy system operator's position
17. Korshun search/attack radar operator's position
18. Navigator's position
19. Pressure bulkhead
20. ECM system waveguide cover
21. Geran active electronic jammer antenna
22. Korshun radar

Tu-142 'Bear-F' technical details

This is the original 12-wheel rough-field undercarriage fitted to the first 12 Tu-142s. These aircraft inevitably became known as 'centipedes' compared with later aircraft with standard four-wheel bogies.

Fuselage
The monocoque fuselage is skinned in smooth aluminium sheets which vary in thickness from 1 to 4 mm. The fuselage is broken down into five major sections: nose glazing; a forward pressurised cabin; a central cylindrical section with diameter of 2.9 m (9 ft 6 in); a conical tail section; and a rear pressurised cabin with a defensive gun installation.

The forward cabin accommodates the crew (there are nine seats with standard seat belts) and various equipment. In case of emergency the crew leaves the aircraft via the nose wheel well. The entrance hatch is opened by compressed air and the nosewheel leg is extended. Three hydraulic actuators start hydraulic motors which move a mobile floor which runs between the crew stations to the exit hatch for a duration of 100 seconds.

The fuselage centre-section is 16 m (52 ft) long. Aft of the cockpit is a compartment containing a PSN-6A life raft. Behind the wing centre section is a cargo bay with two pairs of doors. Aft of this bay is one of the fuel tanks. The tail section has a compartment for two crew members, with windows and blisters. The fairing for the Krypton rear-facing radar is mounted above the rear turret. In case of an emergency, two crew members bale out through the entrance hatch. The cover of the hatch acts as a shield to protect the crew from the slipstream.

This view shows the radome for the Korshun radar, refuelling conduit, cooling air intake and main stores bay. Aft of the main bay is another set of doors for an auxiliary bay.

Wings
The wings of the 'Bear' are swept 35° at the quarter-chord. Swept wings were adopted as air compression begins to have an adverse effect on the aerodynamic characteristics in the Mach 0.715-0.74 (770-820 km/h/478-509 mph) speed regime, which could lead to boundary layer separation and a large drag rise. For the same reasons the wing is of rather thin section.

Compared to earlier Tu-95s, the Tu-142 has a new wing, modified to improve aerodynamic efficiency by 4-5 per cent. The leading edge was extended forward and then angled downward. Along the top surface of the wing there are three pairs of aerodynamic fences, each 190 mm (7.5 in) in height. The aircraft's wing is set at 2.5° anhedral, which improves controllability in the rolling plane.

With the exception of the centre section, almost the whole wing is an integral fuel tank, with eight compartments within the torsion box formed by spars, panels and ribs. As a swept wing generates less lift than an unswept one, it was necessary to install powerful double-slotted flaps with electromechanical control.

The outer wing panel consists of a central torsion box, detachable leading edge, trailing edge and aileron (suspended on nine hinges). The starboard aileron is equipped with a trim tab, and the port aileron has a balance tab.

Tail
By comparison with the Tu-95RTs, the Tu-142's tail unit was also modified. The stabilisers have been made adjustable for trim, while rudder and elevator areas were increased, as was their maximum deflection angles. The stabiliser tips of Tu-142s from the first series were used to mount the Arfa system and Kristall transmitter in streamlined fairings. The fairings had a metal container and dielectric panels. Starting with aircraft No. 4232, the Lira synchronisation device became standard. Leading edges of both fin and stabilisers were equipped with electrical anti-icing, as were those of the wings.

Differences in tail configuration were significant between the Tu-142MK (above) and Tu-142MZ (below). Quite apart from all the additional antennas and lack of glazed blisters, the MZ has a new turret with four, rather than two, 23-mm guns.

Control system
Flight control is effected through the two control columns and two pairs of pedals, which are connected with the control surfaces by rigid rods. GU-54A hydraulic actuators (boosters) are installed in the control system for rudders and ailerons. An artificial feel system is also installed for imitation of loads on rudders and ailerons. Spring loading devices and trimming mechanisms are installed, and with the help of cables they are connected to the AP-15RZ autopilot. A reversionary mode, with unassisted flight control, is also possible. Stability and controllability were enhanced by the Ka-142 unit.

In order to reduce the aircraft's asymmetric tendencies in the event of an outer engine failure, the MP-AF mechanism in the rudder control system will automatically deflect the rudder 11° to oppose the yawing effect. The mechanism works on a signal from the propeller auto-feathering system when the engine's throttle position indicator is at an angle greater than 85°.

Powerplant and fuel
The Tu-142's powerplant consists of four NK-12MP or NK-12MV turboprops with axial-flow compressors, annular combustion chambers, and a gas turbine of a jet type (the expansion of gas occurs in the exhaust nozzle). The engines are installed in nacelles at -1° to the wing chord.

Two four-bladed, variable-pitch AV-60K propellers are mounted co-axially on each engine. The pitch angle of the forward propeller varies between 5° and 91°, while the rear propeller moves between 7° and 92°. In flight the pitch angles change according to mode of flight, while the rotational speed remains constant at 735 rpm.

The forward propeller generates 54 per cent of the power and the rear one contributes 46 per cent. Failure of an engine in flight can lead to an enormous rise in drag, especially as the propeller changes to fine pitch as the engine spools down. If unchecked, this can lead to a loss of control. Consequently, propellers are equipped with systems for automatic and manual feathering (to maximum pitch angles of 91°/92°). Another mechanism also prevents the blades from changing to fine pitch and windmilling.

Transition of the blades to fine pitch after landing (achieved by unlatching the stop to permit the blades to move below the inflight fine-pitch stop with the engine decelerated below flight idle power) creates considerable drag, which provides an essential reduction of the landing run.

Fuel (kerosene -1, -2, TC-1) is distributed in eight integral tanks in the wing, and in three soft tanks in the wing centre section and rear section of the fuselage. Four independent fuel systems feed the engines (one system per engine). Emergency fuel dump is provided and takes 25-30 minutes. Refuelling on the ground is normally undertaken through four central points and two overwing fillers on each wing. Refuelling through 10 overwing fillers increases the quantity of fuel by 4000 kg (8,818 lb). The aircraft is capable of inflight refuelling using the Konus system.

Engines are controlled by throttles (RUD – Rychag Upravleniya Dvigatelem) and fuel shut-off valves located among pilots' and flight engineer's controls.

Aircraft systems
The aircraft has two hydraulic systems. One 150-kg/cm^2 (2,133 lb/sq in) system is used for basic and emergency wheel braking, release and retraction of the nosewheel leg, front wheel steering, operation of the emergency mobile floor, and the windshield wipers. A 75-kg/cm^2 (1,067-lb/sq in) system provides power to hydraulic actuators.

Electric circuits comprise a 27-V DC system, single-phase 400-Hz AC systems of 115-V and 200-V, and a three-phase 400-Hz circuit of 36 V. For high-altitude flights there are five KPZh-30 containers with liquid oxygen.

Defences
Defensive armament of the aircraft consists of a DK-12 tail turret with two 23-mm GSh-23 cannon, each with 600 rounds. The PS-253K sight and PRS-4 Krypton radar are used to aim the cannon. For detection of fighter radars the aircraft has an SPO-10 passive receiver, and SPS-100 ECM equipment is fitted.

In the rear of the undercarriage fairing are numerous three-round launchers for IR decoy flares.

Tu-142 'Bear-F' ASW stores

This Tu-142MK is displayed with five pre-packed sonobuoy cassettes and standard PLAB-250 depth charges.

Weapons

Tu-142s have carried a wide range of maritime stores, usually depth charges (conventional or nuclear) and torpedoes. Mines are also available.

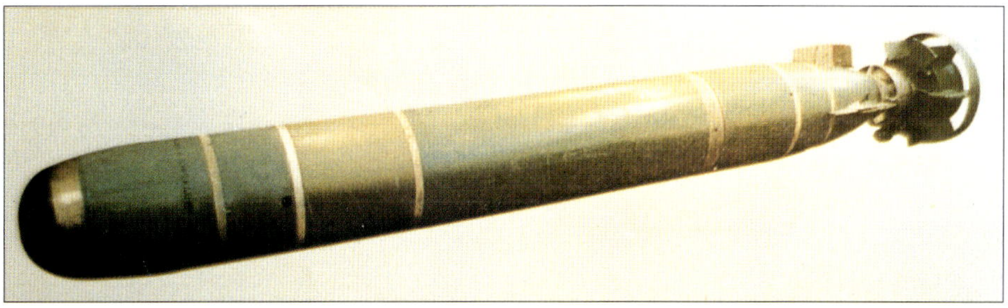

Right: The UMGT-1M Orlan torpedo is an option for the Tu-142MZ, along with the older APR-2. From nose to tail, the UMGT-1's section comprise: homing head, auxiliary equipment section, warhead, battery section, electric motor, steering section, propulsion unit and the braking unit.

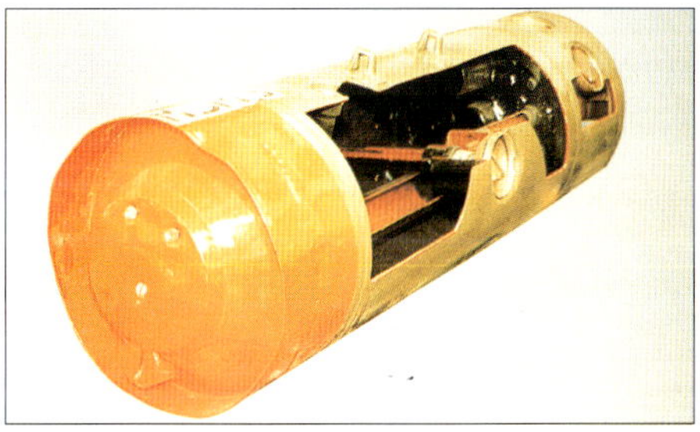

MDM-3 air-dropped mine

MDM-5 air-dropped mine

KAB-250PL (S3V) guided depth charge

Sonobuoys

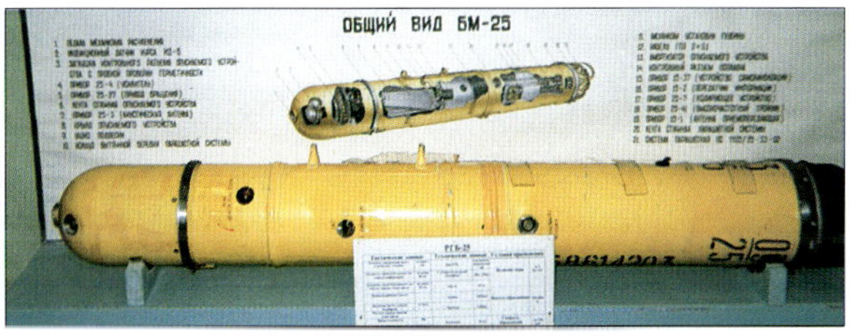

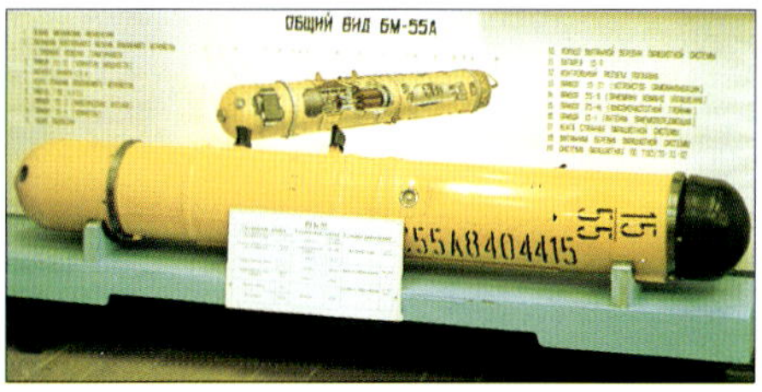

The Berkut generation of sonobuoys for the Tu-142 was based on the RGB-1 (below left), RGB-2 and RGB-3. An example of the latter is seen in the water (left) – like all buoys, the hydrophone submerges beneath the floating transmitter portion. With the Tu-142K/Korshun came new buoys in the form of the passive non-directional RGB-15 (below, painted yellow) and RGB-75 (below, silver), the passive directional RGB-25 (above) and active RGB-55A (above right). For loading into the stores bay the buoys are carried in cassettes (right), these being RGB-75s.

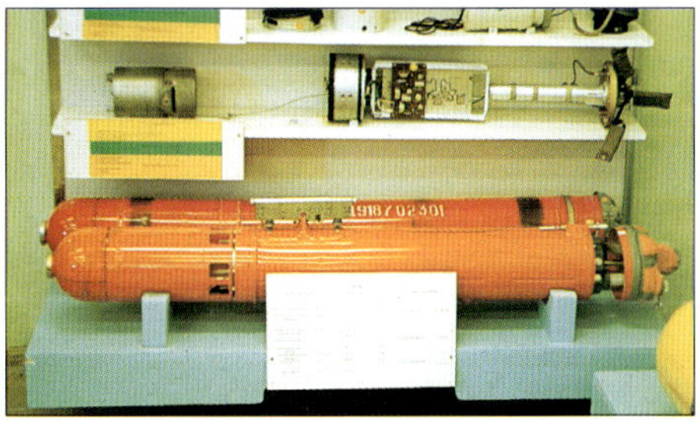

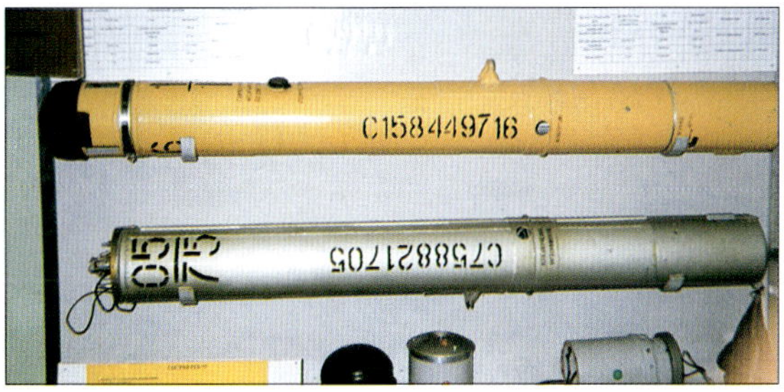

Mission system
Housed in the massive radome is the antenna for the Leninets/St Petersburg Obzor-MS navigation/attack radar (NATO: 'Clam Pipe'), above which is a small radome for a weather radar. The K-016 Sprut missile initialisation and launch system (older Osina system for early aircraft) is installed to provide accurate navigational data to the inertial systems of the missiles before launch.

Powerplant history
The most powerful turboprop engine ever to have been put into production, the NK-12M originated from a German design – the Junkers Jumo 022. Many of the original team, led by Austrian Dipl.-Ing. Ferdinand Brandner, continued to work on the engine after internment, under the supervision of Nikolay D. Kuznetsov at his Kuibyshev (now Samara) OKB. The initial TV-2 engine ran in about 1950 and led directly to successively more powerful engines known as the TV-10 and TV-12. Meanwhile, an alternative approach was being taken with the 2TV-2F, which coupled two of the earlier, less powerful TV-2s to drive a common gearbox. Both 2TV-2F and TV-12 were tested on a Tu-4LL testbed before flying in the first and second Tu-95 prototypes, respectively. The crash of the first prototype was attributed to a failure in the 2TV-2F's reduction gearbox, and all subsequent 'Bears' were powered by the TV-12 (later redesignated NK-12).

Self-protection system
The Tu-95MS has a capable defensive electronics suite centred around the Meteor-NM self-defence system. The passive elements are SPO-10 Beryoza radar warning receivers at the nose and tail, and Mak-UT infra-red missile launch and approach sensors under the nose and on the rear fuselage. Countermeasures are the SPS-100 Geran active jammers under the nose and in the rear fuselage, and in pods under the tail turret, and three-round 50-mm chaff/flare dispensers in the rear of the inboard undercarriage fairings. A 'sting in the tail' is provided by two twin-barrelled 23-mm GSh-23 cannons in the tail turret, controlled by the PRS-4 Krypton (NATO: 'Box Tail') radar sight.

Airframe and control surfaces
The wings are constructed from four-spar inner panels (35° quarter-chord sweep) and three-spar outer panels (33.5° quarter-chord sweep), with 2.5° anhedral. Each wing has two-section double-slotted trailing-edge flaps and three-section ailerons (with tabs). Spoilers and three wing fences are mounted on each upper wing surface. Conventional tail control surfaces include variable-incidence tailplanes, adjustable in-flight according to the fuel used (between 1° down and 3° up); elevators and rudder are all tabbed. The semi-monocoque, circular-section fuselage has a bomb bay in the centre about 8.1 m (26 ft 7 in) long . Flight control is mechanical, hydraulically boosted and controlled by the SAU-3-021 autopilot.

Tupolev Tu-95MS 'Bear-H'

184th Gvardeyskiy Tyazhelo-Bombardirovochnyi Aviatsionnyi Polk 37th Air Army, Engels air base

This 'Bear-H' is the commander's aircraft of the 184th GvTBAP, and wears the name of the city (Saratov) local to Engels air base. It is adorned with the blue fin leading edge traditionally applied to the aircraft assigned to unit commanders. In company with another Tu-95MS, this aircraft made one of the type's rare long-range flights – to the North Pole – on 11 September 2001, the date of the Al Qaeda terrorist attack on New York and Washington.

Tu-95MS 'Bear-H' specification

Dimensions: length 49.13 m (161 ft 2 in); wing span 50.04 m (164 ft 2 in); wing area 289.9 m^2 (3,120 sq ft); height 13.30 m (43 ft 7.5 in); maximum fuselage diameter 2.90 m (9 ft 6 in)
Undercarriage: retractable, with steerable twin nose wheels. Four wheel bogies on each main unit, retracting into large fairings on the wing trailing edges, in line with the inner engines. Main wheel tyre size 1450 x 450 mm (57 x 17.7 in); nose wheel tyre size 1140 x 350 mm (44.9 x 13.8 in); wheel base 14.83 m (48 ft 8 in); wheel track 12.55 m (41 ft 2 in)
Weights: empty 94400 kg (208,113 lb); maximum take-off 185000 kg (407,848 lb); maximum in flight, after in-flight refuelling 187000 kg (412,257 lb); fuel 81670 kg (180,048 lb); maximum landing 135000 kg (297,619 lb)
Performance: maximum speed 830 km/h (516 mph); maximum speed at sea level 550 km/h (342 mph); cruise speed 735 km/h (457 mph); take-off distance at MTOW 2540 m (8,333 ft); *g*-limit +2; ceiling 10500 m (34,450 ft); range with nominal combat load (6 x Kh-55) 10520 km (6,537 miles); range with maximum combat load (16 x Kh-55) 8500 km (5,282 miles); endurance without inflight refuelling 14 hours; range with single inflight refuelling 14100 km (8,762 miles)
Powerplant: four Trud/Samara NK-12MP turboprops, each rated at 11190 ekW (15,000 ehp). Each eight-bladed AV-60K propeller unit comprises two four-bladed co-axial contra-rotating reversible-pitch propellers. Flight refuelling probe ahead of the cockpit, on the nose.

Weapons
The normal load of the Tu-95MS16 consists of six Kh-55SM (AS-15B 'Kent') long-range nuclear cruise missiles on an MKU-6-5 rotary launcher inside the fuselage bomb bay. Additionally, 10 more Kh-55SMs can be suspended under the wings in four clusters (two on each inner pylon and three on each pylon between the engines), although these pylons were removed under the START treaty. Early Tu-95MS6 aircraft carry the older Kh-55 (AS-15A 'Kent') missiles in the fuselage bay only.

Crew
The Tu-95MS is operated by a crew of seven, comprising two pilots, navigator (with access to the astrodome), navigation/offensive weapons operator, defence systems operator, and flight engineer in the forward pressurised compartment. The tail gunner has a separate cabin. All crew (except tail gunner) enter the cockpit via the nosewheel bay.

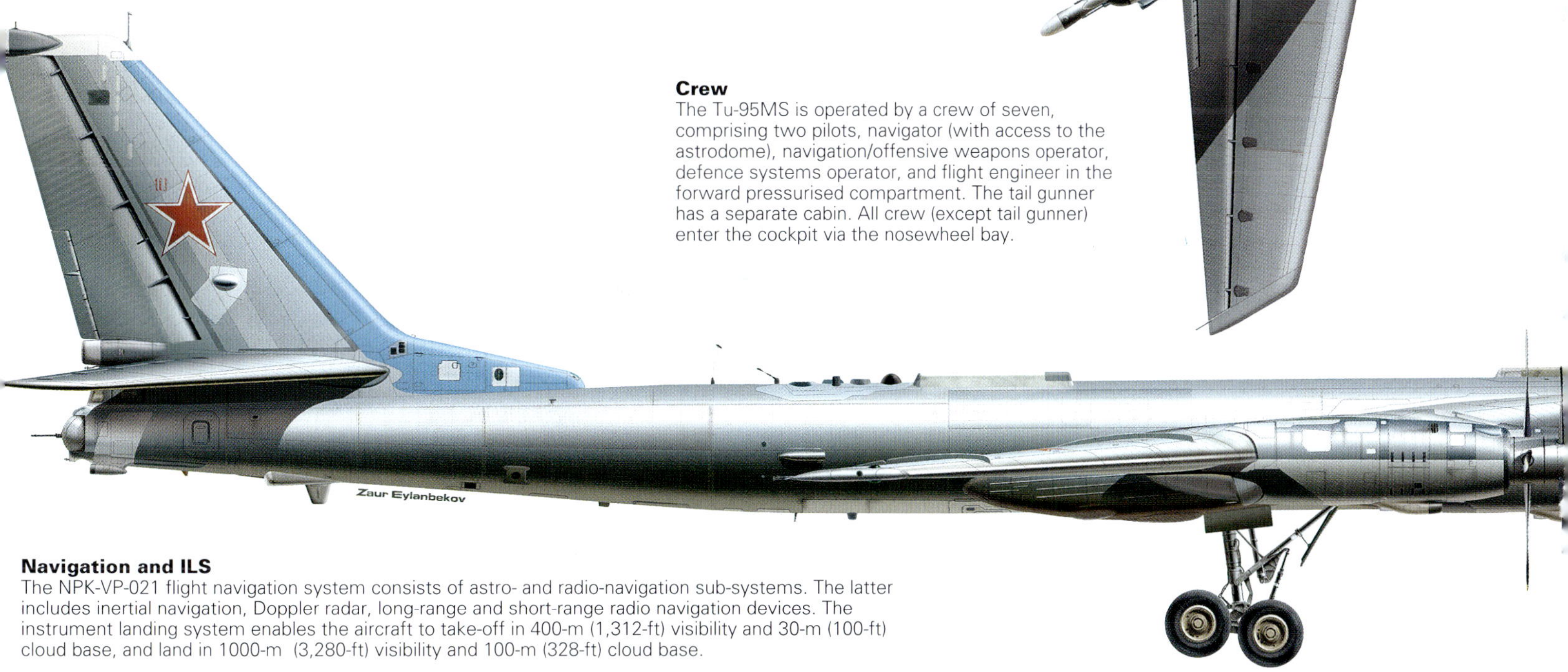

Navigation and ILS
The NPK-VP-021 flight navigation system consists of astro- and radio-navigation sub-systems. The latter includes inertial navigation, Doppler radar, long-range and short-range radio navigation devices. The instrument landing system enables the aircraft to take-off in 400-m (1,312-ft) visibility and 30-m (100-ft) cloud base, and land in 1000-m (3,280-ft) visibility and 100-m (328-ft) cloud base.

Above: As with the Tu-95RTs 'Bear-Ds' of the 392nd ODRAP before them, 76th OPLAP DD 'Bear-Fs' deployed from Kipelovo to San Antonio air base, near Havana, for operations in the west Atlantic. Here a Tu-142M is escorted along the eastern seaboard of the United States by an F-15A from the Langley-based 48th Fighter Interceptor Squadron.

Far right: This view highlights the prominent wing fences on the broad wings of the Tu-142MK. The conduit along the starboard side of the forward fuselage routes fuel from the probe to the main tanks.

per minute central computer. The back-up sub-system consisted of a VNPK-154 computer, DISS-7 Doppler system, Rumb-1B inertial heading and attitude reference system and a number of self-contained systems, for example, the long- and short-range radio navigation system, radar altimeters, automatic radio compasses and the on-board portion of the Os-1 landing system. Redundancy was believed to enhance the system's reliability.

According to calculations, the main subsystem was to maintain an accuracy in the region of 5-10 km (3-6 miles) per flight hour. The backup subsystem's expected accuracy was 1.5-2.5 per cent of the range. It took the NPK-142M at least an hour to be prepared for a mission. When switched on, the system fed the aircraft's initial position data to the surveillance and targeting system, and then complemented them with data on its speed, heading, altitude, roll and pitch.

The flight system included the Bort-42 flight director control system and AP-14PS autopilot, providing semi-automatic and automatic flight based on various input signals fed by the navigation system and 2Kn-K ASW surveillance and targeting system. The Strela-142M communications system was designed to provide interphone communications, land command post and ship communications, and also to record all incoming and outgoing voice and data traffic. The weapons suite was augmented with 16 multiple racks in the form of pods housing RGB-75 and RGB-15 buoy holders. Also, the weapons suite included another three types of multiple racks to mount RGB-25s, RGB-55As, bombs and torpedoes. The new second bay housed a container for various ESGs.

Elements of the 2Kn-K

Unlike the Berkut ASW system, the 2Kn-K system was supposed to detect submarines in both the acoustic and infrasonic wavebands. It comprises onboard equipment, radio-sonobuoys and MGAB small-size air bombs. Onboard equipment is built around four subsystems – radar (RLP), radiohydroacoustic (RGP), computer (BVP) and tactical data display (POTO). Also, it includes the A-081 automatic radio compass and a container housing the system's control panels. The MMS-106 Ladoga magnetometer and Nerchinsk sonic speed measurement system, as well as communications, release control and NPK-142M systems, operate in conjunction with the 2Kn-K.

The radar subsystem's transmitter emits sounding and coded signals, with variable frequency. The radar's output is 80 kW with a horizontal beam width of 1°, polarisation is both vertical and horizontal, and surveillance can be undertaken either in a specific sector to one side or across 360°. The radar can contact the beacons of sonobuoys at a range of 40-50 km (25-31 miles).

The Kayra-P (Kr-P) radiohydroacoustic subsystem provides the reception, visual and audio display and processing of data uplinked by sonobuoys and the Nerchinsk system, and feeds it to the POTO and BVP subsystems. It comprises units served by two radiohydroacoustic subsystem (RGP) operators, each of which has two Kr-4 television displays. One of the screens displays data from four buoys, while the other can show data from the POTO or adjacent operator's workplace. Displayed imagery depends on the type of buoys and their operating modes – manual or automatic.

If the automatic mode fails to establish communications with the buoys, the Kr-4 displays 24 noise boxes. With a buoy transmitting its carrier frequency signal, a horizontal line appears in the relevant box, to be crossed by a vertical line following the arrival of the 'target' signal. The buoy's signal is fed to the processor, which operates in conjunction with the operative recorder for power processing or spectral analysis. Power processing allows an evaluation of the signal and a pass curve target blip on the display. Fed by the operative recorder, the Kr-4 displays show incoming signals processed in to the form of spectrograms of the noises detected, echograms indicating the submarine's position, and oscillograms showing the alteration of the level and heading of hydroacoustic signals.

Kipelovo air base, situated roughly halfway between Moscow and Archangelsk (the nearest major town is Vologda), is home to the 76th OPLAP DD, which has three squadrons assigned. Two fly the Tu-142MK/MZ and one operates the Tu-142MR. Thirty-two 'Bears' – almost the entire 76th OPLAP DD fleet – are visible in this line-up. In the middle of the line is a single Be-12, next to which is one of the Tu-142MR 'Bear-J' communications relay aircraft. Two more 'Bears' are visible on a dispersal in the woods, including another Tu-142MR.

Thus, the operators are presented with processed data from the buoys and can deduce whether a submarine has been detected or not, based on its typical signatures, and, if so, get a fix on it and gauge the nature of its operations.

Kayra-P also incorporates the 15A and 15AK modes for distinguishing the direct signal from an ESG explosion and an echo from the submarine. The 15A mode is designed for receiving signals from MGAB-OZ ESGs, when the depth in the designated area exceeds 5000 m (16,400 ft) and the range to the submarine is less than that. The 15AK mode is intended for processing data generated by detonating MGAB-LZ and MGAB-SZ ESGs. Communication with buoys is handled using 56 communications frequencies at a range of up to 80 km (50 miles) and at a flight altitude of 1500 m (4,920 ft).

Data fed by control panels and all 2Kn-K devices is processed by the BVP subsystem, which comprises an Argon-15 computer with a 208,000 short, addition-type operation capacity. The computer performs dead reckoning of the aircraft's position, buoy impact points and the target's movement, swaps data with other aircraft via the BSS datalink and issues aircraft control commands to the PNK autopilot. Control of the BVP is usually performed by the navigator, although sometimes the RGP operators and nav/TAOs work the system.

A tactical display is provided by the POTO subsystem, which also has an Argon-15 computer. The navigator's and pilot's displays, as well as those of the two RGP operators, show the aircraft's position, ground vector, buoy impact points, position of other assets, and possible submarine locations represented by two-digit signs, vectors and circles. The maximum area displayed by the POTO covers 200 km (124 miles) radius. Having compiled the tactical display, the POTO operates as a one-address computer, swapping data with connected equipment with the exception of the BVP, which exchanges data with the POTO on cue from the BVP itself. An onboard command transmitter is used to prompt RGB-55A sonobuoys to transmit.

Water characteristics, notably the speed of sound through it, are determined using the Nerchinsk measuring system, which operates in conjunction with Kr-P receivers. It comprises an RTB-91 four-buoy bathythermograph device, with the buoys measuring water temperature 200 m (656 ft) down. The Strela-142M communications system, in conjunction with the 2Kn-K, provides tactical data exchange in the area between aircraft, and between aircraft and shore-based command posts.

The 2Kn-K system includes four types of buoys as search assets: RGB-75, RGB-15, RGB-25 and RGB-55A. The Berkut system's sonobuoys could be used as well.

■ **RGB-75 – passive non-directional:** designed for initial submarine search via the submarines' 2-60 GHz infrasonic waveband signatures. The kit comprises 24 RGB-75 sonobuoys whose transmitters operate continuously.

***Below:** Photographed from a NATO fighter, a 'Mod 3 Foxtrot' has its main stores bay open. In the 1970s and 1980s, when SLBMs formed the principal line of strategic deterrence, no game was more serious than underwater warfare, and chances like this for intelligence-gathering were eagerly taken. Fighter crews were ordered to get in close with hand-held cameras to try and get photos of what the 'Bear' held in its belly. During several tense intercepts, fighters were reportedly 'bombed' with sonobuoys as they passed underneath.*

In 1976 the Indian Navy received five radar-equipped Lockheed Constellations from the Air Force to provide a long-range maritime patrol capability over the Arabian Sea and Indian Ocean. The 'Connies' soldiered on until 1984 when they could no longer operate effectively. Their withdrawal left a considerable gap in India's maritime surveillance effort, which needed to be filled with some urgency. In 1986 the Tu-142 was chosen to restore the lost capability, and from 30 March 1988 to October that year eight 'Bear-F Mod 3s' were delivered. The aircraft were essentially similar to the standard Tu-142MK, but had some sensitive items of equipment either deleted or downgraded for export. They were designated Tu-142ME, or Tu-142MK-E in some sources.

Indian Naval Air Squadron (INAS) 312 'Albatross' had operated the Constellations, and was chosen to fly the 'Bears'. The Tu-142MEs wear a depiction of the ocean-going bird on the nose, as displayed by this aircraft during a visit to Russia, probably for overhaul.

Approach to the marker is performed by the A-081 radio compass. The buoys mount no beacons, so their impact points are recorded by the BVP. Submarine acquisition range was supposed to equal 20-30 km (12.5-18.5 miles). Each buoy lasts four to five hours, weighs 9.5 kg (21 lb) and its hydrophone submerges to 150 m (492 ft).

■ **RGB-15 – passive non-directional:** designed to spot submarines by their acoustic signature in the infrasonic and sonic wavebands (2-5,000 Hz), as well as search and pinpoint submarines in conjunction with ESGs. The kit consists of 16 continuous-operation sonobuoys. The airborne operator sets the threshold level of data reception, based on the average noise level from all sonobuoys deployed. Again, the buoys have no beacons. Prompted by the radar, buoys perform negative modulation of the data transmitter's carrier frequency (a similar design was used by the US Navy's AN/SSQ-23 sonobuoys) and the radar scope displays a blip similar to that fed by the beacon. On board the aircraft the Kr-P equipment is used to analyse the spectrum of the signals in the 2-60 Hz waveband visually, and up to 5,000 Hz waveband by ear. When ESGs are used, RGB-15s receive direct signals and echoes from submarines, at a range of 5-10 km (3-6 miles). When operating without ESGs, the RGB-15's range is three to four times shorter. The RGB-15 weighs 9.5 kg (21 lb) and operates for 2 hours, with its hydrophone dropping down to 20, 150 or 400 m (66, 492 or 1,312 ft).

■ **RGB-25 – passive directional:** used to pinpoint a submarine and gauge its movement, the RGB-25 operates at 11 kHz. The buoy submerges down to 20 or 150 m (66 or 492 ft) where it is rotated at 6-12 rpm by an electro-mechanical drive to scan the area. Its bearing detection accuracy is 3°, endurance amounts to at least 40 minutes and weight is 45 kg (99 lb). The radio-frequency kit consists of 10 buoys fitted with self-contained beacons.

■ **RGB-55A – active non-directional:** designed to pinpoint low-noise submarines and gauge their operating patterns before they attack, or to track them briefly. It is also effective in measuring the radial component of a submarine's speed. Following its splashdown, the buoy operates in the passive mode. Only after the aircraft issues a relevant command do the buoys begin 'pinging' with 150, 300 or 600-microsecond acoustic pulses. Echoes are relayed back to the aircraft. Data from two or three sonobuoys provide the fix on the submarine, and describe its pattern of movement. The kit includes 16 sonobuoys, split into four subkits. The 55-kg (121-lb) buoy lasts for up to an hour and deploys its acoustic system down to 20 or 200 m (66 or 656 ft).

When searching for submarines with RGB-15 buoys, three types of ESG can be used: the MGAB-OZ, MGAB-LZ and MGAB-SZ. The -OZ weighs 200 or 800 g (7 or 28 oz) and is used in deep-water areas to pinpoint submarines at ranges from the buoy equal to the distance to the bottom. It detonates at 25, 150 or 400 m (82, 492 or 1,312 ft). The -LZ is employed in areas of ocean with a flat bed and features a 2-m (6-ft 7-in) long, 100-g (3.5-oz) string charge. The charge is flushed out of the bomb casing by water pressure and is detonated at a predesignated depth. The MGAB-SZ bomb is designed for use in challenging hydrologic environments and houses a 200-g (7-oz) 40-coil spiral charge. When set off, the charge generates a series of 4,000-Hz pulses whose number depends on that of the coils. The detonation depth includes 20, 150 and 400 m (82, 492 or 1,312 ft). The aircraft carries 11 such charges.

Magnetic anomaly detection is undertaken by the MS-106 Ladoga system. The magnetometer is a ferrosound design consisting of a magnetic sensor in the upper vertical fin, an orienting system, a measuring device and other support devices. It differs from its predecessors in the two filters it utilises, one of which is used for search and the other for compensating for the aircraft's magnetic fields. The compensation system consists of three coils positioned along the aircraft's axes. The magnetometer's recorder receives signals for the previous 10 seconds, which facilitates their identification but causes a delay in their registration. The acquisition signal is displayed by the 'Signal Signature' panel, and by the magnetometer recorder's chart strip. The magnetometer acquisition range depends on a number of variable quantities but is roughly 600-700 m (656 to 765 yards). To operate the MS-106 in the early days, the crew had to switch off the generators of the first and fourth engines, the hydraulic pump, the autopilot and some other equipment. Later, the electric circuitry was improved.

Operational loadout

Tu-142Ms can operate in both the patrol, and seek-and-destroy configurations. For the latter they normally carry 66 RGB-75s, 44 RGB-15s, 10 RGB-25s, 15 RGB-55As, three AT-2M torpedoes and APR-1 or APR-2 ASW missiles. When RGB-15s are used for initial search with the use of ESGs, the K-142M container is loaded with up to 240 MGAB-LZ or MGAB-OZ charges, and 11 MGAB-SZs. Air-launched mines can be carried too.

The APR-2 anti-submarine torpedo boasts a sophisticated guidance package. After dropping it stabilises its roll and descends on a parachute with a vertical speed of 50 m (164 ft) per second. Following splashdown, the first-stage lock switches off, the command unit kicks in and the missile begins searching in a spiral manner with its engine off at an angular speed of 20° per second. The homer operates in the active/passive mode. With a target acquired, the engine starts and the missile homes in on its target. If no

Right: INAS 312 initially flew from INS Hansa *at Goa-Dabolim, from where it was ideally placed to cover the seas to the west and south of India. This base ('DAB' tailcode) also housed the Il-38 'Mays' of INAS 315.*

target is detected, the missile continues its search on a preset programme.

Naturally, the Tu-142M's ASW techniques underwent some radical enough modifications, which only added workload for the aircrew. Unlike the Tu-142, the Tu-142M employs the NPK-142M for flying to the designated zone. The 2Kn-K system in this case is not used. Entering a 1200 x 1200-km (745 x 745-mile) tactical zone, the BVP and POTO subsystems are switched on. The crew can use the BSS-142 to get situation updates received by the BVP and displayed by the POTO. Prior to that, the search area and its approaches are scanned with the radar for ships, whose noise would interfere with the sub hunt. In hunting for submarines in a designated area, sonobuoys are used, the magnetometer only employed on rare occasion.

'Swinging Tacks' and 'Converging (Diverging) Box' techniques are performed to deploy sonobuoys. Linear barriers are deployed by approaching a predesignated point from either a predetermined or spontaneous direction. To monitor the buoys, the 'Fixed Dimension Circuit Flight' technique is used, as with the Berkut system. If a submarine has to be pinpointed following its detection by RGB-75s and before tracking begins, RGB-15s or RGB-1As are dropped in a closed line via the 'Envelopment' technique. Further tracking is performed via the 'Semi-envelopment' or 'Linear Barrier' techniques, with the use of whatever buoys are available. To search and track stealthy submarines, RGB-15 sonobuoys were supposed to be used in concert with ESGs, or RGB-55As in the active mode. However, given the limited number of the latter, tracking with them was often not feasible.

At all stages of searching and tracking, RGP operators and the navigator process situational data through the use of the BVP and POTO, and perform visual and audio target evaluation and classification. Depending on the tactical situation, targets are engaged following their tracking or immediately after they have been detected and pinpointed. With the mission in the tactical area accomplished, the aircraft returns to base the way it does after a flight to a predesignated area. In combat, elements of the mission would have been accomplished in a different fashion.

Tu-142M/MK in service

In production at Taganrog from 1978, Tu-142Ms began entering the inventory of the 76th OPLAP DD of the Northern Fleet in 1979. The new system, with its numerous attendant 'bugs', slowed the conversion tempo. Snags were encountered by both air and ground crews. Trials had failed to reveal all the flaws of the surveillance and targeting system, which was justified officially by insufficient funds for testing. However, the real reasons were something else. It had become a general practice that even relatively unsophisticated systems would be fielded with loads of deficiencies.

When the Zarechye system development began, the Tu-142M with its Kn-K system was redesignated as the Tu-142MK, while the new variant with the Zarechye system was designated Tu-142MZ.

Above: The Indian 'Bear' population moved out of Dabolim to Arrakonam, resulting in a change of tailcode to 'ARK'. The fleet has been widely used in support of Indian Navy surface operations, especially during times of tension with Pakistan.

Above left: A Tu-142ME from INAS 312 is escorted by a pair of Sea Harriers from INAS 300, shore-based at INS* Hansa*. The 'Bear' squadron is thought to have lost one aircraft in an accident.

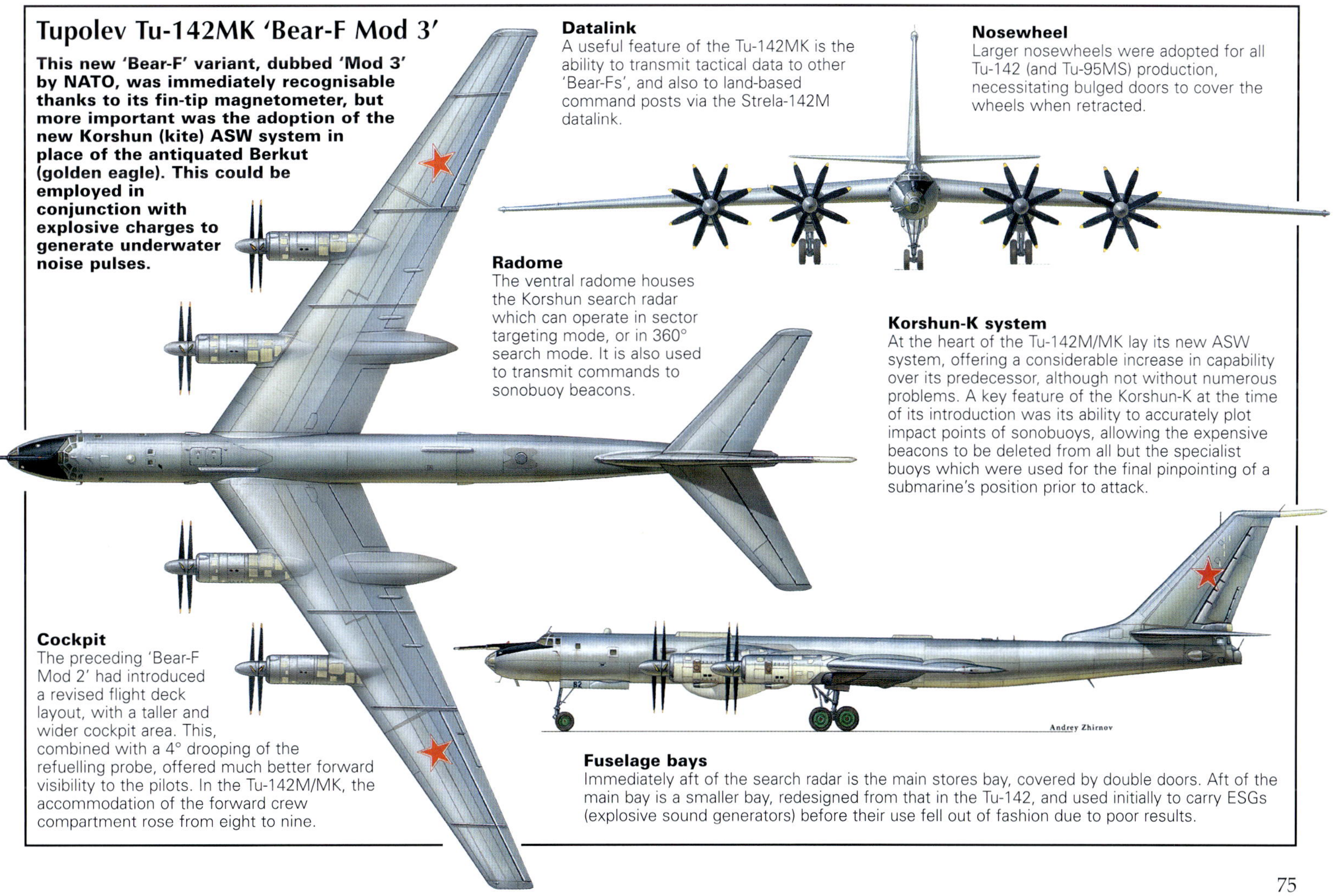

Tupolev Tu-142MK 'Bear-F Mod 3'

This new 'Bear-F' variant, dubbed 'Mod 3' by NATO, was immediately recognisable thanks to its fin-tip magnetometer, but more important was the adoption of the new Korshun (kite) ASW system in place of the antiquated Berkut (golden eagle). This could be employed in conjunction with explosive charges to generate underwater noise pulses.

Datalink
A useful feature of the Tu-142MK is the ability to transmit tactical data to other 'Bear-Fs', and also to land-based command posts via the Strela-142M datalink.

Nosewheel
Larger nosewheels were adopted for all Tu-142 (and Tu-95MS) production, necessitating bulged doors to cover the wheels when retracted.

Radome
The ventral radome houses the Korshun search radar which can operate in sector targeting mode, or in 360° search mode. It is also used to transmit commands to sonobuoy beacons.

Korshun-K system
At the heart of the Tu-142M/MK lay its new ASW system, offering a considerable increase in capability over its predecessor, although not without numerous problems. A key feature of the Korshun-K at the time of its introduction was its ability to accurately plot impact points of sonobuoys, allowing the expensive beacons to be deleted from all but the specialist buoys which were used for the final pinpointing of a submarine's position prior to attack.

Cockpit
The preceding 'Bear-F Mod 2' had introduced a revised flight deck layout, with a taller and wider cockpit area. This, combined with a 4° drooping of the refuelling probe, offered much better forward visibility to the pilots. In the Tu-142M/MK, the accommodation of the forward crew compartment rose from eight to nine.

Fuselage bays
Immediately aft of the search radar is the main stores bay, covered by double doors. Aft of the main bay is a smaller bay, redesigned from that in the Tu-142, and used initially to carry ESGs (explosive sound generators) before their use fell out of fashion due to poor results.

Causing a minor stir when it was first seen by the West in 1985, the Tu-142MZ introduced a wealth of new antennas grouped under the tail and around the nose. The small 'pimple' on the extreme tip of the nose, an antenna for the Geran jamming system, drew particular interest as a similar unit had also appeared on Tu-95K-22 missile-carriers. Speculation at the time – at least in the public arena – suggested it might even be a terrain avoidance/ following radar, although quite why a 'blue-water' ASW platform would need such equipment defied logic. This MZ was intercepted in 1988, while undergoing state acceptance trials.

With some experience in operating the Tu-142MK gained, it became possible to evaluate its capability. As was expected, submarine detection in medium and heavy shipping environments became much more difficult. At least 70 per cent of the submarines spotted by RGB-75 sonobuoys were acquired through spectral analysis of distinguishing signatures by means of discrete components. Problems emerged as far as automatic contact classification was concerned, with operators having to make drawings of noise spectra seen on displays, and then compare them with new displays during another cycle. A decision on a target's presence would be made on the basis of stable emergence of new spectral components from different buoys plotted against the earlier background. This technique was time-consuming and prone to errors. Operations revealed that in 20-30 per cent of contacts, signal processing by different Kr-P channels was not identical. The developer recommended the signals be controlled by means of a single RGB-75 and eliminate the channels providing different results. However, in such a case one or more channels would be precluded from data processing.

Since the RGB-75 range proved to be less than advertised, search techniques had to be revised: following an RGB-75 buoy kit deployment, the RGB-15 kit would be deployed to avoid data channel interference. In a couple of hours it had to be updated. This necessitated the expansion of the RGB-75's frequency channels at least twofold. When an RGB-15 sonobuoy contacted a submarine, its reaction time was 24 seconds: that of an RGB-75 was 400-600 seconds.

Other snags were hit during conversion to the RGB-15 and ESGs. To expand the areas being scanned, a buoy and an ESG had to be dropped at the same place, which reduced bottom interference but caused ESGs to splashdown and detonate before the buoys could kick in. Hence, to deploy an ESG close to a buoy, the aircraft had to make another pass at least 4-6 minutes later. In shallows less than 1000 m (3,280 ft), ESGs were recommended to be detonated at 2-3 km (1.24-1.86 miles) away from the sonobuoy to produce an elliptical coverage area, whose focuses were the buoy and the ESG.

Echoes were normally received against a background of interference, therefore two to five ESGs had to dropped to the same point to get a reliable picture. For an operator to single out an echo via the audio channel, he had to be properly trained and have an ear for music. Low reliability, problems with echo selection, search signatures and other reasons led the air branches of the Fleets to virtually abandon ESG-equipped sonobuoys.

There were other problems. For instance, the RGB-55A's capabilities failed to be used to the fullest. Duration of an echo has to be at least 350-450 milliseconds for a human to be able to discern a difference in its tone. Such a period provides a chance to spot the Doppler effect. However, this duration was unavailable. Flying crews noted a considerable flaw by the developer – the navigator was deprived of a Kr-P display and was unable to participate in evaluating the hydroacoustic situation. The RLP subsystem got its share of flak for the size of the gate in the radar scope, which prevented efficient targeting when attacking pinpoint targets.

First operational forays

Tu-142Ms launched routine operations under combat duty plans in early 1979, with the first unidentified submarine being detected and tracked for 70 minutes as early as April the same year. In early 1981, search operations spanned 4200 km (2,610 miles) but the duration of flights had to be reduced owing to unpredictable weather in the north. The sorties were given thorough after-action reviews to gauge actual fuel consumption and train crews for transatlantic flights. In doing so, account was taken of the fact that the Tu-142M's fuel consumption per hour had grown by 500 kg (1,102 lb) from the Tu-142's 5500 kg (12,125 lb).

On 14 March 1983 a pair of Tu-142Ms made the type's first sortie to Cuba's capital city of Havana. Deployed to Cuban airfields, aircrews were searching for submarines in the Sargasso Sea and reported six submarines detected and tracked for a total of 11 hours 40 minutes. Sorties to Cuban airfields lasted till 1984, with variable frequency, and were terminated due to a decision made to have sonobuoys improved.

In 1984, several Tu-142s were outfitted with Uzor-5V 16-channel FM data recorders. Analysis of results provided the grounds to reproach the buoy developer over its product's poor information capability and efficiency. A 1985 exercise proved that crews were often unable to correctly evaluate the contact authenticity for reasons beyond their control. In the next year, a meeting held by the minister of aircraft industry took a decision to conduct additional trials.

Naval aviation takes a bow: a Tu-142MZ leads a pair of Su-33s over the Zhukovskiy air show. A feature of the 'Mod 4 Foxtrot' was a new conduit fairing on the port side, running for much of the length of the fuselage. This houses a waveguide linking fore and aft elements of the electronic warfare suite.

These trials proved that the RGB-25 and RGB-75 were inefficient in sea states 3 to 5, and technical requirements had to be worked out for their replacement. The Tu-142M fleet was scheduled for improvement by 1989, following preliminary tests in 1988. At the same meeting a decision was made to replace the RGB-75 and RGB-15 with the common buoy RGB-16, operating in both continuous emission and autostart modes.

In other words, it was admitted that the 2Kn-K surveillance and targeting system failed to meet expectations. It, however, was a quantum leap compared with the Berkut, since it incorporated low-frequency and infrasonic buoys, and ESGs. The RGB-75 and RGB-15 designs had become cheaper and simpler by abandoning the beacon and autostart circuitry, there were more advanced data processing techniques, and the bulky RGB-3 had been replaced by the long-range RGB-55A. 2Kn-K gave the crews an opportunity to measure the speed of sound in water, while the tactical situation was shown on displays and could be datalinked automatically.

Of course, the main criticisms should have been taken by the research institutes and the customer organisations, which erroneously portrayed foreign-made submarines as very noisy (it is possible that it was done in coordination with the industry). Naturally, there was nobody to lay the blame for that on.

This was not the only problem. When long-range missions began, the air complex's deficiencies became obvious. Aircrews reported that the operation of the computer within the NPK-142M was below par seven or eight times out of 10. Navigators had to continuously enter corrections, set various measuring devices and correlate their readings. Over a brief period of operation, all three channels of the inertial attitude and reference system in the backup navigational loop failed together over two dozen times, leaving the pilots' and navigators' instrument unable to display the heading. This prompted numerous requests for a backup compass system. Crews asked for the ANP-3V, noting caustically enough that it would be a good replacement for the whole NPK-142M in terms of accuracy. Quite reasonable was the suggestion that the aircraft be outfitted with satellite navigation aids.

As the strategic scenario evolved, so the ASW aircraft fleet's task was modified. Along with developing new foreign submarine patrol areas, supporting the deployment and combat patrols of friendly 'boomers' and some other missions climbed high on the agenda.

The directive of the chief of naval operations, dated 30 October 1983, ordered the 35th ASW Long-Range Air Division activated before 31 December 1983 to comprise the 76th and 135th ASW Air Regiments. Its command was assumed by Colonel Akporisov, who later became the Naval Aviation's chief of staff as a lieutenant-general. There was no need at all to set up such a fancy unit, the reason for its activation was never revealed and the outfit itself was disbanded soon.

'Bears' for India

In 1988, eight Tu-142MKs, redesignated as Tu-142ME, were delivered to the Indian Navy as the only 'Bears' to be exported. They differed from the baseline variant in minor modifications to their avionics suites. They were delivered

Having served in a natural metal finish for most of its career, the naval 'Bear' fleet began to adopt an all-over anti-corrosion grey scheme in the late 1990s. Dielectric panels for antennas and the Ladoga MAD remained in white. A feature of all but the earliest Tu-142s is a white dome above the forward cabin, which houses a remotely controlled sextant for the astro-navigation system. In the Tu-95MS bomber this is replaced by a glazed dome.

Tu-142 'Bear-F' losses

The accident rate of the Tu-142 aircraft has turned out to be very low: for 100,000 hours of flying time, there would only be 3.5 accidents, including two crashes, as detailed below.

6 August 1976: occurred during a wide-approach traffic pattern training flight by Lieutenant Khazagerov, a first pilot candidate, and Major Morozov, a pilot instructor in the 76th independent long-range ASW air regiment of the Soviet Air Force. The report states that 15 minutes after take-off the crew detected a malfunction and started to go through the landing pattern sequence. The aircraft landed at a speed exceeding that required for the given flight weight by 30-40 km/h (19-25 mph). Having rolled straight for 740 m (2,428 ft), the aircraft turned right, rolled off the runway and drove into a deep ditch. The aircraft broke apart with three men dying on the spot.

20 April 1984: two aircraft from the 310th independent long-range ASW air regiment of the Pacific Fleet's air branch were flying in combat formation to their patrol area. The emergency occurred an hour and a quarter after take-off at an altitude of 7200 m (23,622 ft) and a speed of 730 km/h (454 mph). At that time, the aircraft were flying over the Sea of Okhotsk. According to the wingman, the aircraft of the leader, Colonel V.I. Zubkov, started to descend following a blast of black smoke and a white plume on the starboard wing. In about a minute, the aircraft caught fire and started spiralling downward, with a bank increasing up to 60-70°, and crashed in the water, without recovering from its dive. The cause was considered to have been the failure of the No. 3 engine (possibly of the propeller blade or of a turbine disc), which resulted in the severing of the aileron control wires and rupturing of the fuel system, hence causing a fire and putting the aircraft into a spiralling dive from which recovery was impossible.

Some eight years elapsed between the Tu-142MZ's first flight in 1985 and its formal acceptance by the AVMF in 1993. Aircraft were built from new at Taganrog, and a few were completed before production was halted in 1994. The variant entered service with the 76th OPLAP DD at Kipelovo, where this example is seen next to Tu-142MKs. Although studied, further Tu-142 upgrades have been put on hold as the Kn-N/ Zarechye system of the MZ remains capable, and the airframes are young.

Above: This Tu-142MZ was put on display at Zhukovskiy. The MZ inherited new defensive armament from the Tu-95MS, with a four-gun barbette in place of the earlier two-gun armament. At the same time, the lateral observation blisters were deleted.

Right: Close-up of the MZ's nose shows the undernose fairing which mounts a variety of antennas. The 'pimple' below the refuelling probe houses an antenna for the Geran active jamming system. Further Geran antennas are mounted either side of the forward fuselage, and in the tail. Radar warning is provided by the Sayany-RT system. The hemispherical antenna at the rear of the undernose fairing, protected by a cover, is the Mak (poppy) infra-red detection system. Note the landing light in the fuselage, which hinges down from its recess for use.

to INAS 312 at INS Hansa at Goa-Dabolim, adding 'DAB' to the fin to signify their base. Today they remain in service, wearing 'ARK' on the fin since they moved to Arrakonam.

Tu-142MZ

On 4 January 1977, the government passed another decree to upgrade the navy's ASW systems. On 17 July 1979, it was followed by the defence industry's decision to develop the Zarechye system, designed to detect and pinpoint submarines using the RGB-1A/-2 mass production buoys of the Berkut system, the RGB-25/-55 sonobuoys of the Korshun system, and the RGB-16/-26 sonobuoys of the Nashatyr-Nefrit system. The Zarechye system prototype was developed by the Leninets Scientific Production Association, in cooperation with the Kiev-based Underwater Instruments Research and Development Institute. They started developing the system on 2 October 1977 to meet joint performance requirements of the air force and navy, which wanted the system to be fitted on the Nashatyr-Nefrit system with follow-on improvements.

The Zarechye sonar system and its next-generation RGB-16, RGB-26, and RGB-36 sonobuoys were designed to process and display hydroacoustic data, measure its parameters, and provide the Kn-N ASW system with data on detected submarines. New computers, a greater number of data channels, introduction of a separate observation channel for all sonobuoys deployed, and more sophisticated data processing techniques considerably increased the system's efficiency.

Flight tests were to have started in the second quarter of 1982, however they failed to meet the timeframe and the tests had to be put off until the third quarter of 1984. A proposal was made to fit the aircraft with the Sayany ECM system to counter enemy air defence radar- and infrared-guided missiles, and the Kh-35 anti-ship missile. The Sayany system had been under development for several years, and many of its designers had left for their native country (Israel). Flight tests got under way in April 1985 without either Sayany system or the Kh-35.

Tu-142M No. 42172 was the Zarechye prototype. In order to get greater amounts of test information, the aircraft was equipped with various devices for controlling and recording its aerodynamics, electrical equipment, trajectory changes, and the new Kn-N ASW system. Flight tests were conducted over the Black, the Barents, the Norwegian and Okhotsk seas. They ended in November 1986 with 71 sorties flown. An effort to carry out official state tests of the aircraft in August 1987 ended in failure due to some shortfalls in the operation of the RGB-26 sonobuoys. The tests were decided to be conducted in two stages: the first stage was to be carried out without RGB-26 sonobuoys, while the second one was to test the whole set of the aircraft equipment. The second stage of the tests, begun on 13 November 1987, was completed on 30 November 1988. However, the decree on fielding the upgraded anti-submarine system was issued only in 1993. Test reports emphasised that the efficiency of the upgraded Kn-N ASW system was 2 to 2.5 times greater than that of the 2Kn-K, while its kill probability had been increased by 20 per cent due to increased accuracy of detecting a submarine's movements.

Crews who had tested the Tu-142MZ stressed that should the NPK-142 fail, it would be possible to control the aircraft with two operating engines. It seemed quite reasonable to fit the aircraft with auxiliary power plants in order to make do without airfield power sources when getting ready for take-off, as well as softer seats and biostimulants to reduce crew fatigue during long flights. Test navigators pointed out that it was necessary to change the layout of workstations, since the NPK-142M no longer featured the VNPK-154 standby system. They also emphasised poor reliability of the navigation system computer (its failure time totalled nine hours out of 22). During the tests 124 depth bombs of various types were dropped, 24 of which failed to operate.

Production halt

The first Tu-142MZ rolled off the assembly lines in 1993 (the Tu-142M was being simultaneously modernised in maintenance units). At the same time, mass production of the Tu-142 aircraft was stopped in 1994. The Kuibyshev and Taganrog plants had manufactured between them a total of 100 Tu-142s in various versions.

Series Tu-142MZ aircraft differed slightly from the prototype: the NK-12MK engines were replaced by the more reliable NK-12MP power plant, the aircraft incorporated some structural features inherited from the Tu-95MS missile-carrier, and featured an upgraded drop control system. AT-2M torpedoes could no longer be carried, while

Below: Production Tu-142MZs benefitted from advances in turbine technology and were fitted with NK-12MP engines, offering much greater reliability.

Tupolev Tu-142MZ 'Bear-F Mod 4'

76th OPLAP DD
Kipelovo air base

An attempt to redress the shortcomings of the Tu-142MK, the MZ introduced a much better ASW system including a new family of sonobuoys. Buoys from the previous two generations could still be used. The prototype (42172) was a converted Tu-142MK.

UMGT-1 torpedoes, APR-2 anti-submarine torpedoes, and UPLAB-50 training bombs could be carried externally. Naturally, the aircraft was equipped to carry the new RGB-16, RGB-2, and RGB-36 sonobuoys. In addition, the aircraft retained the ability to use the sonobuoys of the 2Kn-K ASW system. Most importantly, the Tu-142MZs were fitted with the modified Kn-N ASW system, in which the Zarechye sonar subsystem, which provides simultaneous processing and analysing of data, transmitted by sonobuoys in audio and subsonic frequencies, replaced the Kayra-P sonar.

Main differences between the Kn-N and the 2Kn-K ASW systems included the following: a TsIKL control mode, which allows the flight to be simulated on the ground; A-311 new radio-navigation system; improved ergonomics of the crew workstations; and the observation channel, which automatically decided whether a sub had been detected when working with the RGB-1, -16 and -26 buoys. In addition, the new system featured Trek (track) and Amplitudnyi Trek (amplitude track) modes, which made target detection considerably easier when processing signals transmitted by the RGB-16 and the RGB-26 sonobuoys.

Other differences included: the frequency range of the RGB-16 sonobuoys was increased from 24 to 64, which lifted limitations on their joint deployment as their data transmitters operate continuously; mode 26 was introduced to carry out passive tracking with direction-finding, based on discrete components of noise signature of surface vessels and submarines; one of the data processing modes, transmitted by the RGB-36 sonobuoys, allowed the crew to simultaneously define the range rate, the range and the direction of a submarine in the active operation mode; the Kr-K system is capable of simultaneously processing data from eight RGB-16 sonobuoys in subsonic and low-sound frequencies, as well as passive detection; and measures were taken to increase the likelihood of detecting the signal against the background of ocean-floor noise with the help of depth bombs.

Kn-N employs three types of sonobuoys:

- **RGB-16 – passive wide-range:** capable of receiving acoustic noise within a frequency range of 2 to 5,000 Hz with a submarine detection range of 12 to 15 km (7.5-9.3 miles). It weighs 9.5 kg (21 lb) and can operate for up to five hours (up to six hours with an automatic start). It boasts a more efficient underwater jamming-proof capability as compared to the RGB-75 and RGB-15, and a lower level of its own background noise within the low frequency range.
- **RGB-26 – passive low-sound:** capable of tracking submarines in the passive mode with direction-finding of their noise signatures by discrete components. Operation time is 1.2 hours, while its weight is 13.9 kg (30.6 lb).

Overall, the ASW complex of the Tu-142MZ is known as Nashatyr-Nefrit (ammonia-jade). While this introduced a new and much improved sonar system in the form of the Zarechye suite, with a far greater degree of automation compared with the preceding 2Kn-K system, much of the avionics equipment was retained from the Tu-142MK, including the Korshun radar and the oft-criticised NPK-142M navigation system. Comprehensive electronic defences were added, resulting in a plethora of new antennas, while auxiliary power units were installed to allow autonomous operations from austere locations. Internally, many of the crew work-stations were improved with new layouts.

Left: Despite the addition of much new equipment, the navigator in the Tu-142MZ retains a station in the nose section, with a swivelling seat to allow the occupant to look forward through the glazed nose, or to the main instrument console to starboard. The navigator's domain is accessed via a crawlway between the instrument panels for the pilot and co-pilot.

***Above:** In its initial form (this is the prototype at Taganrog) the Tu-142MR was based on the Tu-142MK and featured a standard glazed nose, underneath which was a large radome housing the Groza-134VR radar. The aircraft lacked the port-side waveguide fairing of later production aircraft. The drogue for the VLF trailing wire antenna is clearly visible.*

***Right:** Existence of the 'Bear-J' was publicly acknowledged in the West in 1986, although the exact nature of its role was not clear at the time. This view highlights the additional communications antennas on the lower rear fuselage sides.*

■ **RGB-36 passive-active directional:** capable of detecting a submarine's range rate, range and direction in the active operation mode. It weighs 58 kg (128 lb).

Tu-142MR relay aircraft

With the development of the Soviet submarine fleet came the need to maintain steady communication with submarines while submerged in remote areas of the ocean. As a result, a *retranslyator* (relay aircraft) version – the Tu-142MR 'Bear-J' – was created, based on the Tu-142M 'Bear-F Mod 3.' The governmental decree on the development of the Tu-142MR was issued on 3 October 1972. It was designed to carry out semi-automatic and automatic relay of messages, transmitted by coastal command posts in HF (high-frequency) and VLF (very low-frequency) ranges, and to relay messages transmitted by submarines, surface vessels, and naval command posts, to submerged subs in VLF frequencies.

In order to facilitate operation of the communications and relay system, the aircraft was equipped with the following autonomous systems: AC power supply system with turbo generators; hydraulic system for aerial extension device actuators; and a cooling system for the automatic waveguide tuner of the VLF transmitter. The structure of the tailfin and the leading edge was also slightly changed to incorporate antennas.

In addition to the pilot and the co-pilot, the crew incorporates the navigator, flight engineer, four communications system operators and the gunner (in the tail cockpit). The nose section houses the Groza-134VR radar and the aircraft instrumentation system (if the aircraft is not fitted with the Sayany-RT system, the radome is situated under the fuselage). The canopy is VLF-proof – the forward silicate canopy and windowpanes boast translucent coating, bonded to the airframe. The Plexiglas of the pilot's and the navigator's windows are provided with detachable frames with a 1-mm thick steel (tinned) wire mesh. Pilots noted that the protective mesh impairs visibility and increases eye fatigue during long flights. The navigator's window mounts four tinned brass wire meshes. The radio operator's cabin is also protected by steel netting. The exterior surfaces are coated with colourless varnish and special backing enamel.

The aerial extension device, equipment and distribution devices of the power supply system, Etyud relay equipment, and transmitter and receiver units are all housed in the cargo bay of the aircraft. The forward part of the extending aerial external unit consists of an air intake, heat exchanger, two turning sections, front flap panel and locking device system. There is a fairing on the spine. A steel zinc-coated 7.5-mm (0.295-in) thick wire represents the aerial. It was determined that, following a certain number of cycles, the cable was likely to become jammed due to accumulation of twists and back-spin.

Maximum extension speed of the aerial is 5 m (16.4 ft) per second, while it retracts at up to 4 m (13.12 ft) per second. It takes 37 minutes to extend it to its full length of 7680 m (25,197 ft) and 48 minutes to reel back in. Flight altitude with a fully extended aerial is between 8000 and 10000 m (26,247 and 32,808 ft). The extending aerial external unit and the air intake increased drag considerably. With the aerial extended at an average flight weight of

Tu-142MRs entered full service in 1982 and were occasionally encountered by NATO interceptors. They were occasionally seen operating with 'Bear-Fs', as in this 1989 intercept by USAF F-15s of a Tu-142MR and Tu-142MK. When communicating with submarines by VLF, the Tu-142MR deploys its wire aerial to full length and flies a circle to keep the aerial as vertical as possible. At high weights this entails the aircraft operating very close to the stall, placing great strain on the pilots.

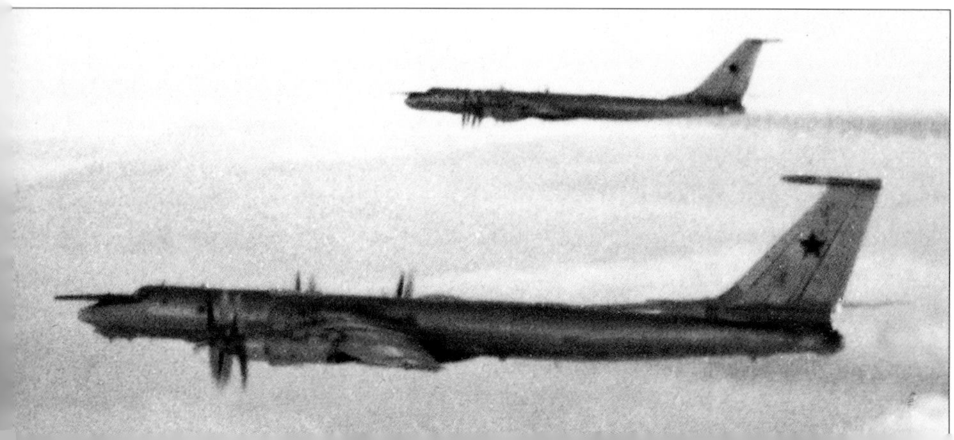

138 tonnes (135.8 tons) and speed of Mach 0.66, fuel consumption rises to 5.3 kg/km (18.8 lb/mile).

When the weight exceeds 160 tonnes (157.5 tons), flight is conducted at angles of attack which are close to critical, and calls for close attention, and may even be dangerous in turbulent air. Under such conditions, it is almost impossible to engage the autopilot as it constantly switches off. The aircraft communications system was developed and manufactured by the Gorkiy-based Radio Communications Research and Development Institute. The system comprises two HF radios, two VHF and UHF radios, a VLF radio transmitter, two sets of VLF receivers, seven HF receivers, relay equipment, alphanumeric communications equipment and a scrambler. Control panels and interface units provide the equipment with a capability of simultaneous operation in VLF and HF frequencies.

Tu-142MRs are fitted with the Groza-134VR pulse radar to detect radar activity areas and carry out surface surveillance. The radar features a horizontally-searching aerial and an indicator of the azimuth-to-range type. All radar units, except for the antenna, are located in the pressurised compartment. The radar's detection range against large targets like industrial sites is 250 km (155 miles), while in storm clouds it is reduced to 140 km (87 miles).

Tu-142MR first-stage flight tests, totalling 45 sorties, lasted from 23 March until 20 October 1978. The second stage began on 3 April 1979 and finished on 5 May 1980. These flights were conducted from Kirovskoye airfield, Crimea, and Kulbakino airfield, Nikolayev. In 1982 the Tu-142MR aircraft entered the inventory of the Russian Armed Forces. The aircraft's cruising range is 12315 km (7,652 miles), its cruising speed is 735 km/h (457 mph), and its flight ceiling is 10700 m (35,105 ft). Take-off run is 2300 m (7,546 ft) and landing roll is 1800 m (5,905 ft).

Conclusion

The Tu-142's fate was similar to that of the Tu-95RTs, which were also constantly upgraded. There is no doubt that the wings of fast heavy aircraft are not designed for low-altitude flights, and from March 1985 a regular series of groundings has affected the fleet. In 1986 operation of the Tu-142M, built in 1978-86, was temporarily suspended due to cracks in the landing gear fairing, and it was only in April 1989 that the joint decision by the Ministry for Aircraft Industry, the air force, and the navy on rectifying the problems was adopted. Three times the navy has appealed to the defence industry and 12 times to the Ministry for Aircraft Industry over the issue of poor reliability of its long-range aircraft, Tu-142 included.

As noted above, aside from some minor airframe/aircraft system improvements, the modernisation of the Tu-142 followed a series of modifications to the mission equipment, as summarised below:

- **Berkut-95:** with audio frequency sonobuoys. Apart from the extra range, the Tu-142 was no better than the Il-38, although the latter did not pose the same levels of maintenance problems as the 'Bear'. The Berkut-95 featured a number of shortfalls: high operation costs due to the necessity of massive employment of expensive sonobuoys, which had a limited submarine detection range; poor reliability of computers; lack of visual picture; and low automation level of processing raw data from transmitters. Almost all Tu-142s, fitted with the Berkut-95, had left the AVMF inventory by 1992.
- **Korshun (2Kn-K):** replaced Berkut-95 in 1980, including the RGB-75 and the RGB-15 low-frequency sonobuoys, which improved initial submarine detection capability. It also incorporated more sophisticated data processing systems, as well as depth bombs. Another important feature of the Korshun system was that its sonobuoys were much cheaper, thanks to the removal of transmitting beacons and automatic starts, and a shift from the bulky RGB-3s to range-finding sonobuoys. The tactical display had its own computer and was capable of displaying and transmitting tactical data from one aircraft to another and from the aircraft to the command post. Navigation and tactical functions were divided. Efficiency of the MMS-106 magnetometers, mounted on the aircraft, turned out to depend on many systems, and it was only in the 1990s that flight crews noted some increase in their efficiency. One of the shortfalls of the Korshun system was the low automation level of initial processing of data, and a great rate of false alarms. The navigation system also had some shortfalls. The aircraft crew received another member.
- **Zarechye (Kn-N):** the shortcomings of sonobuoys of the 2Kn-K led to a replacement of the hydroacoustic subsystem, including a new family of buoys. New weapon systems were developed simultaneously with detection systems.

By mid-1996 the air branches of the two fleets operated 15 Tu-142s, 20 Tu-142MKs and 12 Tu-142MZs, more than half of which were in reserve storage. The Tupolev OKB offered to upgrade the Tu-142Ms, replacing their avionics, fitting them with new anti-submarine torpedoes, improving the airframe, and equipping them with the Morskoy Zmey (Sea Dragon) avionics. However, for a number of reasons, including financial restraints, no positive decision has yet been taken.

Lt Col Anatoliy Artemyev (Soviet/Russian navy, Retd); translation by Zaur Eylanbekov

Most, if not all, production Tu-142MRs were much closer to the Tu-142MZ in antenna configuration than the prototype. They had the Sayany-RT warning system installed in an undernose fairing and under the tail, linked by waveguide fairing on the port side, while the tail section incorporated the MZ's four-gun turret and lacked observation blisters. Addition of the Sayany-RT nose fairing in turn required the relocation of the Groza-134VR radar to a large thimble radome on the extreme nose in place of the glazed section.

Production of the Tu-142MR is thought to have numbered around 10 aircraft, with operational aircraft probably divided between Northern and Pacific Fleets. This view of a 76th OPLAP DD 'Bear-J' on the Kipelovo flightline shows details of the Groza installation, as well as the three most obvious external communications features: underbelly fairing for VLF trailing wire antenna, fin-tip probe antenna for HF communications, and spine bulge for an undisclosed, but probably satellite, communications system.

In later years free-fall bombers such as this Tu-95 'Bear-A' posed little threat to NATO or the US, and were used mainly for training. Nevertheless, they were watched closely by NATO fighters. Such missions could be dangerous, often requiring the interceptor to make an 'ID run' at close quarters under the 'Bear' to read its nosewheel numbers or to photograph unusual antennas. Captured by the crew of another 'Bear', this remarkable photo shows an F-4E from the Iceland-based 57th FIS. At such close quarters the Phantom crew would be almost deafened by the growl from the Tu-95's thrashing blades.

Tu-95 - missile-carriers

At the beginning of the 1970s there were five large air bases within Soviet strategic aviation: Uzin, Semipalatinsk, Mozdok, Engels and Ukrainka. Myasishchev M-4 and 3M 'Bison' bombers were stationed at Engels and Ukrainka, whereas the three remaining bases accommodated Tu-95 'Bears'. Two regiments – the 409th TBAP (Tyazhelobombardirovochnyi Aviatsionnyi Polk, Heavy Bomber Air Regiment) and the 1006th TBAP – were stationed at Uzin, Ukraine, and two more were stationed at Semipalatinsk, Kazakhstan: the 1223rd TBAP and the 1226th TBAP. The fifth Tu-95 unit – the 182nd TBAP – was stationed at Mozdok in Northern Ossetia. The main targets for Soviet strategic bombers were in the USA and, for the air division stationed at Ukrainka, China. The location of fixed air bases for heavy bombers in the southern part of the Soviet Union should not be misleading: strategic aviation uses forward 'jump' airfields at Anadyr, Tiksi, Magadan and Vorkuta, which are located much closer to the US.

Apart from special versions such as the Tu-95RTs maritime reconnaissance aircraft and Tu-142 anti-submarine warfare aircraft, production of the original 'Bear' came to an end in 1965. Powered by temporary 2TV-2 engines, the Tu-95 (or *izdeliye* V) prototype undertook its maiden flight on 12 November 1952 with Alexey Perelyot at the controls. It crashed on 11 May 1953, killing the crew. The prototype of the final version, powered by four powerful Kuznetsov NK-12 turboprops, took-off only two years later on 16 February 1955, piloted by Mikhail A. Nyukhtikov. In October 1955, production plant No. 18 at Kuibyshev (formerly, till 1935, and presently, since 1991, the town's name is Samara) built the first two series aircraft. From 1955 to 1965 the factory made 49 'Bear-A' bombers (30 Tu-95s and 19 Tu-95Ms) and 71 missile-carriers (48 Tu-95K 'Bear-Bs' and 23 Tu-95KM 'Bear-Cs'). Several other variants which saw service were only small modifications of the above-mentioned versions: the Tu-95A and Tu-95MA were adapted for dropping nuclear bombs, the Tu-95KD received an inflight refuelling system, and the Tu-95MR 'Bear-E' had additional reconnaissance equipment. The Tu-95U and KU were trainer conversions.

Since the basic airframe has remained unchanged for many years, the history of the Tu-95 is, in essence, the history of its armament. The heavy Kh-20 missile carried under the fuselage of the Tu-95K (or Kh-20M of the Tu-95KM) became obsolete after only a few years: it was too slow to avoid anti-aircraft defences, and its seeker needed the target to be illuminated by the aircraft's radar, so the aircraft had to approach dangerously close to the target. The Kh-20M (AS-3 'Kangaroo') missile weighed 12000 kg (26,455 lb) and carried a 3-MT nuclear warhead over a distance of 450-650 km (280-405 miles) at a speed of 2120 km/h (1,317 mph). The 'simple' Tu-95 and Tu-95M were proven to be even less useful – they had no missile armament so that they had to directly overfly the target to drop their bombs.

Second-generation missiles

Dramatic changes in aircraft armament came about at the beginning of the 1970s. It was decided then to transfer on to the Tu-95 the much more modern missile systems developed for the Tu-22M 'Backfire' and Tu-16 'Badger' bombers. According to a government resolution of 13 February 1973, modernisation was planned in two separate programmes. Fifty recently made Tu-95KMs (from series production of 1961-1965) and Tu-95Ks (from production of 1958-1962) were to be equipped with the recent and most powerful K-22 missile system, which employed the Kh-22 missiles developed for the Tu-22M. The less advanced (but cheaper) K-95-26 missile system, with KSR-5 missiles from the Tu-16K-26 'Badger-G' bomber, was to be provided for 33 older Tu-95s and Tu-95Ms.

Right: Giving the Tu-95K-22 a conventional bombing capability was the use of external pods, each holding 15 250-kg (551-lb) bombs. These were carried, as here, on the three existing missile pylons. A Tu-95KD 'Bear-B' had earlier been tested with two containers on a centreline attachment. The fins on the rear ensured that the pods separated cleanly from the aircraft if jettisoned. The intended role for this configuration was anti-airfield attack, saturating the target with all 45 bombs dropped in a single pass.

Above and left: Primary armament for the Tu-95K-22 was the Kh-22 missile (AS-4 'Kitchen'), which could be carried singly on the centreline or on two underwing pylons. This aircraft was fitted with three, although this was not an operational load. 'Bear-Gs' and their Kh-22s were mostly targeted against US Navy carrier battle groups.

This K-95-26 air-missile complex was to include the Tu-95M-5 aircraft (*izdeliye* VM-5), two KSR-5 missiles, and Volga missile initialisation and launching system. The KSR-5 (AS-6 'Kingfish') missile, developed by the OKB-2-155 design bureau (now MKB Raduga) at Dubna near Moscow, was in series production from 1966. This was a simplified version of the Kh-22 (see below) designed for the modernisation of Tu-16s. The KSR-5 (*izdeliye* D-5) was equipped with an active radar seeker and was intended for attacking targets offering a high radar contrast, such as ships, bridges and power stations. The missile weighed 3900 kg (8,598 lb), including a 700-kg (1,543-lb) warhead. Dimensions were: length 10.52 m (34 ft 6 in), diameter 0.90 m (2 ft 11 in) and wing span 2.61 m (8 ft 7 in). The KSR-5 flew at up to 3200 km/h (1,988 mph) speed and its range was 240 km (149 miles).

KSR-5s were carried by the Tu-95M-5 on two beams under the wing root, freeing the fuselage bomb bay for other armament. The Rubin-1KV radar was identical to that fitted in the Tu-16K-26. A Tu-95M-5 prototype was converted from series Tu-95M No. 601 from the 409th TBAP at Uzin. The conversion process was protracted: the aircraft arrived at the Kuibyshev factory in January 1973 and the upgrade was not completed until October 1976, when the Tu-95M-5 performed its first flight. On 21 May 1977, after 32 test flights, development of the Tu-95M-5 was halted. Work on the Tu-95M-5 had progressed at a snail's pace because it was obvious from the start that the more advanced upgrade of the Tu-95K/KM into the Tu-95K-22 was more promising. Moreover, a new age of Soviet strategic aviation development began in the summer of 1976 when work started on the Kh-55 strategic cruise missile.

A converted Tu-95KM, No. 2608, became the prototype for the Tu-95K-22 (*izdeliye* VK-22) 'Bear-G'. Rework of this aircraft began at Kuibyshev in May 1974 and the modernised aircraft made its maiden flight on 30 October 1975. The air-missile complex K-95-22 included – as well as the aircraft – Kh-22M missiles, PNA-B (NATO 'Down Beat') radar, as well as the Kama missile initialisation and launch system. Two missiles were carried under the wing roots on BD-45 pylons (which, like the missiles, radar and other equipment, was borrowed from the Tu-22M); a third missile was suspended under the fuselage on the former BD-206 pylon, equipped with an adapter. In spite of three available pylons, the aircraft only carried a maximum of two missiles. Other equipment was also improved, including the large Rezeda active jammer in the aircraft's tail and the smaller Siren jammer in the nose.

At 3600 km/h (2,237 mph), Raduga's Kh-22M (AS-4 'Kitchen') missile is 50 per cent faster than the Kh-20M, while it is half the weight at 5800 kg (12,786 lb). There are two variants: one with an active radar seeker and a 900-kg (1,984-lb) conventional warhead, and one with a Doppler-inertial navigation system and 150-kT nuclear warhead (the standard variant was the nuclear weapon). The range of the first variant is 310 km (193 miles), being limited by the seeker's capability, while the nuclear missile can reach 510 km (317 miles). Length of the Kh-22M is 11.65 m (38 ft 3 in), diameter 0.92 m (3 ft) and wing span is 3.0 m (9 ft 10 in).

Tests of the Tu-95K-22 continued for a very long time. The official commissioning was no earlier than 1987, which was after the commissioning of the Tu-95MS with cruise missiles. However, some Tu-95K/KMs from front-line units were converted into Tu-95K-22s long before the official commissioning, so that the first Kh-22M missiles were launched from a military Tu-95K-22 as early as 1981.

Another development was the adaptation of the Tu-95 for the tactical task of destroying enemy airfields. The importance of such missions was fully realised after the 1967-1973 Middle East wars. On the initiative of the Long-Range Aviation command, some Tu-95K-22s were modified to carry containers, each with 15 250-kg (551-lb) bombs. Two containers could be suspended on a double beam under the fuselage, or a container on each of the three missile-carrying pylons. A single-pass salvo of 30-45 bombs was intended to completely destroy the airstrip.

Rebirth

In August 1976 came a message which brought about a revolution in Soviet strategic aviation: under development in the US was the new AGM-86 Air-Launched Cruise Missile. The ALCM was equipped with a terrain-reference navigation system which correlated the nature of the ground over which it was flying with a map stored in the missile computer memory. The Russians were also working on such a system, but as they were not overly optimistic as to its eventual success, they intended to abandon the programme. The information from the United States triggered off a storm.

The final incarnation of the first-generation 'Bear' was the Tu-95K-22 'Bear-G', modified from the earlier Tu-95K 'Bear-B' and KM 'Bear-C'. The most obvious new feature was the adoption of Rezeda jamming equipment in an extended tailcone in place of the tail turret (also fitted to some Tu-95RTs 'Bear-Ds'), while a Siren jammer was mounted in a nose 'pimple'. Additional equipment (such as the Kurs missile datalink) was housed in fuselage pods, while the wings, as here, often mounted air sampling pods for monitoring nuclear particles in the atmosphere. Tu-95K-22s were concentrated at Ukrainka with the 1223rd and 1226th TBAP, but have since been replaced by the Tu-95MS.

The Tu-95MS began operations in early 1983. Training sorties often involved pairs or groups of aircraft flying long-range missile launch profiles.

Above: The massive Obzor-MS ('Clam Pipe') nav/bomb radar takes up much of the space in the nose, suspended on a rotating pedestal mounted on the roof of the radar bay. Above it is a much smaller scanning antenna, either side of which are the mountings for EW antennas. The yellow pipe takes fuel from the nose-mounted probe and feeds it along the starboard side of the fuselage to the main tanks. The strut-mounted antenna is noteworthy, a feature possibly associated with the Tu-95MSM upgrade.

Right: Close-up of the nose shows the hemispherical multi-faceted Mak-UT infra-red detection/missile warning system and, alongside, an antenna for the active jamming system.

Using 'jump' airfields in the Russian Far East, Tu-95MSs have sporadically tested US air defences around Alaska. This intercept was undertaken in 1985 by F-15s of the 21st TFW.

Over the course of several months, the design team of the Ramenskoye PKB (Priborostroitelnoye Konstruktorskoye Byuro, instrument design bureau) received enormous financial aid for continuation of its work on 'correlation-extreme' navigation. On 9 December 1976, the government took an important resolution in ordering from three design teams new missiles in airborne, sea-launched (from surface ships and submarines) and land-based versions. The Raduga missile design bureau at Dubna submitted a project for the Kh-55 – a small, subsonic airborne missile similar to the AGM-86 (early proposals included also sea- and land-based versions). The Novator design bureau at Sverdlovsk (now Yekaterinburg) proposed the similar naval complex 3K10 Granat, with the 3M10 missile and a land-based version. Lastly, NPO (Nauchno-Proizvodstvennoye Obyedineniye, Scientific-Manufacturing Corporation) Mashinostroyeniya at Reutov near Moscow suggested a quite different design: the all-purpose Meteorit complex with the 3M25 heavy supersonic missile in a naval version Meteorit-M (Morskoy, sea), land version Meteorit-N (Nazemnyi, land-based) and airborne version Meteorit-A (Aviatsionnyi, airborne).

At the beginning of the programme the Meteorit offered the greatest expectation, offering the tempting speed of 3000 km/h (1,864 mph) at an altitude of 22000-24000 m (72,178-78,740 ft) over a promised range of 5000 km (3,107 miles). The Meteorit-A was 12.8 m (42 ft) long and weighed 6380 kg (14,065 lb) at launch. The missile had folding swept wings and folding tailplanes, as well as small canards; it was powered by a ram-jet cruise engine with ventral air intake. Unfortunately, only one poor-quality photograph of this missile is known and only scant information is available.

Meteorit was tested for a very long time with poor results. The first one was launched on 20 May 1980 from a land-based launcher at the Kapustin Yar exercise ground; the test was a failure. Only the fifth launch, on 16 December 1981, was considered a success, although the missile flew only 50 km (31 miles). The airborne Meteorit-A (AS-X-19 'Koala') was launched for the first time from the experimental aircraft Tu-95MA on 11 January 1984 – and again without success. The second launch on 24 May also failed. The Tu-95MA (converted from the fourth series Tu-95MS) carried two Meteorits suspended under the wing roots, and had the option for two more missiles carried between the engines. Since at the time the Tu-95MS with Kh-55 missiles was entering service in military units, work on the Meteorit-A airborne missile was halted at the end of 1984. Until the late 1980s the Russians still hoped to bring the Meteorit into service, continuing tests of the seaborne version, but eventually the whole programme was terminated at the end of 1989 after more than 30, mostly abortive, tests.

In December 1983, the Raduga Kh-55 (AS-15 'Kent') missile was commissioned into service in the air forces, while in April 1984, the Novator Granat (SS-N-21 'Sampson') system for submarines was commissioned into service in the navy. The ground forces were attempting to introduce the RK-55 (SSC-X-4 'Slingshot') into service, but all 84 ready missiles were scrapped at the test stage in accordance with the Soviet-US treaty of December 1987 covering the liquidation of medium-range missiles.

Cruise missile tests

Tests with the Kh-55 missile were much more successful than those of the Meteorit. The Tu-95M-55, or *izdeliye* VM-021, test bed aircraft intended for testing Kh-55 missiles took off for the first time on 31 July 1978 (it was the aircraft formerly used for testing the KSR-5 missile, i.e. Tu-95M-5). Conversion for its new role was undertaken from July 1977 to June 1978. Former equipment and some fuel tanks were replaced by the APP-95 Dub (Oak) missile preparation system (APP = Aparatura Podgotovki i Puska, initialisation and launch equipment), and by DISS-7 and MIS-45 navigation systems. Initially, a single launcher was planned inside the Tu-95M-55's bomb bay, but later a six-round MKU-6-5 rotating launcher was installed, as planned for the series production aircraft. Between 1978 and 1982 the aircraft made 107 experimental flights and launched 10 Kh-55s. On 28 January 1982, the Tu-95M-55 crashed during take-off from Zhukovskiy airfield, killing all 10 on board.

In 1978, the DMZ (Dubnenskiy Mashinostroitelnyi Zavod) factory in Dubna built a small batch of Kh-55 missiles intended for tests, and soon afterwards series production began in a factory at Kharkiv (Ukraine). On 14 December 1980 the factory delivered the first series missile to the armed forces. This weapon was ceremonially launched on the Day of the Soviet Army – 23 February 1981 – at Semipalatinsk (the identity of the launching aircraft remains unknown: at that time only two aircraft were adapted for this task, the experimental Tu-95M-55 and the Tu-142MS prototype).

Production of Kh-55s was accorded the highest priority. In 1981, the Kharkiv factory produced 40 missiles, and in the ensuing years the production rate increased dramatically. In September 1982 the department manufacturing the Kh-55 in Kharkiv adopted 24-hour shiftwork. At the same time, the factory in Smolensk started production of the new Kh-55SM version. In December 1986, production of Kh-55s at Kharkiv was stopped, although in the first quarter of 1987 the factory built 16 missiles with additional fuel tanks for tests with the Tu-160, carried out by the Tupolev design bureau. From that time Kh-55 missiles were only built at Smolensk. In total, about 1,500 missiles had been manufactured by the early 1990s.

The strategic cruise missile Kh-55 (*izdeliye* 120) is used against targets of known co-ordinates. The Osina (Aspen) missile initialisation and launch system prepares the missile launch, which consists among other things in bringing the inertial navigation axes of the missile exactly into line with those of the aircraft, and in transferring a digital map of the terrain from the aircraft's onboard computer into the missile's memory. The Kh-55 missile is guided to the target by its inertial navigation system. In some sections of the route the correction system is switched on which, by means of Doppler navigation radar and radio altimeter, sees the terrain over which it is flying and compares it with the three-dimensional map stored in the computer memory

For a period in the early 1990s, the 'Bear-H' became more visible with select airshow appearances, during which it impressed onlookers with its sheer size and brutish power, as demonstrated above at the 1993 Zhukovskiy show. Following the appearance of a single aircraft in 1993, the 1994 IAT show at RAF Fairford drew two 'Bears', a VVS Tu-95MS being accompanied by a Tu-142MK from the AVMF. The 'Bear-H' (left) was from the 182nd TBAP at Mozdok. Below a Tu-95MS practises for an airshow appearance with the MiG-29s of the 'Swifts' team.

Right: Having observed the Tu-95MS for several years from the cockpits of its interceptors, the USAF got the chance for a detailed look in May 1992 when two aircraft flew to Barksdale AFB, Louisiana, on an exchange visit. The B-52 and Tu-95MS later came face to face again at the 1993 IAT airshow at RAF Fairford.

Above: Tu-95MS bombers sporadically probed and prodded at US defences, approaching from various directions. In July 1988 this bomber was one of two intercepted off the Canadian coast by F-15s from the Massachusetts ANG, which had been on alert at Loring AFB, Maine.

(in the West such a system is called Tercom, Terrain Contour Matching). If necessary, a course correction is made. Thanks to this system, the missile accuracy can be below 100 m (328 ft), which is more than sufficient for a missile with a nuclear warhead. The BSU-55 (Bortovaya Sistema Upravleniya, Board Control System) missile autopilot allows for obstacle avoidance at low altitude and horizontal manoeuvring for avoiding interception.

Of conventional aerodynamic configuration, the Kh-55 missile has unswept folded wings and folding control planes. The wing and control planes are unfolded after launch, and the Soyuz R95-300 cruise turbofan engine is lowered from the bottom of the missile's aft end. Flight to the target is typically between Mach 0.48 and Mach 0.77 at low altitude. To extend its range, the missile can be fitted with two conformal fuel tanks scabbed on to the side. The missile is used only with a 200-kT nuclear warhead.

Two-type missile force

As in the US, which adapted the new B-1B and older B-52 to carry ALCMs, the Russians prepared two platforms for the Kh-55. The Tu-160 'Blackjack' supersonic bomber was then in the early stages of development and eventually achieved initial operational capability (IOC) in 1987. The 'Bear' was a cheaper and readily available carrier for the new missiles. This time, however, the Russians did not confine themselves to a mid-life upgrade of existing aircraft, but resumed production of the Tu-95, expecting future development of the weapons. Since the most modern variant of the 'Bear' was then the Tu-142M anti-submarine warfare aircraft made by the factory at Taganrog, its airframe was chosen as the platform for the new armament. In July 1977, Tupolev was ordered to redesign the Tu-142M into a Kh-55 missile-carrier, designated Tu-142MS.

According to the initial design, the aircraft was to carry 12 Kh-55 missiles, six on each of two tandem MKU-6-5 revolving launchers. However, proper fore-and-aft balancing of the aircraft with two launchers turned out to be impossible, and one of the armament chambers had to be dispensed with. Thus, the basic version of the aircraft carries six Kh-55 missiles on a single drum inside the fuselage. An overloaded variant, with shorter range and slower speed, can carry 10 more missiles under the aircraft's wings (each wing carries one cluster with two missiles between the fuselage and inner engine, and one cluster with three missiles between inner and outer engines).

At Taganrog, the conversion of Tu-142MK airframe No. 42105 into the first Tu-142MS began in the second quarter of 1978 and was completed in September 1979. In the same month the aircraft made its maiden flight. When compared with the Tu-142MK the aircraft has a shorter front fuselage with a new, wider cockpit and Leninets Obzor-MS radar in the nose. The bomb bay is located on the aircraft's centre of gravity. New electronic warfare and communication systems were fitted, as were modernised NK-12MP turboprop engine offering longer service life. A new electric system was equipped with a more powerful generator.

Into production and service

When series production began in 1981, the aircraft was redesignated as the Tu-95MS (*izdeliye* VP-021) 'Bear-H'. On 3 September 1981, the first series aircraft Tu-95MS, No. 01, performed the first successful launch of a Kh-55 missile. On 26 March 1982, Tu-95MS No. 2 flew to Akhtubinsk to begin state acceptance tests. On 31 December 1983, upon introduction of adjustments resulting from the state acceptance tests, the Tu-95MS aircraft with Kh-55 missiles was officially commissioned into service. The factory at Taganrog also built the Tu-142 and converted Il-76MDs into A-50 'Mainstay' early warning aircraft, and the planned Tu-95MS production run unquestionably exceeded its capacity. Accordingly, at the beginning of 1983, Tu-95MS production was moved to Kuibyshev.

On 17 December 1982, the first two aircraft (Nos 01 and 02) were delivered to the 1023rd TBAP at Semipalatinsk. A month later, the next Tu-95MS arrived at Semipalatinsk, and in February 1983 pilots from the 1023rd TBAP launched Kh-55 missiles for the first time during exercises. In 1984 the next unit – the 1126th TBAP stationed on the same airfield at Semipalatinsk began receiving its new aircraft. In 1985 the 1006th TBAP at Uzin acquired its first Tu-95MS, and in 1987 the 182nd TBAP at Mozdok began to convert to the 'Bear-H'. Older Tu-95K-22 aircraft from these air bases were moved to Ukrainka, to replace Myasishchev M-4 and 3M bombers. The displaced 'Bisons' were either converted into tankers or withdrawn.

Above: The Tu-95MS16 introduced uprated armament with two twin-barrelled GSh-23 cannon replacing the single-barrelled AM-23s. The gun-aiming radar remained the PRS-4 Krypton unit, known to NATO as 'Box Tail'. The four-gun tail was derived from that fitted to the Tu-22M2, and was also applied to the Tu-142MZ.

Right: At Tushino in 1995 the flypast included this Tu-95MS simulating refuelling from an Il-78, flanked by a pair of Su-27s from the 'Russian Knights' team.

There are two production versions of Tu-95MS. The difference between these two versions is commonly related to the number of missiles carried: older aircraft designated conventionally as Tu-95MS6 carry six missiles inside the weapon bay, whereas the newer machines (Tu-95MS16) carry 10 additional missiles under the wings. However, the difference between the two is much more significant.

In June 1983, just as series production of the Tu-95MS and its Kh-55 missiles was beginning, a government resolution was issued concerning their modernisation. Both the Tu-95MS16 and Kh-55SM missile arose from this programme. The most important alteration was replacing the Osina missile launch system with the much more modern Sprut (Octopus) system, as used in the Tu-160, enabling the simultaneous launch of 12 or more missiles. The aircraft's self-defence and communications systems were also improved: two double-barrelled GSh-23s cannon in the rear turret replaced the original AM-23 single-barrelled guns. The missile's onboard sensors were improved by increasing their resistance to jamming. In 1986, the modernised Tu-95MS16 aircraft and Kh-55SM (*izdeliye* 125) missile entered production, and were officially commissioned into service with the Soviet air forces. During START negotiations, Russia presented the Kh-55 and Kh-55SM missiles with the 'official' designations RKV-500A and RKV-500B, corresponding to their Western designations AS-15A and AS-15B.

According to official data submitted on 31 July 1991 during signature of the START I treaty, in September 1990 the Soviet Union had 84 Tu-95MS aircraft with 672 'RKV-500' missiles, as well as 63 older combat versions (Tu-95K-22, K and M), and 11 Tu-95U training aircraft at the Ryazan-Dyagilevo training centre. Several aircraft built in the next year were delivered to Uzin, Ukraine. In February 1992 Russia's President Boris Yeltsin announced the decision to cease Tu-95MS production, at a total of 88.

Tu-95 in Soviet air forces, September 1990, by variant

Air Base	MS16	MS6	K-22	K	M	U	Total
Uzin-Shepelovka	21	-	-	1	1	-	23
Mozdok	22	-	-	-	-	-	22
Semipalatinsk	13	27	-	-	-	-	40
Ukrainka	-	-	46	15	-	-	61
Ryazan-Dyagilevo	-	-	-	-	-	11	11
Total	56*	27	46	16	1	11	157

** One Tu-95MS16 was awaiting delivery at the Kuibyshev factory, thus the total number of aircraft according to the START treaty was greater by one than the number given in the table. Additionally, several aircraft of various versions were assigned to the LII flight-test centre at Zhukovskiy.*

Above and left: The art of mid-air refuelling is never easy, but the central position of the probe and the Tu-95's stability make the process straightforward. When refuelling from Il-78s 'Bears' use the UPAZ pod mounted on the rear fuselage rather than the wing-mounted pods.

Bottom: Eight Tu-95MSs line up on the Engels ramp, with Tu-160 'Blackjacks' in the distance. The 'Bears' were previously assigned to the 182nd TBAP at Mozdok, but have been withdrawn from their previous base, close to Chechnya, and now serve with the 184th GvTBAP.

***Below: 184th GvTBAP aircraft are adopting the names of cities. This Tu-95MS is named* Kaluga.**

In service now for 20 years, the Tu-95MS gave the elderly 'Bear' a new lease of life, and benefitted from the improvements made to the basic design, many of which came from Tu-142 production. While production of the maritime 'Bears' remained at Taganrog, Tu-95MS construction was moved back to Kuibyshev (Samara), where the first generation of 'Bears' had been built. Service entry of the Tu-95MS was undertaken swiftly by Soviet standards, allowing the retirement of the Myasishchev 'Bison' from the bomber role and the transfer of the older Tu-95K-22 'Bear-G' bombers away from the centre and west of the USSR to the less strategically important base at Ukrainka near Lake Baikal in the Far East. From this base the principal targets were Chinese rather than from NATO.

After the disintegration of the USSR most of the Tu-95MS aircraft remained beyond the borders of the Russian Federation. Kazakhstan inherited 40 bombers at Semipalatinsk, whereas the Ukraine had 25 bombers at Uzin. Russia quickly reached agreement with Kazakhstan and all 40 aircraft flew to Ukrainka air base in Russia's Far East (the last four bombers departed for Russia on 19 February 1994). In exchange for the bombers, Kazakhstan obtained tactical aircraft. The agreement with Ukraine turned out to be much more difficult. After several years of negotiations, concerning mainly the Tu-160s of the 184th GvTBAP at Priluki, in November-December 1999 Ukraine sold to Russia only three Tu-95MS 'Bears', the remaining aircraft being scrapped. In 1998, the aircraft from Mozdok air base, located dangerously close to the Chechnya trouble spot, were transferred to Engels.

At present (according to official data for July 2001) Russia has 63 Tu-95MS aircraft with 504 Kh-55 missiles, as well as two Tu-95s of older versions. These aircraft are distributed between three air regiments: the 184th Guards TBAP at Engels, and the 1223rd and 1226th TBAPs at Ukrainka. About 60 per cent of these aircraft are recent versions with the Sprut system; the remainder being from the first series with the Osina system. From 1999 on, Tu-95MS bombers have been named after towns, examples being *Saratov*, *Smolensk*, *Blagoveshchensk*, *Kaluga* and *Ryazan*.

Like all Russian military aviation, Tu-95s do very little flying. Average flying time of a pilot in the 37th Air Army (long-range air forces) amounted in 2001 to a mere 20 hours (10 hours in 2000). Nevertheless, for the first time in several years, Russian strategic aircraft began venturing out on long-range flights. During the Zapad-99 exercise in June 1999 a pair of Tu-95MSs made a 15-hour flight from Engels to the Iceland region. On the way back one of the aircraft launched a Kh-55 missile on an exercise ground near the Caspian Sea. In September 1999, two pairs of Tu-95MSs took off from Anadyr and Tiksi airfields in Chukotka, Russian Far East, and approached the Canadian coast. In November and December 2000 two Tu-95MSs were temporarily ferried to Anadyr airfield, three to Tiksi and two to Vorkuta. In 2001 Russia planned another series of long-distance sorties, including a flight to Cam Ranh Bay in Vietnam, but these missions were not carried out. In late April 2002, two 'Bear-Hs' again ventured out into the Pacific during a Far East military exercise, and were intercepted by USAF F-15Cs from Alaska.

Current modernisation programme

As already noted, the history of the Tu-95 mirrors the history of Russian airborne strategic missiles, and this continues today. The basic function of the Tu-95 until 1990 was nuclear deterrence (the only weapon of the Tu-95MS – the Kh-55 cruise missile – exists only with a 200-kT nuclear warhead). However, as early as the late 1980s Russia began considering the transformation of the Tu-95MS into a dual-purpose weapon for both nuclear and conventional missions. This marked a significant change from the military-political point of view. Previously, Russian strategic aviation armed only with nuclear missiles could be used only as a deterrent in times of peace or as the ultimate weapon in a global conflict. By adopting non-nuclear weapons, strategic aviation could also be used in smaller conflicts of varying scale, beginning, for instance, with the use of single missiles against terrorist training camps at the other end of the world. This trend was abruptly amplified in the early 1990s when the Cold War was over, and the US demonstrated the great capability of strategic aviation armed with smart conventional weapons in Desert Storm.

In 1992 design work began on the Tu-95MSM aircraft with a new Sigma missile launch system and new Kh-101 conventional strategic missiles (and later also with the modernised Kh-555, new medium-range Kh-SD and strategic nuclear Kh-102 missiles, all derived from Kh-101 technologies). From autumn 1999 conversion of air force

Above and right: Both Tu-95MS6 and MS16 carry six Raduga Kh-55 (AS-15 'Kent') missiles internally, but the MS16 can also carry 10 further missiles on underwing pylons. However, this capability was removed to comply with START treaty regulations. However, these pylons can also be used to each carry two Kh-101s, giving an external total of eight missiles.

Demonstrating a spirited take-off during the Zhukovskiy show, this Tu-95MS is assigned to the LII test centre for various trials, hence the calibration markings on the rear fuselage. As the massive mainwheels go through their somersaulting manoeuvre during retraction, drag increases markedly. Another peculiarity during gear retraction is the nose-up pitch change resulting from so much weight being moved aft, and its consequent effect on the aircraft's centre of gravity.

Inside the Tu-95MS

Above: In the rear of the forward crew compartment, facing aft, is the first navigator's position, dominated by a large screen for the Obzor-MS radar.

Right: The flight deck of the initial production Tu-95MS differed only in detail from earlier 'Bears'. Later aircraft featured a more modern flight deck with central console mounting the throttles.

Below: The engineer's position is also only subtly different from earlier aircraft. Again, in Tu-95MS16s the station was considerably revised.

Above: The communications station, opposite that of the flight engineer, has the cryptographic terminal installed, albeit hidden from prying eyes.

Below: The second navigator's station is on the port side, next to the first navigator. Defensive electronic warfare equipment is also controlled from here, although the necessary equipment has been removed in this case. It normally occupies the space below the screen.

Missiles carried by Tu-95MS/MSM

	Kh-55	Kh-55SM	Kh-101	Kh-SD	Kh-65SE
Launch weight	1210 kg	1500 kg	2200-2400 kg	1600 kg	1250 kg
Warhead	nuclear 200 kT	nuclear 200 kT	conventional 400 kg	conventional 400 kg	conventional 410 kg
Length	6.04 m	6.04 m	7.45 m	6.04 m	
Body diameter	514 mm	770 mm			514 mm
Wing span	3.1 m	3.1 m			3.1 m
Minimum flight alt.			30-70 m	30-70 m	40-110 m
Ceiling			6000 m	6000 m	
Cruise speed			190-200 m/s	190-200 m/s	160 m/s
Maximum speed	260 m/s	250 m/s	250-260 m/s	260-270 m/s	260 m/s
Accuracy	100 m	100 m	12-20 m	12-20 m	
Maximum range	1500 km	2000 km	5000 km		250-280 km

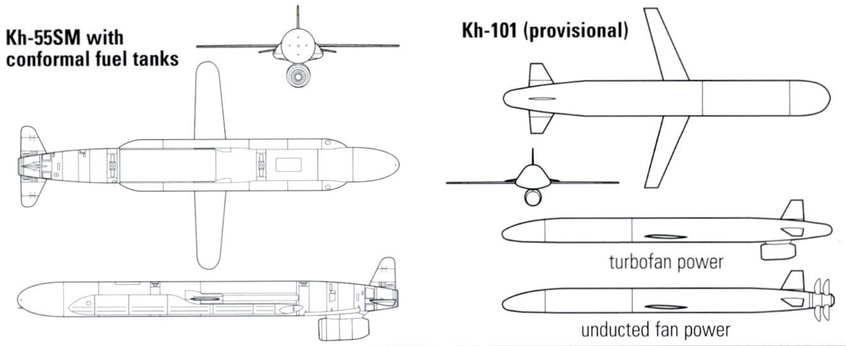

aircraft was carried out by the Air Force's 360th ARZ (Aviaremontnyi Zavod, repair plant) at Ryazan. The aircraft are upgraded on the occasion of major overhauls, which also prolong service life to 30 years. The Tu-95MSM should remain in service until around 2020 (unless further modernisation and service life extension is undertaken in the future). All 35 or so Tu-95MS aircraft with the Sprut system are being upgraded, the last due in 2005; older aircraft with the Osina system will not be modernised.

There is little information about tests of new armament for the Tu-95MSM. The Kh-101 missile, the most important item of the whole programme, is now being introduced into series production. According to unconfirmed information, the air force's Tu-95s fired Kh-101 missiles for the first time during exercises on 6-8 October 1998. In January 2002, the then commander-in-chief of the Russian Air Force, Anatoliy Kornukov, informed about his decision on starting series production of "new accurate missiles for the strategic aviation" (Kornukov did not mention which type of missile, but it may be guessed from the description that it was the Kh-101). According to Kornukov, the missile will be supplied to military units in 2003.

Work on the Kh-101 (*izdeliye* 111) strategic missile began as early as the late 1980s as a replacement for the Kh-55. The fundamental innovation was installing a much more accurate homing system, enabling the use of a conventional warhead instead of a nuclear one. The Kh-101 missile has a classic configuration with slightly swept fixed wings, and fixed control fins. The propulsion system of the Kh-101 remains unknown. According to one report, the missile is propelled by a turboprop engine, with pusher airscrews installed in the rear part of the fuselage, but another propulsion system with a turbofan engine, similar to that of Kh-55, is also possible.

Guidance to the target is by inertial navigation system with electro-optical correction system. Television terminal guidance system is activated in the direct vicinity of the target. This system compares the image seen by the camera against the target picture stored in the missile's memory. Probably, although this has not been confirmed, the Kh-101 is also equipped with a GPS receiver for accurate positioning. The missile warhead contains a penetrating charge with combined high explosive and incendiary effects. The Kh-102 missile with nuclear warhead (and with simpler homing system) will be developed from the Kh-101 in the future. The Tu-95MSM is capable of carrying up to eight Kh-101 missiles in four clusters, two under each wing, while six additional missiles of another type (Kh-55, Kh-555 or Kh-SD) can be carried in the bomb bay.

Rework and overhaul of the Tu-95MS fleet is undertaken at the 360th ARZ at Ryazan, where two aircraft are seen stripped down. As Tu-95MS16 aircraft are cycled through the plant for overhaul and life extension, they are being upgraded to Tu-95MSM standard with the Sigma launch system, compatible with the new family of conventionally armed missiles such as the Kh-101 and Kh-555. The series MSM upgrade began in 1999 following tests with a trials aircraft which involved Kh-101 firings. Older Tu-95MS6 aircraft are not due to receive the MSM upgrade.

Cruise missile conversion

In 2001, the overhaul of Kh-55 missiles and their upgrade to Kh-555 standard were ordered from the DMZ factory at Dubna (according to unconfirmed information, the Kh-555 missile was launched on 12 January 2000 for the first time). The most important alteration is installing the terminal guidance system of the Kh-101, as well as a conventional, instead of nuclear, warhead. It is very probable that the Kh-555 missile will be fitted with a new engine. The R-95-300 turbofan of the Kh-55 is almost 20 years old and was manufactured in series at Zaporozhye, Ukraine. Russia will certainly prefer a Russian-made engine. There are several new engines of similar class, which can be taken into consideration. The most likely option is a new engine developed by the Lyulka-Saturn company in Moscow, designated *izdeliye* 36MT, which is now undergoing static tests (this engine was installed in the new version of the Kh-59MK tactical missile, which was formerly also powered by the R-95-300). Other engines that may be considered are the TRDD-50 designed by OMKB at Omsk, and Moscow Soyuz's modernised RDKF1-300 or new R-125-300.

Having been developed to launch just one type of weapon (the nuclear Kh-55), the 'Bear-H' is entering the conventional arena and is in the process of acquiring the ability to carry several types of weapon, both nuclear and conventional. Blessed with a capacious bomb bay, it is possible that the Tu-95MS might also be used for free-fall bombing in benign threat environments – as is the case with the B-52 – turning the story of the 'Bear' full-circle.

In the long run, Russia plans to use equipment prepared for the Kh-101 (Sigma missile launch system, inertial navigation system with electro-optical correction, as well as TV terminal seeker) in the new medium-range missile Kh-SD (Sredney Dalnosti, medium-range). There are two types of Kh-SD warhead: penetrating (e.g. against deep fortifications, command posts) or cassette (e.g. against airfields and other surface targets). No information is available on the appearance, construction and propulsion system of the missile. The Tu-95MSM will carry six Kh-SD missiles inside the bomb bay plus – optionally – eight missiles under the wings.

Information about Kh-SD partly tallies with data for the Kh-65 missile, already known from summer 1992 when, during Mosaeroshow at Zhukovskiy, a small brochure about it was issued. Next year, a full-size mock-up of its anti-ship version, designated Kh-65SE, was shown (in Abu Dhabi in February 1993 and then in September at Zhukovskiy and at Nizhnyi Novgorod). Some parameters of the Kh-65 are identical or very close to those of the Kh-SD, including length, flight speed and warhead weight. The Kh-65's weight is smaller (which may be a result of additional fuel tanks in the Kh-SD) and the guidance system is different. The external appearance of the Kh-65 is very similar to that of the Kh-55, the only conspicuous difference being the wide, flat forward fuselage. The missile is powered by a TRDD-50 turbofan engine installed under the rear part of fuselage.

According to the initial information from 1992, the Kh-65 was intended for attacks against stationary targets of known coordinates and its homing system was to be similar to that of the Kh-101 (inertial-Doppler navigation with electro-optical position correction), but without a terminal homing system. The range of Kh-65 was quoted at 580-600 km (360-373 miles). The Kh-65SE anti-ship version is intended for attacks against big radar-contrasting targets with radar cross section exceeding 300 m^2, particularly naval ships in a strong jamming environment. The missile has an active radar seeker for terminal guidance. The range of this version of the missile is 250-280 km (155-174 miles). The Kh-65 has been displayed as part of the armament of some tactical aircraft, such as the Su-35 and Su-27IB. It is possible that the Kh-65 missile is one of the variants of the Kh-SD programme. The general designation Kh-SD would suggest a development programme for a new family of medium-range air-to-surface missiles, rather than of a specific type.

Piotr Butowski

Right: With production of the Tu-160 proceeding with occasional single aircraft and development of a new bomber hampered by funding difficulties, the Tu-95MS/MSM will remain the backbone of Russia's strategic bomber fleet for some years. Young airframes, capable systems and the current vogue for stand-off cruise missiles ensure a bright future for the 'Bear'.

Left: Engels is the focal point of Russian strategic aviation, housing a regiment of 'Bear-Hs' and one of 'Blackjacks'. On the banks of the Volga in central Russia, Saratov is the closest major city to the base, and the 184th GvTBAP's lead aircraft was named in its honour. The 'Blackjacks' are assigned to the 121st GvTBAP, which absorbed the aircraft returned from Priluki, Ukraine, where the 184th GvTBAP was formerly based.

Special Report

Brazilian Navy

Força Aéronaval da Marinha do Brasil

Brazil's long coastline places it in a commanding position over the south Atlantic, and it has a large navy to match. Although the service has operated a carrier since 1960, it was used in the ASW role, and did not embark any fighter/attack aircraft. With the purchase of a new, larger carrier and Skyhawks to go on it, all that has changed.

The Tracker and Skyhawk represent the past and present of Brazil's carrier aviation. A total of 21 Trackers (13 S-2As and eight S-2Es) was delivered in 1961 for service with 1º Grupo de Aviação Embarcada aboard Minas Gerais*, receiving the local designations P-16A and P-16E (above). From delivery until their retirement in 1996 they were operated by the air force. The Tracker had been Brazil's only truly operational fixed-wing carrierborne aircraft, and its withdrawal left carrier aviation in the doldrums. All changed in 1998 when the Skyhawks were purchased, followed by the acquisition of the French carrier* Foch*. A change of law meant that it would be the navy, rather than the air force, which operated the new arrivals.*

Wing tanks are usually carried by Brazil's AF-1s, occasionally augmented by centreline tanks, as here. With the delivery of buddy pods, from late 2002 the fleet will gain an important refuelling capability, greatly extending its reach.

Carrier operations by the diminutive A-4 Skyhawk, once a common sight on US Navy (not forgetting Argentine and Australian) aircraft-carriers in the 1960s and 1970s, has again become reality on the new Brazilian aircraft-carrier A-12, NAe (Navio Aeródromo) *São Paulo*. In March 2002 the vessel (formerly the French carrier FNS *Foch*) made a two-week cruise along the Brazilian coastline. As well as a number of exercises for all the carrier's crew, Skyhawk pilots of VF-1 had the opportunity to work on their carrier qualifications to achieve the necessary 30 arrested landings.

Brazilian carrier operations

In 1956 the Marinha do Brasil (Brazilian Navy) expanded its capability by purchasing the retired aircraft-carrier HMS *Vengeance* from the Royal Navy at a price of $9 million. The carrier, which between 1952 and 1955 had been in use by the Royal Australian Navy, was renamed A-11 NAeL ('L' meaning *Ligeiro* = light) *Minas Gerais* and underwent a lengthy refit in the Netherlands. Finally, on 6 December 1960, the carrier departed for Brazil carrying three Westland Whirlwind helicopters from the UK and three Grumman TBM-3 Avengers, which were obtained from the Royal Netherlands Navy. The Avengers were the first fixed-wing aircraft to operate from the new carrier. Soon after, three more Avengers were added to the inventory of the Marinha, although they did not operate from the carrier on a permanent basis. More fixed-wing aircraft to be operated from the NAeL *Minas Gerais* were bought during 1963: six North American T-28R-1 Trojans were a start and, of course, plans were laid for new aircraft purchases.

In November 1963 the *Minas Gerais* sailed for the first time to conduct carrier operations with Brazilian naval aircraft. This exacerbated the long existing rivalry between the Navy and Air Force. During November 1958 the Força Aérea Brasiliera (FAB) had already formed a

Right: Although the Skyhawks could operate from Minas Gerais*, as demonstrated in January 2001, the larger size and fully angled deck of* São Paulo *provides a far greater safety margin, as well as offering much greater capacity in terms of aircraft numbers. Following its initial two years of work-up operations, the vessel is to receive an update at the end of 2002 which significantly enhances its defences. The vessel had received a major engine/boiler overhaul in 1986/87.*

The stalky undercarriage and long tailhook are apparent as an AF-1 pilot approaches the moment of truth, heading for one of* São Paulo*'s four arrester wires which span the 8° angled deck. The P-408-engined Skyhawk is very responsive on the approach.

squadron specially focused on carrier operations. The 1° Grupo de Aviaçao Embarcada (1° GAE), established at Base Aérea (BA) Santa Cruz, started to train its pilots and prepare them for the arrival of S-2A Trackers and SH-34J Seabat helicopters. During 1965 the Marinha lost the rights to fly fixed-wing aircraft by Presidential Decree No. 55.627. The FAB had finally convinced the government that it should control all fixed-wing naval aviation. All naval aircraft had to be transferred to the air force, leaving just helicopters in the navy inventory.

Skyhawks for Brazil

On 8 April 1988, the current President of Brazil, Fernando Henrique Cardoso, ended this absurd situation by signing Presidential Decree No. 2598. Since taking office in 1994, Cardoso had been lobbied by Naval department staff, explaining to him the importance of fixed-wing aircraft in the Navy inventory. The law returned the rights of the Marinha do Brasil to operate fixed-wing aircraft.

Studies for new aircraft to be operated from the NAeL *Minas Gerais* began in the mid-1990s. Aircraft evaluated by the navy were the Sea Harrier, Super Etendard and A-4 Skyhawk. Any plans to acquire the Sea Harrier were quickly dropped due to excessive costs of acquisition, operations and maintenance. The Super Etendard option was abandoned, simply because these aircraft were not available for sale. By far the best candidate was the McDonnell Douglas A-4. The Skyhawk's dimensions also contributed to this choice, because of the operational limitations of the NAeL *Minas Gerais* (which was launched in February 1944). By the beginning of 1997 an active search for 'Scooters' had begun.

Former US Navy/Marines Skyhawks at AMARC, Davis Monthan AFB, were deemed too expensive to refurbish due to high airframe hours. Naturally, the Brazilian search soon led to the Middle East, where former Kuwaiti Air Force (T)A-4KUs were offered for sale. A big advantage of these aircraft was the fact that, on average, each airframe had logged only 1,700 flying hours. A letter of intent to buy was signed in December 1997. However the final signature was only to be placed after the Presidential Decree had been signed. For the amount of $79 million, 23 Skyhawks (20 A-4KUs and three TA-4KUs) were purchased. The deal also included around 65,000 spare parts and armament, including AIM-9H Sidewinders and 127-mm rockets. Twenty spare Pratt & Whitney J52-P-408 engines were delivered as well.

A deck crew member signals to a VF-1 pilot to cut power after a successful arrest (above), while three AF-1s rest between launches on the* São Paulo *(left). The Brazilian Skyhawk fleet represents the 23 Kuwaiti aircraft which had survived the Gulf War. Before the conflict they were due to be passed back to McDonnell Douglas as a trade-in as part of Kuwait's Hornet purchase. This had yet to be implemented when Iraq invaded in August 1990, and the A-4KUs of Nos 9 and 25 Squadrons flew many missions in an attempt to halt the aggression. As Iraqi forces moved in, the Skyhawks operated briefly from highways before pulling back to Dhahran in Saudi Arabia, from where they continued the fight as the 'Free Kuwait Air Force'.

The AF-1s are armed with AIM-9 Sidewinders for their principal air defence role, although they can carry free-fall bombs for surface attack. The trusty 20-mm Colt-Browning Mk 12 cannon in the wing roots are useful in either role. A forthcoming modification programme may include a radar – a vital addition for the primary task, better defences and enhanced attack systems.

Especially for the official hand-over ceremony in Kuwait, one A-4KU – BuNo 160180 (former KAF 801) – was painted in Brazilian navy colours and was given the serial N-1001. This scheme consisted of an overall single-tone grey with large titles and a full-colour (green, yellow, blue) rudder. This was, however, not to be the scheme adopted in service. Before their shipment to South America, the aircraft were overhauled by Boeing Maintenance in Kuwait to prepare the airframes for their journey.

Finally, on 5 September 1998 the vessel MV *Clipper Ipanema*, with the twenty-three 'Scooters' aboard, docked at Arraial do Cabo harbour, close to Base Aérea Naval (BAN) São Pedro da Aldeia. At the port of arrival the aircraft were prepared for transport by road to the airfield, a large-scale operation executed during the hours of darkness over five nights. Due to a lack of space, all aircraft were parked on the large helo apron.

Single-seat A-4KU Skyhawks were given the local designation AF-1, while the two-seat TA-4KUs became AF-1As. In September 1998 a competition was opened to find the most suitable company to maintain the airframes and train new groundcrew locally. The Marinha assigned the contract to the Chicago-based company Kay & Associates Inc., for a maintenance and training programme which began in October 1999. A section of the HS-1 (Sea King squadron) hangar was made available for this purpose.

VF-1 'Falcões'

To prepare for the entry into service of the Skyhawks, construction began at BAN São Pedro da Aldeia of office building and hangars for the new squadron. Skyhawk operations also required the enlargement of the runway from 1800 to 2400m (5,905 to 7,875 ft). On 2 October 1998, the 1° Esquadrão de Aviões de Interceptação e Ataque (First Interception and Attack Squadron) – VF-1 – was officially established at BAN São Pedro da Aldeia with Captain Carlos Augusto Marcondes as its first commanding officer.

Despite having no Skyhawk flight experience, he was still considered the most suitable person to command the new squadron, as he had been closely involved in the acquisition of the new aircraft as part of the Development Team of the Directory of Aeronautics, based in Rio de Janeiro. Besides, he had flown Sea Kings with HS-1 for more then 10 years. According to Commander Marcondes it was, and still is, a great experience to build a new operational fighter squadron from scratch, with many challenges to be faced during this work-up period. The squadron had started with only 30 sailors.

Training for the first pilots was already under way at Naval Air Station Meridian in the United States. One of the pilots, Lt Alvarrenga, was among the last Naval Aviators to be trained on the TA-4Js of VT-7. Other pilots did their train-

Hook down and nose high, an AF-1 approaches São Paulo in classic pose. When built, the A-4KUs had been based closely on the A-4M Skyhawk II and, as such, featured powerful P-408 engines, squared-off fintip and the deep nose associated with the angle-rate bombing system. All were fitted with the bolt-on refuelling probe which ducts fuel back to the main tanks in the centre.

The two Mitchell-Brown steam catapults are easily powerful enough to launch Skyhawks and Super Etendards at maximum weight. One is located in the waist position and one in the bow, both with a stroke of 50 m (164 ft).

ing course at NAS Meridian on the T-45 Goshawk. The squadron started operations with three Brazilian pilots and an Argentinian instructor pilot. Lt Cdr Vignale was a very experienced Super Etendard pilot who had also instructed on the TA-4J at NAS Meridian. His presence at São Pedro helped VF-1 to set up squadron operations.

In April 2000 the first of the 23 AF-1s took to the air in a new colour scheme designed by VF-1. All the early test flights were flown by former US Navy Commander Curt Francis, by then employed by Kay & Associates Inc.. After his retirement from the US Navy, Curt had flown for Lockheed Martin, assigned to the (O)A-4AR Fightinghawk programme for the

Despite its age the Skyhawk remains a tricky opponent in air combat – the principal role for the Brazilian AF-1s. The Skyhawks have the most powerful engine fitted to the breed: the 49.82-kN (11,200-lb) thrust J52-P-408, which was installed at virtually no cost in weight or fuel consumption compared to earlier engines.

Fuerza Aérea Argentina, initially at the 'Skunk Works' and later at Base Aérea Villa Reynolds. When this programme came to an end, only office work could be offered to him, which was not to his interest. Now working for Kay & Associates Inc., he trains pilots and performs all post-maintenance test flights. Another involvement is the LSO training programme.

During September 2000 the first flight to the *Minas Gerais* was made. After a couple of 'touch-and-goes' the Skyhawk returned to São Pedro. Four months later, on 18 January 2001, the first trap landing on the NAeL *Minas Gerais* was undertaken. The opportunity for further approach training to be undertaken was seized in October 2001 when USS *Nimitz* arrived in Brazilian waters. In two days a total of 45 approaches was made. Each AF-1 flight to the *Nimitz* included two or three bolters, giving the three VF-1 pilots a great opportunity to simulate carrier landings.

Later in the same month, the first air-to-air refuelling experience was gained, when AF-1s were refuelled by a Brazilian air force KC-130H tanker aircraft from BA Galeão. In December, an AF-1 fired the first AIM-9H Sidewinder. Throughout 2001 further AF-1s completed their overhaul and became operational. VF-1 expects to be fully operational with a full aircraft complement in 2004.

Skyhawk pilot training

Although advanced training for navy pilots had been standard, basic training had involved students receiving instruction at BAN Punta Indio in Argentina, or at BA 2 Capitán Curbelo near Punta del Este, Uruguay. However, all future AF-1 pilots will now receive their basic training in Brazil by the air force. The first part of this three-year training course begins at BA Pirassununga, where Marinha students fly over 100 hours in the T-27 (EMB-312) Tucano. The second phase – combat jet training – is undertaken at BA Natal, where another 100 hours are flown on the AT-26 (EMB-326) Xavante.

Advanced and carrier training will still take place in the US, where the student pilots are trained on the Goshawk. Again, around 100 hours are being flown. During this phase the students will make their first arrested landing on an aircraft-carrier, as well as their first catapult launch. Normally this occurs after 70 flying hours on the T-45, but it can also begin at an earlier stage of the course, dependent on the availability of a training carrier. Back in Brazil, a 50-hour AF-1 conversion course will be flown. After mastering the Skyhawk, the course progresses to various combat exercises, including air-to-ground missions with bombs and 20-mm cannon, and air combat manoeuvring (ACM) against other AF-1s. The training syllabus also includes FCLP (Field Carrier Landing Practice) at São Pedro.

After the 50-hour conversion course the AF-1 pilot still needs to become carrier-qualified. A minimum of 50 hours on type is required before the pilot can attempt a trap. After 30 traps pilots receives their qualifications to operate from the NAe *São Paulo*. At a later stage, night landings will be added to the syllabus, as soon as a special landing system on the carrier becomes available.

Currently VF-1 has eight pilots, who are all carrier qualified, and around 10 more are expected to join the squadron this year. By 2003 VF-1 will have around 30 pilots available. Two of the first VF-1 Skyhawk pilots are training to become LSOs (Landing Signaling Officers), both having already gone through the

Brazil's Skyhawk fleet

AF-1 serial	US Navy BuNo.	KAF A-4KU serial
N-1001	160180	801
N-1002	160181	802
N-1003	160183	804
N-1004	160186	807
N-1005	160188	809
N-1006	160189	810
N-1007	160190	811
N-1008	160192	813
N-1009	160193	814
N-1010	160195	816
N-1011	160196	817
N-1012	160197	818
N-1013	160198	819
N-1014	160199	820
N-1015	160201	822
N-1016	160202	823
N-1017	160203	824
N-1018	160204	825
N-1019	160205	826
N-1020	160206	827
AF-1A serial	**US Navy BuNo.**	**KAF TA-4KU serial**
N-1021	160212	883
N-1022	160213	884
N-1023	160215	886

Above: As well as national and naval markings, VF-1's Skyhawks wear a stylised falcon below the cockpit as a unit marking.

Right: Thanks to its compact size the A-4 does not require folding wings. São Paulo's two deck lifts – one forward of the island in the deck-centre and one abaft the island on the deck edge – each measure 16 x 10.97 m (52.5 x 36 ft) and can easily accommodate the AF-1. The carrier has retained the distinctive white triangular markings at the end of the landing deck applied while in French navy service.

LSO school, in the US. One of them, Lt Alvarenga, is currently receiving additional training on the French carrier *Charles de Gaulle*, while the other is being trained on *São Paulo* by Commander Curt Francis.

2002 and beyond

Although nearly all the A-4 airframes have received an intensive overhaul, not all aircraft are currently used. Due to the present small number of pilots, about one third of the total fleet is used for flight operations. Stored airframes are parked in VF-1's new hangars, but could be made fully operational again at short notice. When more pilots join the unit, further AF-1s will be brought out of storage. It is planned to operate up to 15 aircraft in the near future (from 2004), leaving eight airframes in flyable storage. If needed, they can also be returned to flight operations.

In late 2002, the Marinha is due to receive six buddy refuelling pods for the Skyhawks, ordered from the United States. The pods will enable the AF-1s to operate over an expanded operational area. Even though many spare parts were delivered with the aircraft, not all of them are easily available. Therefore, the Marinha do Brasil has contracted several Brazilian companies to copy and manufacture rare parts for the AF-1s. An important supplier is the 'Gespi' company in São Paulo State.

Right: Skyhawks currently surplus to requirements are stored in excellent condition, all panels and openings being covered with protective tape and the cockpits firmly sealed. The pylons are also specially protected to keep the connections in working order. If and when required, the aircraft can be returned to operational status at short notice.

Below: Twelve AF-1s are seen in one of VF-1's hangars at São Pedra de Aldeia, where they are held in storage. The active fleet will grow through 2002/03 to allow the deployment of up to 14 aircraft to the carrier.

Although VF-1 is in the middle of its working-up period, plans are in a final stage to upgrade the fleet of Skyhawks. The primary air defence task of VF-1 means the upgrade will see an improvement in that specific area of the AF-1's operational capability. Although the final package has yet to be confirmed, according to VF-1 commander Marcondes it is likely to comprise a radar, cockpit upgrade with multi-function displays and HUD, mission computer, FLIR, datalink and radar warning receivers. These Argentine Fightinghawk-like upgrades will enable the Skyhawk to use a modern array of armament, set to include the Brazilian MEE-1 air-to-air missile. Several international and national companies will bid for this contract. The contents of the programme largely depends on the budget to be made available by the Ministry of Defence. After receiving the upgrade, the AF-1(A) Skyhawk should offer the Marinha do Brasil at least 6,000 hours of service, allowing it serve well into the century.

Plans for other fixed-wing aircraft are currently under review, including the offer from EMBRAER for an upgraded version of the Grumman S-2 Tracker. After the Argentine navy showed the capabilities of its Turbo-Trackers on the NAe *São Paulo* in May 2002, it now seems likely that EMBRAER will receive more interest in its offer.

NAe *São Paulo*

In September 2001 the first Skyhawk landed on Brazil's new carrier, the A-12 NAe *São Paulo*. This vessel replaced the aging NAeL *Minas Gerais* on its arrival on 17 February 2001. The 'Clémenceau'-class carrier was bought from the French navy, in whose service it had carried the name FNS *Foch* (R99), and had seen action off Lebanon and in the Adriatic. With the commissioning of the nuclear-powered *Charles de Gaulle*, *Foch* became surplus to requirements. With the Skyhawks in mind, the Brazilian government decided to buy *Foch* for FF80 million (US$10.5 million). This price tag only included the carrier with all its non-offensive communication systems. All defensive armament was removed, consisting of eight 12.7-mm bow-guns, two 30-mm cannon, two Crotale EDIR missile launchers and two Sadral systems. Together with *Foch*, the French also tried to sell some Breguet Br1050M Alizé aircraft, but the Marinha was not interested.

On 4 September 2000, the first 300 Brazilian crew arrived in Toulon to start their bi-national training onboard *Foch*. After completing this successful 10-week training period, an additional 150 Brazilians arrived to transfer the aircraft-carrier, together with 450 French naval crew, to the naval shipyard in Brest on the French west coast, where it arrived on 10 November. Only five days later, on the 15th, FNS *Foch* was officially handed over to the Marinha do Brasil and named A-12, Navio Aeródromo *São Paulo*. The new Brazilian carrier remained in Brest for another two months while its operational capabilities were tested, and while it received maintenance of the propulsion and aviation systems.

For conversion training VF-1 has three AF-1A two-seaters. Six were originally supplied to Kuwait as TA-4KUs but only three survived the Gulf War. They are unique among two-seat Skyhawks in being fitted with the Pratt & Whitney J52-P-408 engine (as used in the single-seat AF-1s).

Above: The landing system and raised blast deflector plate are visible in this scene of an AF-1 being readied for launch, while an Esquilo hovers in the plane-guard role in the background.

Right:* São Paulo *leaves Santos harbour in April 2002 with three AF-1s, two ASH-3s and a single UH-14 aboard. The vessel is 265 m (869 ft 5 in) long and displaces 33300 tonnes (32,780 tons) in deep load. This compares with the corresponding figures for* Minas Gerais *of 214.1 m (702 ft 5 in) and 19655 tonnes (19,345 tons).* São Paulo *is also considerably wider, measuring 45.72 m (150 ft) at maximum beam. It can reach a top speed of 32 kt (59 km/h; 37 mph) and has a range of 7,500 nm (13890 km; 8,630 miles) at a cruising speed of 18 kt (33 km/h; 21 mph).

On 25 January 2001, the anniversary of when the State and City of São Paolo were founded, the carrier set sail for the first time. According to the current ship's commander, Captain Nigro, "the aircraft-carrier was born with a very strong spirit", referring to the prosperous and religious activities of the Bandeirantes who had used São Paulo city as their starting point, and the current economic importance to the country of both the city and state. On 1 February the carrier left port to point its bow south, heading for one of the most scenic coastal cities in the world – Rio de Janeiro. After a brief refuelling stop in the Senegalese port of Dakar, the NAe *São Paulo* arrived on 17 February.

On arrival, the complete crew of the A-11 NAeL *Minas Gerais* transferred to the A-12, and aviation safety and crew training started. The carrier also needed adjustment to the catapult system, as new parameters had to be set to launch the AF-1 Skyhawk. By the beginning of September the carrier received the first Skyhawk on its flight deck, but had to return for maintenance in the naval shipyard in Rio the same month. A more successful sea trial was undertaken in February 2001.

On March 10 *São Paulo* left harbour on its first operational cruise. Aboard were one HI-1 IH-6B (Bell 206B) Jet Ranger III, one HU-1 UH-12 (HB-350BA) Esquilo, one HU-2 UH-14 (AS 332M) Super Puma (all three types in the utility role), two HS-1 SH-3A/B (ASH-3D/SH-3H) Sea Kings (in the ASW role) and three AF-1 (A-4KU) Skyhawks (in the air defence role). The primary task of the carrier is air defence, and in the future it will be equipped with a maximum of 14 aircraft, when more pilots and Skyhawks become available to VF-1.

For its next cruise *São Paulo* headed south on 22 April for exercises with the Uruguayan and Argentine navies, landing on Argentine Navy Super Etendards and S-2T Turbo-Trackers from 2 to 5 May (as described in the 'ARAEX' sidebar). More trips were planned for 2002, although by the end of the year the NAe *São Paulo* will return to the naval shipyard in Rio de Janeiro to have Italian 'Aspide' surface-to-air missiles installed.

Naval helicopters

During 1952 a law was passed to allow the Brazilian Navy to reorganise and to create a Navy Aeronautic Division, which occurred in 1954. In 1955 the Naval Aviation Centre and Training facility (CIAAN) was founded in Rio de Janeiro. Shortly after that the first officers were sent to the United States to be taught about naval aviation operations. Together with the purchase of two survey vessels from Japan in 1957, two Kawasaki-Bell 47Js were acquired to become the first helicopters for the Marinha do Brasil. The helicopter fleet was soon expanded with the arrival of three new-built Westland Widgeon helicopters and a few SH-34J Seabat helicopters, which were transferred from the Brazilian air force. These helicopters began operations from the newly-acquired 'Brooklyn'-class cruisers.

More helicopters were added to the inventory in the shape of three Westland Whirlwinds, which arrived on the new Brazilian carrier NAeL *Minais Gerais* together with three Grumman Avengers. With carrier operations to be started in the near future the Brazilian navy founded the Primeiro Grupo de Aviação Embarcada, with all flying components of the Naval Air Force included. The 1° GAE consisted of the 1st Anti-submarine Aircraft Squadron equipped with the fixed-wing Avengers, 1st and 2nd Anti-submarine Helicopter Squadrons and the one General-Purpose Helicopter Squadron (HU-1), which were equipped with Widgeons as well as a couple of Bell 47s. With the arrival of more Bell 47s the first Instruction Helicopter Squadron (HI-1) was formed in 1962.

In the meantime the Força Aéra Brasiliera (Brazilian air force) had also founded an air wing which was focused on future carrier operations. The FAB had ordered Grumman Trackers and six Sikorsky SH-34J Seabat helicopters to be operated in its 1° GAE. When, in 1965, the presidential decree gave the air force the exclusive rights to operate fixed-wing aircraft, the Brazilian Naval Arm was forced to

HU-2 flies the AS 332F Super Puma, locally designated UH-14 and employed mainly on marine assault duties. Other tasks include long-range SAR, utility and VIP transport, and also patrol. The helicopters are fully navalised, with flotation equipment in the nose and wheel sponsons, and have (non-automatic) folding tail and main rotor blades to reduce space when stowed on ships. However, they can only operate from three of the navy's larger vessels.

transfer all its fixed-wing aircraft to the FAB. From that period on the Naval Air Arm was only allowed to conduct operations with helicopters. In return, the navy took over the recently delivered Seabats from the Air Force. With the arrival of these helicopters another new squadron was formed, HS-1 (Escuadrão de Helicópteros anti-submarino). In 1975 the Seabats of HS-1 were retired, having been replaced by four Sikorsky S-61D-3 Sea Kings which had been delivered from 1970.

The fleet grows

Between 1965 and 1968 more new helicopters were acquired for utility tasks within HU-1: the Westland Wasp Mk 1, Westland Whirlwind Srs 3 and the Hiller FH-1100. Also, the training syllabus had seen some changes when the Bell 47s in HI-1 were replaced by 20 Hughes 269As from 1963. These helicopters were gradually replaced by 18 IH-6A (Bell 206A) JetRanger IIs in 1974. In 1978 the arrival of a new type of helicopter led again to the foundation of a new squadron; HA-1 started operations with nine Westland SAH-11A Sea Lynx (Mk 21s). The Sea Lynx was tasked in anti-submarine, attack and reconnaissance roles. The old S-61D-3s of HS-1 were augmented by Agusta AS-61D (ASH-3A) Sea Kings in 1984.

In 1979 HU-1 received the locally built Helibras UH-12 (HB-350BA) Esquilo. These were augmented in 1987 by the twin-engined UH-13 (HB-355F) Esquilo. Four other utility squadrons were established outside the main naval base at São Pedro da Aldeia to operate the Esquilo helicopter, but only in small numbers. The second utility squadron (HU-2) was established in 1986 and, to begin, received six UH-14 (AS 332F) Super Pumas, followed by another two in 1993.

Brazilian Naval Aviation saw some more changes in the 1980s when the IH-6As of HI-1 were replaced by the IH-6B (AB 206B JetRanger III) version between 1985 and 1987. Also six more Sea King helicopters were taken into the inventory of HS-1 in 1996, this time former USN SH-3H (SH-3B in Brazil) Sea Kings. The 1990s also saw the introduction of the AH-11A Super Lynx (Mk 21A) to HA-1 from 1996, as well as the modernisation of the five remaining Mk 21s to 'A' standard.

Currently the Marinha do Brasil has around 78 helicopters in the inventory assigned to eight squadrons. All rotary-wing aviators receive around 18 months' training in HI-1 'Garça' on the IH-6B. On successful completion of the course, pilots then transfer to the operational units of the navy for conversion to type.

Utility helicopters

In the Marinha do Brasil five units are assigned to the utility role: HU-1 'Aquia' at BAN São Pedro da Aldeia, HU-3 at BAN Manaus, HU-4 at BAN Ladário and HU-5 at BAN Rio Grande. Currently these units use a mixed fleet of 17 single-engined UH-12 (HB-350) and nine twin-engined UH-13 (HB-355) Esquilo helicopters, all locally manufactured by Helibras under licence from Eurocopter. The largest squadron, HU-1, operates 14 Esquilos, on a wide variety of tasks including liaison, support and SAR missions. The Esquilo can be equipped with small arms such as the MAG 7.62-mm or 12.7-mm machine-gun, or heavier armament like a seven-round 70-mm rocket launcher.

Two of HU-1's twin-engined Esquilos are used in support of Antarctica expeditions. Before the Ocean Supply Ship H-44 *Ary Rongel* sails off at the beginning of November, the crews assigned to fly the helicopters during the voyage undergo special training. The training spans two months and is conducted in the Itatiaia Mountains, north of Rio de Janeiro. The aim is to familiarise the crews with cold

JetRangers have served the Brazilian navy since 1974, initially in IH-6A JetRanger II form. Today the rotary-wing training function is performed by the IH-6B JetRanger III, which flies exclusively with the 1º Esquadrão de Helicopteros Instrução (HI-1). The JetRangers can be fitted with light armament and gunsights for weapons training. They are occasionally detached for liaison work to the carrier, this aircraft carrying an 'A-12' marking on the nose.

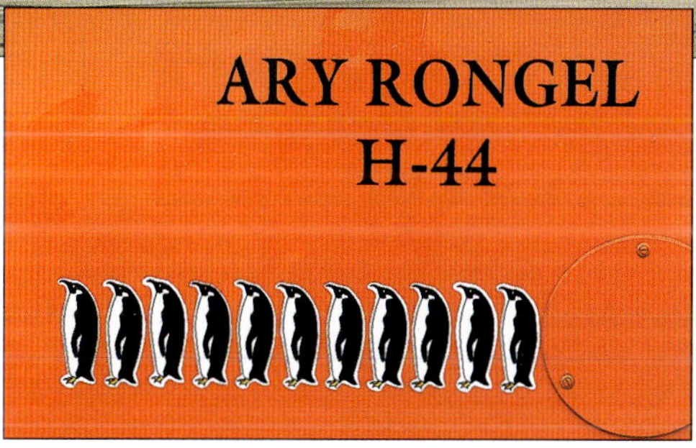

HU-1 operates a mixed fleet of single- and twin-engined Esquilos (licence-built Ecureuil), locally designated as UH-12 and UH-13, respectively. Below is a UH-12 (HB 350BA, equivalent to AS 550U2) operating from the carrier on SAR duty, while above are the two UH-13s (HB-355F, equivalent to AS 555U2) given a high-conspicuity scheme for service aboard the polar survey ship Ary Rongel and from Brazil's Teniente Rodolfo Marsh research station on King George Island, Antarctica. Each penguin symbol (left) represents an Antarctic cruise.

weather and severe conditions. To enhance conspicuity, parts of the tailbooms are painted orange, as well as the doors on either side. During their stay in the south, the helicopters mainly provide a shuttle service between the Ocean Supply Ship and the base camp in Antarctica. During their five-month stay, each Esquilo logs about 100 flight hours. On their return to São Pedro the orange paint is removed. Each expedition to the new continent is marked on the cabin door by a penguin.

The other utility squadron stationed at São Pedro, HU-2 'Pegasus', operates seven UH-14 (AS 332F) Super Pumas. In addition to the roles already described for HU-1, these larger, long-range helicopters are used for VIP and troop transport, and in the assault role, the latter conducted in close co-operation with the marines. Due to the UH-14's large size, it is only able to operate from two 'Mattoso Maia'-class assault ships and the NAe *São Paulo*. The UH-14s do not have an automatic rotor blade-folding system, which makes them harder to handle on a flight deck or in a ship's hangar. They are equipped with a self-inflating flotation system, which activates on contact with water,

The oldest helicopter in the Brazilian Naval rotary-wing fleet is the Sea King of HS-1 'Guerreiro'. Locally designated SH-3A (ASH-3A) and SH-3B (SH-3H), it forms the backbone of

Below: Veteran of the rotary-wing fleet is the Sea King, which currently serves in two forms. This is an SH-3A, one of the original aircraft which were upgraded to virtual SH-3H standard by Agusta.

HS-1's night-capable Sea King fleet is based at São Pedro de Aldeia, but is regularly detached to the carrier to provide ASW coverage. The SH-3A (above) and SH-3B (below) have broadly similar capabilities, although the latter has a better dunking sonar, and can be identified by the larger housing for it. The Bs previously served with the US Navy as SH-3Hs, and retain the extended starboard sponson which formerly housed a towed MAD 'bird'. The unofficial owl squadron badge on the hangar celebrates the squadron's night capabilities.

the airborne ASW fleet, providing the carrier air group with over-the-horizon surface ship and submarine surveillance. Seven survivors of the original SH-3A fleet (Sikorsky- and Agusta-built), which entered service during the 1970s, were extensively modified by Agusta in 1987. One of the most significant parts of this upgrade included the installation of equipment which enables the SH-3A to launch the Aérospatiale AM.39 Exocet anti-ship missile.

Replacing the Trackers

Six SH-3Bs (and two spares) arrived in 1996 from the US on board the A-11 NAeL *Minas Gerais*. These former USN SH-3Hs were bought by the Marinha do Brasil to compensate for the grounded Grumman S-2 Trackers of the FAB. The 'new' Sea Kings were equipped with AQS-18V sonar, which is more powerful than the AQS-13B fitted to the SH-3A. Armament of the SH-3B includes Mk 9 depth charges and Mk 46 torpedoes. Both types of Sea Kings in HS-1 are night vision-capable, the unit's unofficial badge featuring an owl's face to symbolise this capability.

The Marinha do Brasil is currently deciding on a life-extension programme for all its remaining Sea Kings. According to the deputy squadron commander of HS-1, the upgrade will cover ESM, FLIR, radar, communication equipment and up-to-date avionics. It is planned to finish this upgrade programme by 2003/2004. This will enable the Sea King to remain operational in Brazil at least until 2010.

The other helicopter with teeth in the Marinha is the AH-11A Super Lynx (Mk 21A) of HA-1 'Lince'. This unit, which also has its home at São Pedro da Aldeia, is unlike all other helicopters squadrons as it operates on board the frigates of the Brazilian navy. However, as frigates are also part of the carrier battle group, the AH-11A can also be employed in protection of the carrier *São Paulo*. The Super Lynx's systems and equipment allow an all-weather capability, with autonomous overwater navigation. The MIR II Mk 2 ESM can identify enemy radar emissions, while the Sea Spray 3000 radar provides 360° surface search. The AH-11A can carry up to four Sea Skua air-to-surface missiles, two Mk 46 torpedoes or two Mk 9 depth charges. HA-1's main tasks are surveillance, surface attack, over-the-horizon targeting (OTHT), vectored submarine attack and electronic warfare. The type's all-weather capability is also put to good use on many other tasks.

Brazil has made big plans for naval aviation, with upgrades scheduled for the Skyhawks, Sea Kings and the NAe *São Paulo*, plus the desire to acquire a fixed-wing patrol aircraft such as the Turbo-Tracker. Huge investments are due to be made very soon. However, the outcome of the 2002 Presidential election is being watched closely, and it is unclear whether the new President will be as much a 'friend' to the Marinha do Brasil as the current one. However, with the A-12 NAe *São Paulo*, Brazil retains its place in the short list of countries which are able to afford an aircraft-carrier.

Cees-Jan van der Ende and Marco Muntz

The Marinha do Brasil's association with the Westland Lynx dates back to 1978, when the 1º Esquadrão de Helicopteros de Esclarecimento e Ataque (HA-1) received nine Mk 21s, essentially similar to the Royal Navy's HAS.Mk 2. Experience with these led to a 1991 order for nine Super Lynx Mk 21As with Gem 42-1 engines, 360° radar and ESM, but lacking the optional overnose FLIR installation. Additionally, the upgrade of the five surviving Mk 21s to this standard was authorised, with new serials assigned (N4010/14). The first aircraft, both new-build and conversions, were handed over in 1996. The Super Lynx is an outstanding small ship's helicopter, operated on numerous utility duties in addition to its primary ASW/ASUW roles. For anti-surface unit attacks the aircraft is armed with the MBDA Sea Skua missile (right), cued by the Seaspray 3000 undernose radar.

ARAEX – Argentina-Brazil joint exercise

Between 1 and 5 May 2002 the Argentine and Brazilian navies held the joint ARAEX IV exercise, the latest in a series of joint operations between the two nations. This exercise was notable for being the first time Argentina's Super Etendards had made arrested landings on a carrier since 1987.

When the Argentine carrier ARA (V-2) *25 de Mayo* was retired in 1987, the Super Etendard pilots of the Segunda Escuadrilla Aeronaval de Caza y Ataque (EA32) were left without a vessel to operate from. Some measure of carrier proficiency was maintained using the dummy deck at the home base of Base Aeronaval Comandante Espora at Bahía Blanca, while occasional bolters were performed on the Brazilian 'Colossus'-class carrier *Minas Gerais* during previous ARAEX exercises, which began in 1992. However, the catapults of the older Brazilian ship were not powerful enough to launch the Super Etendards, so they could not perform arrested landings. However, Argentine S-2T Turbo-Trackers from the Escuadrilla Aeronaval Antisubmarina (EA2S) had been able to operate from *Minas Gerais*, and did so during the ARAEX series. The retirement of the old carrier and arrival in 2001 of the larger and more capable *São Paulo* finally provided the opportunity for the Super Etendards to return to full carrier operations.

NAe A-12 *São Paulo* had left its berth in Rio de Janeiro on 22 April, accompanied by the tanker G27 *Marajó* and the Type 22 Batch 1 frigate F49 *Rademaker* (ex-HMS *Battleaxe*). The latter carried a Lynx from HA-1. On 30 April the carrier group conducted joint operations in the first URUEX exercise to be held with the Uruguayan navy, which supplied the 'Commandant Rivière'-class frigate ARU *Artigas* carrying a Wessex HC.Mk 2.

As the *São Paulo* group sailed south the following day, it was joined by two Argentine vessels: the Meko 360 destroyer ARA *La Argentina* (D-11) carrying a single AS 555SN Fennec from 1º Escuadrilla Aeronaval de Helicópteros (EAH1), and the supply ship ARA *Patagonia* (B-1, ex-FNS *Durance*), which transferred fuel to the carrier. The two vessels were no strangers to this process, as they had performed it many times when in French navy service. The ARAEX group took up station around 100 miles off the Puerto Belgrano naval base to prepare for flight operations.

Assigned to the Argentine Grupo Aeronaval Embarcado (GAE) were three Super Etendards from EA32 and three S-2Ts from EA2S, which operated alongside four AF-1s from Brazil's VF-1. Argentine rotary-wing participation consisted of an Alouette III from EAH1 in the plane-guard role, and three SH/ASH-3Ds from EAH2 for logistic transport flights between vessels. The Brazilian navy operated two HS-1 SH-3s, one UH-12 and two UH-13s from HU-1, and two UH-14s from HU-2.

Fixed-wing flight operations got under way on 2 May, the 20th anniversary of the sinking of the ARA *General Belgrano* during the Falklands/Malvinas campaign. The occasion was solemnly marked by both Argentine and Brazilian crews. Between 2 and 5 May the Super Etendard, AF-1 and S-2T made 10 flights each, during the course of which four 'Sue' and four S-2T pilots were certified for carrier operations. The Argentine GAE had its own landing signals officer (LSO). Brazilian officers were particularly impressed by the performance of the S-2T Turbo-Tracker, ironic considering that the navy had retired the radial-engined S-2E in the 1990s and had cancelled a re-engining programme.

Juan Carlos Cicalesi, Santiago Rivas and César Del Gaizo

F-51 Mustangs in Korea

USAF, RoKAF, RAAF and SAAF service, 1950-1952

Thrust into battle within days of North Korea crossing the 38th Parallel, supposedly outdated F-51 Mustangs took on a front-line fighter-bomber role as the only aircraft suitable – and available in any numbers – to operate on long missions from Korean airfields. It would be two-and-a-half years before Fifth Air Force could finally retire the last of its 'Fifty-Ones'.

Very few fighter-type aircraft survive from one war to the next. Technology sees to it that most military weapons are outdated within a very short time of becoming operational. The North American F-51 Mustang was one of those rare birds that bridged the gap, and it did so during a period when piston-engined aircraft were rapidly becoming extinct because of the new jet age. For two years after World War II ended, the Mustangs were the hot rods of US Air Force and Air National Guard units.

In 1948, though, the primary interceptor and fighter-bomber roles began to go to the new Lockheed F-80 Shooting Star. This was especially true in the Far East. By early 1950, the Mustang had disappeared from squadron inventories at key air bases in the Philippines and Japan. Far East Air Forces (FEAF) HQ was touting the fact that its armada of fighter aircraft was now an all-jet force equipped to handle any situation that might occur. There were five key fighter wings in the Far East in 1950: the 18th at Clark AB in the Philippines, the 51st on Okinawa, the 8th at Itazuke AB, the 35th at Johnson AB near Tokyo and the 49th at Misawa AB. All had previous experience in the Mustang and had recently converted to the F-80.

Crossing the 38th Parallel

On 25 June 1950, the North Korean Army crossed the 38th Parallel into South Korea. For the next 37 months, these five fighter/fighter-bomber organisations were involved in the first major conflict of the Cold War. It hurled them into combat against hundreds of thousands of Chinese and North Korean troops. Most air units, sooner or later, would have to face the threat of one of the most advanced jet fighters in the world...the MiG-15.

FEAF faced several major problems in the first few weeks of the war. The South Korean airfields were in no shape to accommodate large numbers of aircraft, so most units resorted to launching their missions from Itazuke AB, flying to the front lines and then returning. The F-80 was the front-line fighter of choice and the only aircraft available in sufficient quantity, but problems with fuel consumption and short range made it incapable of the required loiter time over the fluid front lines. The only aircraft able to spend long periods over the lines and carry a wide variety of ordnance was the F-82G Twin Mustang; they, however, were not available in anywhere near sufficient numbers to have a chance of slowing down the North Koreans' march to the south.

The Mustang had all but disappeared from the USAF inventory in the Far East; fewer than three dozen Mustangs were in Japan, most in storage. Only 10 operational examples had survived and they had been given to the fledgling Republic of Korea Air Force (RoKAF) to

Above: A 40th FIS Mustang returns to Pohang (K-3) after a strike on North Korean positions in September 1950.

Left: A red/white spinner was the trademark of the 40th Fighter-Interceptor Squadron, part of the 35th Fighter Group at Johnson AB, responsible for the air defence of Tokyo during the immediate post-war years. The 40th FIS was the first regular FEAF squadron to convert back to F-51s and was joined shortly afterwards by the 39th, while the 41st FIS remained in Japan. F-51D-30-NA 44-74383* Bugs Bunny *was photographed at Pohang in 1950, laden with typical ground attack weapons – a pair of yellow napalm tanks and half-a-dozen 5-in HVARs.

equip a special training programme. Five days into the war, nine USAF pilots along with the small group of RoKAF pilots moved their Mustang force to Taegu Air Base (K-2) in South Korea. This small force of warriors, led by Major Dean Hess, introduced the North Koreans to the Mustang's lethal capabilities. The Mustangs were able to inflict some damage on North Korean equipment and personnel, but a much greater number of fighter-bombers was needed to put a dent in the NKPA's advances.

Urgent request

With the war only 10 days old, the decision-makers at FEAF requested an emergency influx of F-51s from the US. This plea immediately received approval, and within five days 145 Mustangs had been flown from various Air National Guard units to NAS Alameda in California, where they were loaded onto the carrier USS *Boxer* along with 70 experienced Mustang pilots.

Right: The badge of the 18th FBW which, by 1952, was responsible for all F-51 fighter-bomber operations with four squadrons, including No. 2 Sqn, South African Air Force.

Below: 1st Lt Mario Prevosti poses with his 12th Fighter-Bomber Squadron F-51D at Chinhae AB in 1951. The 12th FBS's Mustangs were among the more colourful examples in the FEAF.

South Korea's air force operated ex-USAF F-51s for the entire war, initially with the aid of USAF pilots under the 'Bout One' programme. Their primary base of operations was to be Kangnung AB (K-18), though early missions were flown from Taegu (K-2). Here Mustang '4' is seen at Taegu in 1950; RoKAF markings appear to have been hastily applied.

In the meantime, the 30 or so Mustangs that had been in storage in Japan had been reactivated, and General George Stratemeyer ordered the formation of a provisional squadron to fly them. The pilots selected for this unit were pulled from squadrons of the 18th Wing based at Clark AB. The unit became known as the 'Dallas' group and comprised mostly 12th Fighter-Bomber Squadron pilots and maintenance personnel. Their first destination was Johnson AB to pick up their F-51s, after which they immediately proceeded to Taegu AB. Each of the FEAF wings committed two of its three squadrons to the war effort, while retaining one, thereby allowing each to cover its original air defence role. The 18th at Clark AB sent the 12th and 67th Squadrons to Korea while the 44th remained behind in the F-80s.

The confusion about specific unit designations was almost as great as that surrounding the fluid front lines. The first two hastily organised Mustang units were known as the 'Bout One' and 'Dallas' groups. In a matter of days – and before the Mustangs could actually get into the routine of flying combat missions – the 51st Fighter Squadron (Provisional) was formed to absorb them. Originally, the former unit had been under the control of 5th Air Force in Japan and the latter under 13th Air Force in the Philippines. Now, the collective Mustang efforts would be controlled by 5th Air Force (Far East Air Forces).

First missions

Finally, on 15 July, the 'Dallas' group flew its first missions from Taegu – the same date that USS *Boxer* sailed from San Diego with its deckload of F-51s, destined for Japan. Keep in mind that only about 30 Mustangs were available for combat at this time. These were split between the 35th Fighter Group's 40th Squadron and the 51st Provisional Squadron.

Major Harry H. Moreland was the commanding officer of the cadre that had come from the Philippines. He recalls one of the very early missions flown from Taegu.

"I was in a two-ship flight that launched from the strip at Taegu. We were going to run the road between Seoul and Taejon and hit anything we could find. Our first target was a couple of North Korean trucks that we strafed and then napalmed. There wasn't much activity on the roads, and right after we had headed back to the south, I spotted a huge haystack in the middle of a field that had tread marks leading up to it. I made a low-level pass and could see that it was a tank that had been hastily covered with hay. Since we had used up our napalm, we decided to see what our 50-calibre rounds could do to it. We knew we couldn't knock out the tank with it, but after a couple of passes, we noticed that the hay had caught fire. We made a few low passes right over the top of the tank to fan the flames with our prop wash. The flames literally cooked the tank, and as we circled, there was a large explosion. We had destroyed the tank with our machine-guns and the help of an intense fire."

The pilots in the early provisional unit flew in such an aggressive manner that South Korean President Syngman Rhee named them the 'Flying Tigers of Korea'. The name stuck. As a result of being compared to the legendary 'Flying Tigers' of early World War II fame, the Mustangs in the 12th Squadron received tiger

Above: Australia was among the first Commonwealth countries to offer personnel and equipment to the UN forces fighting in Korea. Mustang-equipped No. 77 Sqn, RAAF, based at Iwakuni AB, had been part of the Allied occupation force in Japan. Committed to FEAF at the end of June 1950, the squadron's 26 aircraft joined the 35th FIG at Pohang. Here No. 77 Sqn aircraft are pictured at Iwakuni in the summer of 1950. The aircraft nearest the camera (A68-809) was the personal mount of the squadron's CO, Wg Cdr L.T. Spence, killed attacking North Korean positions at Angang-ni in September 1950. No. 77's Mustangs were replaced by Gloster Meteor F.Mk 8s in April 1951.

Left: By late November 1950, the South African Air Force had Mustang-equipped No. 2 Sqn in place at Pyongyang East, North Korea. Attached to the 18th FBG, No. 2 flew Mustangs until early 1953, when it joined the 18th Group in re-equipping with F-86F Sabres.

Above: A sometimes overlooked aspect of F-51 operations in Korea were the activities of the 45th Tactical Reconnaissance Squadron ('The Polka Dots', hence the spinner and wingtip markings) – the only operator of the RF-51D during the war. The appropriately named 44-84843* Ridge Runner *(a factory-converted former F-6D) is pictured returning from a mission during 1952. The 'photo Mustangs' very seldom ventured as far as the Yalu River because of their vulnerability to ground fire and the MiG-15. RF-80s generally handled photo assignments north of Pyongyang. Inset: The badges of the 45th TRS each graphically represented the squadron's armed reconnaissance role.

Right: Servicing is carried out on an RF-51D at Kimpo (K-14) during November 1951. The squadron's Mustang complement actually included more 'straight' F-51Ds than 'RF's. These undertook armed visual reconnaissance missions, strafing targets of opportunity.

teeth paint jobs, which became the unit's trademark for the duration of their Mustang operations in Korea.

Also during this early period, the handful of young RoKAF pilots were getting their hands-on training under heavy fire. FEAF records show that the first Mustang lost in the war, on 4 July 1950, had South Korean markings (tail number unknown). The RoKAF pilot was Lt Col Geunseok Lee, who was killed when his aircraft plunged into the ground after attacking a North Korean tank near Shihung, south of Seoul.

Meanwhile, the fast-moving *Boxer* pulled into port in Japan on 23 July, after an eight-day trip across the Pacific. The unloading had top priority and the ground crews were out in force to get the Mustangs combat-ready. Initial orders stated that all aircraft had to be ready to move into Korea within two weeks of their arrival from the US.

About the same time, FEAF got more good news regarding the number of available Mustangs committed to the war effort. The Australian No. 77 Squadron was pledged to the United Nations' efforts and would bring a full complement of Mustangs. Initially they would be attached to the 35th Fighter Group, which brought it up to full strength with three squadrons. Although all efforts to halt the North Korean juggernaut had failed, they were about

Armed only with .50-calibre ammunition, the 45th's F-51Ds on armed reconnaissance missions usually worked in pairs, generally heavy small arms fire forcing them to fly in 'hi-lo' formation, one aircraft watching for ground fire while the other operated at low level. F-51D-25-NT* Big Bertha *was pictured during April 1951.

***Right:** A hot summer's day at Hoengsong (K-46) during 1951 provides little relief for ordnance crews preparing 39th FIS and 67th FBS aircraft for a midday mission. Each aircraft carries a typical load comprising 5-in HVARs and napalm.*

***Below:** This red-trimmed F-51 of the 36th FBS, 8th FBG is in the process of being 'bombed-up' at Taegu AB in August 1950. The stockpile of 500-lb GP bombs in the foreground represented far less than a day's work for the squadron's aircraft. The North Korean army was pushing UN forces deeper into the Pusan Perimeter; the front line was just minutes from Taegu, allowing the Mustangs to fly several missions daily.*

***Below:** The 5-in HVAR assembly area at an unknown 18th FBG Mustang base, pictured during 1951, is strewn with empty crates and HVARs in various states of assembly. F-51s routinely carried six rockets for use against vehicular traffic; modified HVARs, with shaped warheads, were employed against armoured vehicles. Typically HVARs were accompanied by pairs of 500- or 1,000-lb GP bombs or napalm tanks.*

to be swarmed by some of the best close support aircraft in the UN inventory.

No. 77 Sqn had the dubious honour of posting the second Mustang loss of the war, on 7 July. The pilot of A68-757 was jumped by a Yak while on a bombing run against the Pukyong railroad station, and was successfully rescued.

As the new wave of F-51s began to enter the forward areas, it was decided to bring another squadron into the war. The 67th Fighter Bomber Squadron had been standing by at Johnson AB after its arrival from the Philippines, and was next in line for the former ANG Mustangs. The 'Fighting Cocks' from the 67th moved to Ashiya AB on 30 July, and four days later were winging their way over the Sea of Japan, headed for Taegu AB. On 4 August, another major move was made on paper. The provisional 51st Squadron that had been formed was changed back to its old 12th Fighter-Bomber Squadron designation. At this time, the foundation was laid for the formidable 18th Fighter Bomber Group to begin its long and illustrious combat tour in Korea.

The remaining squadron of the 35th Fighter Group committed to action in Korea – the 39th – finally received its allotment of Mustangs during the first week of August. On 7 August, all aircraft from Ashiya AB launched with full ordnance loads and headed for targets along the MLR (Main Line of Resistance) around the Pusan Perimeter. Once they had expended their loads, they recovered at their new base of operations (Pohang AB) on South Korean soil. Some of the aircraft were reloaded and sent out again within an hour of landing. Squadron records show that the 39th logged 26 combat sorties that day. The 35th Fighter Group now had three squadrons operating inside the Perimeter: the 39th, the 40th and Australia's No. 77 Squadron.

Five squadrons

The number of sorties the F-51s could generate in a single day dealt a heavy blow to North Korean efforts. At this point in the war, five fully-complemented squadrons were flying the Mustang in combat. (The RoKAF unit was still not a fully-trained/fully-outfitted force and No. 2 Sqn, South African Air Force, did not enter the war until November.) The final group to utilise the *Boxer* F-51s was the 8th, based at Itazuke, Japan. In the spring of 1950, this outfit was one of the most experienced in the USAF at flying the F-80 Shooting Star. On 11 August, the 8th Group followed the lead of the 18th and 35th in converting two of its three squadrons back to the Mustang (35th and 36th Squadrons). The 80th 'Headhunters' continued to fly the F-80 and split its time between air defence of its assigned sector in Japan and flying bombing missions over the Korean peninsula. With the conflict slightly over six weeks old, Fifth Air Force could now utilise five Mustang-equipped fighter-bomber squadrons in the war effort.

During November 1950 the 35th FBS at Kimpo AB (K-14) experimented with the use of naval 11.75-in 'Tiny Tim' rockets against bridges on the Yalu River. Exactly how many of these reputedly erratic weapons were acquired from the neighbouring Marine Corps Corsair unit, and whether they were effective, appears to have gone unrecorded, though the experiment seems to have been short-lived.

Left: While ground crews toiled to keep their charges operational, they rarely got the opportunity to witness the fruits of their efforts. This demonstration of the effectiveness of a napalm strike was executed for 18th FBG personnel on a small island close to Chinhae AB.

Below: During a six-week period in July/August 1950 the 12th FBS dropped over 2,500 napalm tanks – on average around 60 per day. Mustangs expended a huge number of tanks, especially when their targets were troop concentrations, against which this much feared weapon was particularly effective. Preparing and filling napalm tanks was a messy, time-consuming process that utilised large numbers of local civilian workers.

Even with the breakout from the Pusan Perimeter, the F-51s were tasked with missions that took them all the way up to the Yalu River from bases on the far southern tip of the peninsula. To say that they were operating in a 'target-rich environment' would be an understatement. The Mustang pilots took a heavy toll on enemy tanks and trucks moving on the myriad roads. By early October, this effectiveness had forced the North Koreans to move by night and hide by day. The diminishing number of targets was also exacerbated by the fact that the F-80s' use of new elongated tip tanks meant they were beginning to be able to operate over the entire Korean peninsula, thereby increasing the number of fighter-bombers looking for something to hit.

At this point, the US Marines were just getting in place with their very capable night-fighter squadrons – VMF(N)-513 with F4U-5N Corsairs and VMF(N)-542 with F7F-3N Tigercats. They teamed with the 3rd Bomb Wing's B-26 Invaders for the night interdiction/intruder missions. Still, this was not enough to stem the flow of supplies and equipment during the hours of darkness. It was inevitable that the Mustangs would get sucked into this role, albeit on a very small scale. Many pre-dawn launches of F-51s were made from several bases in South Korea. Their strategy was to catch the North Koreans either settling in or still moving at first light; this they achieved, though not to an extent that was decisive.

There were recorded instances when the Mustangs were sent north on true night-time missions. A notable example of just such a mission took place long before the Inchon

Below: Prior to the Inchon landings of 15 September 1950, US ground troops were wedged in by the NKPA in the Pusan Perimeter. With C-54 transports flying wounded out and bringing in munitions through Pohang and Pusan, the 35th and 36th FBSs moved to Tsuiki AB, Japan, where this F-51D was photographed.

The weapon of choice

1st Lieutenant Marvin Wallace
39th Fighter-Interceptor Squadron

"During those early days, we flew a lot of missions out of Pohang AB and also some from Tsuiki AB in Japan. One of the most memorable was one when we were pushing the North Koreans back north of the Naktong River. At this particular time, we were flying out of Tsuiki AB in Japan and I was flying wing for the 35th Wing Commander, Colonel Fred Grey. We had already flown one mission that morning and had recovered at Taegu AB, where we rearmed and refuelled. Just moments after taking off for our second mission, we received a frantic call from BUTTER 1-4, a forward controller. It seemed that a North Korean anti-tank gun was blocking a column of friendly tanks on a mountain road about 20 miles [32 km] from Taegu.

"The situation was deteriorating rapidly and the controller was desperate, requesting at least one Mustang to come in immediately. I peeled off from the flight and went straight to the area. It only took a few seconds to spot the anti-tank gun in front of the column and the lead tank was only about 100 yd [91 m] in front of it, but was protected from being knocked out by a curve in the road. In another minute, he would have been in full view to the enemy gun. By this time, I was skimming just a few feet above the ground, headed right at the target. I cleared the lead tank and placed my two napalm tanks right into the middle of the gun and its crew. It must have been a thrill for the guys in the tank because I couldn't have been more than 10 ft [3 m] above them as I went by. I really felt good about that!

"But that isn't the end of my story. The mountains were full of retreating North Koreans and they were trying to get across the river near the town. So, I went to work on them by myself, as it was a good 15 minutes before the rest of my flight came into the area. Colonel Grey wanted to know if I had a target for his napalm. Since most of the enemy's vehicles had been destroyed earlier, I directed him in on a ridge where about 400 troops were cut off by the two tank columns. This was a tough way for a troop to be taken out, but this was war! The Colonel killed a lot of them.

"You need to know there is a weird thing about the effects of napalm. When the area was checked by friendly troops, they reported that there were a large number of dead and many were not burned at all. The napalm burns so furiously that it sucks the oxygen from the air and creates a smothering effect. This was what killed a lot of the North Korean soldiers.

"For the rest of the mission, we strafed troops in the hills and those going across the river, which was shallow enough to be waded. Later that day, we learned that two more flights of Mustangs hit the same area with napalm and 50-calibre rounds, and that when the smoke cleared there were over 1,200 dead enemy troops. Our flight did not pick up a single bullet hole from our low-level passes.

"It was always my theory to turn as hard as possible, going in and coming out with rapid altitude changes, keeping the straight-in firing portion of the run as short as possible. Some of us were lucky and others weren't. The small arms fire was extremely heavy and the Mustang was vulnerable to getting hit in the coolant lines."

Right: Conditions on the Korean airstrips were pretty basic, especially for hard-pressed maintenance personnel. Apart from major overhauls, performed in Japan, all necessary work was carried out at squadron or group level or at Field Maintenance Units. The Packard Merlin powerplant of this 12th FBS Mustang receives attention at the 18th FBG's rear base at Chinhae (K-10) during 1951.

Below: Damaged aircraft deemed beyond repair were stripped of useable parts for use on serviceable machines. This 39th FIS Mustang is pictured at Chinhae during the winter of 1951/52.

Heavy-calibre North Korean AAA exacted a heavy toll on the Mustang fighter-bombers; this 39th FIS F-51 was one of the luckier ones. Had there been sufficient numbers available, F-47N Thunderbolts may have been employed in the role in Korea, instead of Mustangs. Certainly serious consideration was given to the use of F-47s, given the reduced vulnerability of its air-cooled engine to the damage caused by ground fire.

Landing, as the NKPA noose was tightening on UN forces wedged in the Pusan Perimeter. Intelligence reports stated that the North Koreans were bringing large numbers of reinforcements down several roads deep into South Korea. The single wing of B-26s was hard-pressed to cover all the roads, prompting Captain William S. Slater, a pilot in the 12th FBS, to volunteer for a nocturnal foray in support of the B-26s.

His take-off was recorded at 23:30 from a base in Japan. It was a moonless night and there was a heavy cloud layer most of the way up the peninsula. As Slater approached the Naktong River, he began letting down through the clouds. His only reference was several small fires that were still burning from daylight strikes. When Slater had descended below the clouds, he was amazed to see rows of truck lights extending far to the north – all southbound. His weapon load consisted of two 500-lb bombs, rockets and 0.50-in rounds. On his first pass he let the bombs go. Circling around and lining up again, he fired a salvo of rockets, which blinded him for a few minutes. After expending both types of heavy ordnance, he had only 0.50-in ammunition left. Lining up carefully and completing several passes, he left the truck convoy in a shambles when he'd finished off his last round. The weather was still poor, so he flew all the way back to Japan on instruments and landed with no problems and no damage to his Mustang. He later was awarded the Distinguished Flying Cross for this mission.

For a brief period, the 35th and 36th Squadrons operated from Tsuiki AB in Japan because the space available in South Korea was stretched beyond capacity and the heavy transports were damaging the few hard runways (PSP or pressed gravel). This situation put both squadrons at a distinct disadvantage, both in terms of wear-and-tear on the aircraft and pilot fatigue due to the long distances required for every mission. There was no place for them to recover in Korea except in an emergency. This condition lasted for only a short while, and three weeks after the Inchon Landing broke the stalemate (7 October) the 35th Squadron 'Black Panthers' moved to Suwon (K-13). The 36th remained in Japan for another three weeks before it was able to get space at Kimpo (K-14). This put them within easy reach of all targets in North Korea.

45th TRS 'recce' Mustangs

On 3 September 1950, another chapter opened in the saga of the F-51 in Korea when Mustangs began to fly in a photo-reconnaissance role. The 45th Tactical Reconnaissance Squadron was accordingly formed at Itazuke AB in Japan, its first foray into Korea beginning at Taegu (K-2) a month later. The 45th was outfitted with standard F-51D models used for armed reconnaissance missions, and in November it also began to receive newly converted RF-51D models. These conversions were made in Japan as aircraft became available, although it was very difficult to find spare Mustangs because the vast majority were rushed into the fighter-bomber mission. The squadron became an integral part of the famous 67th Tactical Reconnaissance Wing later based at Kimpo AB.

The primary job assigned to the 45th was photographic and visual reconnaissance over the front lines and a limited distance north of the MLR. On numerous occasions they were known to penetrate deep into North Korea, but the arrival of the MiG-15 severely curtailed

Some aircraft were not repairable. On 6 March 1952, SAAF Lieutenant Topper Van der Spuy suffered an engine failure on take-off from Hoengsong (K-46) on his fifth mission of the day. Van der Spuy jettisoned his bombs as his aircraft tore into a rice paddy and broke in half, rupturing the fuel tank. Fortunately, the spilt fuel did not ignite and the pilot was able to escape once the aircraft had come to a halt.

'Dogpatchers' – 18th Fighter-Bomber Wing

As Mustang numbers dwindled and units were pulled back to Japan to re-equip, once again, with F-80s, the 18th Fighter-Bomber Wing, and its combat element, the 18th Fighter-Bomber Group, became the main operator of the type in Korea. Its original squadrons – the 12th and 67th Fighter-Bomber Sqns (the 44th FBS having remained at Clark AB with F-80s) – and the attached No. 2 Sqn, SAAF, were joined by the 35th FIW's remaining Mustang unit in Korea, the 39th FIS, in mid-1951. The strengthened 18th FBW fought on with the F-51 until the end of 1952.

Most combat wings in Korea were issued with a C-47 transport for the movement of personnel and equipment; the 18th FBW was no exception.

forays north of Pyongyang by propeller-driven aircraft without adequate protection. With the majority of the 45th's assigned aircraft being the straight F-51Ds, they worked in pairs and 'ran the roads' looking for targets of opportunity that could be taken out or slowed down with their 0.50-in guns. The RF-51s were particularly vulnerable to ground fire because they had to fly straight and level at low altitude in order to get the required pictures. The loss rate jumped so high that none of the squadron's Mustangs was allowed to go out alone. A 'hi-lo' tactic was adopted wherein one aircraft was always about 3,000 ft (914 m) higher, spotting for the one at low level. This proved to be a productive tactic.

FEAF records show that the first RF-51D loss in combat occurred on 1 December 1950. It was flying a photo mission slightly north of the front lines when a hail of ground fire raked the Mustang, hitting the coolant system. The pilot bailed out over enemy lines and was immediately captured. Two months later, on 1 February 1951, the second 45th Squadron RF-51 (44-84844) was shot down north of the MLR. The pilot, Second Lieutenant James Rice, bailed out, and after some evasive efforts he was picked up by helicopter. While the loss of two Mustangs might not seem important, they were specialised aircraft and few in number, so the losses were significant.

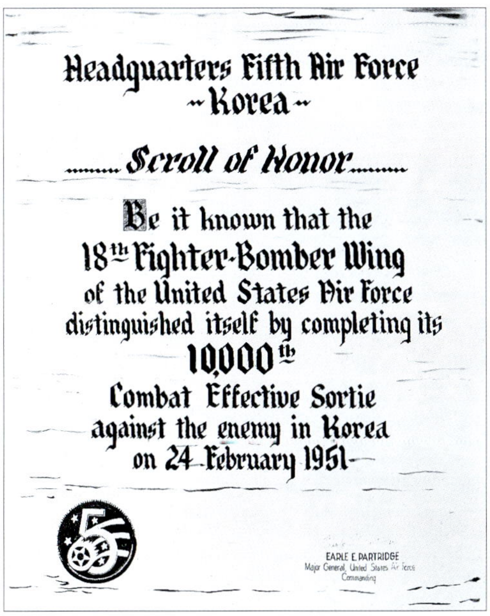

Headquarters Fifth Air Force
– Korea –

Scroll of Honor

Be it known that the
18th Fighter-Bomber Wing
of the United States Air Force
distinguished itself by completing its
10,000th
Combat Effective Sortie
against the enemy in Korea
on 24 February 1951

EARLE E. PARTRIDGE
Major General, United States Air Force
Commanding

Above: Major-General Earle Partridge of 5th Air Force presented the 18th Fighter-Bomber Wing with this 'scroll of honor', to mark its 10,000th sortie, in late February 1951. The 18th continued to fly the F-51 until the end of 1952; by then it had flown close to 50,000 sorties.

Bases in the north

As the cool Korean autumn weather began to move in, the number of bases on the peninsula increased. Some of the Mustang units and Marine Corsair squadrons used former North Korean bases in and around the capital city of Pyongyang. The three primary bases in that area were Pyongyang Main (K-23), Pyongyang East Airdrome (K-24) and Hamhung West Airdrome (K-28). These bases gave the aggressive F-51 pilots almost point-blank access to targets along the Yalu River. On these particular missions, they came under heavy anti-aircraft fire from gun emplacements on the north side of the river (in Manchuria). The 18th Fighter-Bomber Wing had more Mustangs working this location than any other unit. The 12th and 67th Squadrons, along with the South African unit, were among the first to see action over the area that later became famous as 'MiG Alley'.

Although the North American P-51 and Boeing B-29 were probably covered more closely by the media in World War II than any other aircraft type, it became evident that public interest was clearly, and understandably, focused on the new jets and the spectacular air battles between the North American F-86 Sabre and the MiG-15. However, the little Mustang was appreciated by all who flew and serviced it, and especially by the troops on the ground. Once the Republic F-84 got into the war in large numbers, the F-51s were used sparingly north of Pyongyang because of the MiG threat. In the late autumn of 1950, though, the Mustang did attract worldwide attention for a few days: when MiG-15s crossed south of the Yalu River and challenged UN aircraft, it was the agile Mustangs they encountered.

The pilots flying the MiGs learned several lessons in just a couple of days. First, they could turn inside an F-51, and second, if they were not careful, the Mustang would hose them

Right: Captain John Horn of the 39th FIS captured this dramatic view on film over Pyongyang on 14 August 1951, showing the 39th's commander, Major Murrit Davis, flying Sexy Sally***, releasing his napalm tanks. That afternoon, the mission was repeated, but claimed the lives of Horn, Davis and Horn's wingman.***

Mustang pilots with No. 2 'Flying Cheetahs' Sqn, SAAF were awarded no less than 55 Distinguished Flying Crosses by the USAF for their efforts during the Korean conflict. This No. 2 Sqn Mustang was photographed at Chinhae AB during 1951.

down when they bottomed out of their firing pass. In addition, the fuel consumption rate for the jets was astronomical at lower altitudes. The first encounter between these two types came while two flights of 18th Fighter-Bomber Group F-51s worked with a forward air controller in a T-6 'Mosquito' operating 10 miles (16 km) south of Sinanju.

No major damage was done to either side in the encounter, but the significance of those few minutes rippled all the way to the Pentagon within hours. A war that had been all but won had suddenly taken on a whole new dimension. The anticipated six-month duration had been extended, indefinitely. Worse was yet to come: the Chinese were about to enter the war in such large numbers that UN ground forces were forced to retreat towards the 38th Parallel.

The November-December 1950 period was unusually tough on Mustang operations, though the weather had very little to do with it. The only USAF fighter-bombers available in quantity were the F-80s and F-51s. The F-84s were just joining the war in early December, so the pressure brought by the Chinese Army moving rapidly – almost unchecked – to the south put the slower, vulnerable Mustangs in a precarious position. Records show that 14 Mustangs were lost (to combat and 'operational' causes) in November, followed in December by another 20. At this rate, the aircraft rushed over on the USS *Boxer* were not going to last long. The emphasis had been shifted to the F-84 and F-86, for obvious reasons, so the older aircraft had to suffer their losses, cannibalise parts from damaged airframes and keep going. The supply of older, prop-type replacement aircraft was inadequate and squadrons flying the F-51 faced great difficulty in

A 67th FBS Mustang, loaded with napalm, cruises at altitude over North Korea, en route to a strike on supply routes from neighbouring Manchuria.

First MiG-15 encounter

Captain A.R. Flake
67th Fighter-Bomber Squadron

"MiG-15s first appeared in the area just north of the Yalu River in the fall of 1950. After months of ground support missions with no major interference from enemy aircraft, we began to see plenty of MiG activity north of the river, and on several occasions they came over on our side. They were usually in two-ship elements at high altitude. They would circle over us to observe our operations and tactics. My first 'almost' aerial combat took place when two MiGs made two passes at the flight I was leading. This attack occurred while we were on a low-altitude search and destroy mission. My element had just climbed up to fly top cover for our other element. We were up around 2,000 ft [610 m] and circling when my wingman, Lieutenant Harold Ausman, called out two MiGs off our right wing and quite high.

"Expecting them to stay high, we made no effort to counter. Suddenly, they inverted, dropped their noses and accelerated toward us. It was an almost vertical pass. As we started a hard turn in hopes that they would come on down and engage, they broke it off. Then, seconds later, after they had made a spiralling climb to the left, they initiated a second pass and this time it was not nearly as steep. We turned into them and I was almost head on against the leader when they broke off again. Although they were out of range, I held down on the trigger in hopes of getting some good gun camera footage of them. When we got back to base, they processed the film immediately and it clearly showed the MiGs on several frames. The film was rushed to HQ FEAF and from there on to the UN in New York. This was the first clear evidence that there were Soviet-built jet aircraft involved over Korea.

"During the following week, my flight encountered several Yak-types, which we engaged, and shot down three of them. The first occurred during a low-altitude mission near the Yalu. Four Yaks, in elements of two, crossed the river heading just east of our position. Their movement was called out by my wingman and I ordered the flight to climb at max power and assume combat formation. Almost immediately, the Yaks began a diving attack on our two elements. I turned my element hard right to meet them head-on. As we came around, the lead Yak fired a short burst and broke to his left and down. I proceeded to engage. Continuing a right turn I lost sight of him, momentarily, then picked him up as he pulled out of the dive and started a left climbing turn.

"He was now well below me. While I manoeuvred to attack, he apparently lost sight of me because he continued in an almost gentle climbing turn. This enabled me to gain a firing position from behind and slightly inside his circle. Having expended part of my ammunition on ground targets, I did not want to fire until within optimum range. At a slight overtake speed and estimating about 600 ft [183 m] range, I pulled the trigger. Almost immediately, the Yak began trailing smoke, and seconds later, the pilot levelled the aircraft out and promptly bailed out. The Yak went in and exploded in a fireball. The pilot had a good chute, so he probably survived.

"The next day, in the same general area, we found ourselves in a very tight spot. This time, we were jumped by four Yaks and the pilots were good and aggressive. The one I went up against was well trained and eager to engage. After passing each other in a head-on pass, I pulled almost vertical to gain altitude while slowing my speed to quicken my turn. At my next sight of the Yak, he was above me and his nose was almost on me, and he was closing so fast he would be firing before I could get in position. Rolling and inverting, I headed straight down. I can't recall all the details, but I knew that I could not turn with him. I used every ounce of power and speed I could get from my Mustang in order to overcome the Yak's superior manoeuvrability.

"The fight ended up in a series of vertical moves, from 9,000-10,000 ft [2743-3048 m] at the top all the way down to the deck and then straight back up again. The only decent shots I was able to get off were when the Yak was at the top of his climb. At that time, he would drop his nose and tumble back down – and it is almost impossible to hit a spinning target. After three or four of these manoeuvres, I began to get a few hits, and on one pass I was slightly above and to the right. As I fired, his right wing appeared to pull away at the wingroot. The Yak rolled abruptly and crashed, spreading flames and debris over a wide area. The pilot never had a chance to get out. He had been one of the better ones that any of us went up against."

Above: For forward operations the 18th FBG used this dusty airstrip at Hoengsong (K-46). Note the partially painted 'sharkmouth' on the Mustang taxiing out with a full load of napalm and HVARs – possibly a newly-arrived aircraft acquired from another squadron.

12th Fighter-Bomber Squadron

The most colourful Mustangs of the 18th FBG were undoubtedly those of the 12th FBS 'Flying Tigers', which adorned its aircraft with a 'sharkmouth' marking much like that of the most famous 'Flying Tigers', the AVG/23rd Fighter Group.

Before settling at Chinhae AB (above) the 18th FBG was based at Pusan (K-9, below), Pyongyang East (K-24) and Suwon (K-13). Eighteen months at Chinhae followed, before the wing moved to a new base at Osan, where it re-equipped with F-86F Sabres.

Above, right and below: Buildings dating from Chinhae's days as a Japanese seaplane base remained and were used by the 18th FBG for maintenance and the like. Known as 'Dogpatch', Chinhae AB was the 18th's main base from 1951.

This late-1951 view shows an ex-40th FIS aircraft reassigned to the 67th FBS after the 40th left Korea to re-equip with F-80C Shooting Stars. As both squadrons used red trim in their markings, few alterations were necessary.

staying at full strength. The only thing to do was to start moving some squadrons back into jets and to shift their Mustang inventory to the remaining units in order to make them effective.

In the first week of December, the Chinese were pushing toward the outskirts of Pyongyang and on 3 December the 35th Squadron was forced to relocate to Seoul City Airport (K-16). Before either the 35th or 36th Squadrons could move, they were hit by an air attack that blanketed the parking area at their Pyongyang airfield with fragmentation bombs, damaging several Mustangs. Three had to be destroyed because there was no time to repair them to flyable condition. The loss of aircraft had become too great and was being felt in all of the Mustang squadrons.

The 8th Fighter-Bomber Group was the first to give up its Mustangs. On 8 December, both its squadrons flew their fighters back to Itazuke to make the change, once again, to jets. The 80th was one of the premier outfits in the Far East and had been one of the first to fly the F-80. This re-equipment was permanent and the unit remained with jets for the duration of the war. The two squadrons effected were the 35th and 36th; their sister squadron, the 80th, had never relinquished its F-80s. This meant they were brought together as a three-squadron group again. The Mustangs that had been flown by these two squadrons were turned over to the 18th and 35th Groups to replace lost aircraft.

The decision to finish concentrating the available F-51s became a reality in May 1951. The previous month (April) had proven to be the worst month for Mustang casualties, a total of 25 being lost to ground fire.

One of the more talented fighter pilots in the 8th FBG was 'Whistling Joe' Rogers of the 36th Fighter-Bomber Squadron, seen here in his F-51 Buckeye Blitz VI, at a forward base in South Korea. Rogers' aircraft was well-known for the steel tubing he mounted under its wings to create a loud whistling sound during a dive attack on enemy positions.

'Dogpatchers' to the end

The first unit to be effected was the 35th Fighter-Interceptor Group. Its 40th Squadron was pulled back to Johnson AB in Japan to begin flying the F-80 again. The 39th Squadron remained in Korea with all of its Group's remaining Mustangs. Since it was not feasible to have a one-squadron group, the 39th was attached to the 18th Fighter Bomber Group. This brought the 'Dogpatchers' up to four-squadron strength (comprising the 12th, 39th and 67th, plus No. 2 Sqn, SAAF). The Aussies of No. 77 Sqn stood down from Mustang operations during the first week of April 1951 to begin conversion to the Gloster Meteor. In early June 1952, the 39th Fighter Squadron pulled away from the 18th Group to begin conversion to the MiG-killing F-86 Sabre. As the various units ceased Mustang operations, the aircraft were turned over to the 18th Group and the ROK unit for continued operations in the war.

From the summer of 1951 through the end of December 1952, the 18th Group flew countless close support and interdiction missions against the Communist forces. Their coverage by the media was minimal, at best. The personal

Duelling with Yak-9s

Captain Howard I. Price
67th Fighter-Bomber Squadron

"We were operating out of Suwon AB (K-13) because we had just been kicked out of Pyongyang by the Chinese. We took off and climbed to an altitude of 5,000-10,000 ft [1524-3048 m]. Our number 4 man aborted just a few minutes into the mission, so the three remaining Mustangs continued north into the Sinuiju area. We were armed with four rockets and a full load of 50-calibre rounds. Our mission was armed reconnaissance, looking for any vehicular traffic that might be moving supplies down to the front lines. Anytime we got this close to the Yalu River, we were always on the lookout for the new MiG-15 fighters that were based in Manchuria. It was easy to spot them from the dust they kicked up right before they took off.

"On this day (6 November 1950), several Yak-9s came at us from the east. First Lieutenant George Olsen, my wingman, spotted them and called out their position. They were climbing up from our 9 o'clock position. By my count, there were six of them. I turned into them so I could identify them positively as enemy. I could see no insignia on the aircraft, but could identify them as Yaks. I called out that they were bogeys and for us to attack. I dove on the nearest one. He saw me and panicked, rolled over and dove down. Immediately, I picked up the one that had been closest to him. He started to turn sharply to the left and away from the other aircraft when he suddenly snap rolled. I followed him down and, while he was in a turn, I clobbered him with a long burst.

"Pieces flew off of the Yak and about that time, his engine quit and the pilot bailed out. What surprised me was that his parachute was square, unlike our round ones. I didn't see where he landed, but swung around to try to get another of the Yaks. All I saw was Lieutenant Reynolds chasing a lone Yak, so the others must have dove down to tree-top level. Their aircraft were much slower than our Mustangs.

"Just as I pulled up alongside Reynolds, he radioed that all of his guns had jammed. The Yak that he was chasing was trying to get across the Yalu to safety. I slid down and across as Reynolds pulled off. Seconds later, I fired each of my rockets, trying to save the 50-calibre ammo. One rocket went right over one of the Yak's wings and the other went right under. Actually, these rockets were worthless, as they had been stored on Guam since 1945. The propellant had solidified on the low side and you never knew if it was going to hook to the left or right.

"I finally lined up the Yak from slightly below and fired a long burst. I hit him from behind the cockpit all the way up to his engine compartment. I must have fired three or four bursts. All at once, a bluish flame started coming out of his engine, and at the same time he nosed over into a very steep dive. I think he

Pictured at K-9 in September 1950, immediately before the Inchon landings, Captain Howard Price shows off the modified 5-in HVARs used to attack armoured vehicles. Fitted with a shaped explosive warhead, these weapons were reputedly capable of disabling a T-34 tank.

was killed because he made no attempt to pull out and he went straight into the ground from 5,000 ft [1524 m]. We watched him impact the ground somewhere close to Sinuiju. No other aircraft were sighted in the area, so we continued our patrol for another 30 minutes. There was no road traffic in our area, so we headed back home. It had been a very successful mission for me."

Pictured during June 1952, six months before the Mustang was withdrawn from front-line units in Korea, this 67th FBS Mustang awaits its next mission from the forward base at Hoengsong.

achievements of the Group's maintenance and ordnance personnel was absolutely outstanding, and was achieved through the leadership and personal direction of experienced squadron and group commanders. They were fighting with an aircraft that had become obsolete, which meant that the source of replacement airframes and spare parts had all but disappeared.

The 18th Fighter Bomber Group's achievements in the Korean War would hold up against any that had been documented in military aviation history. Unit records show that the 18th officially began operations on 29 August 1950. On 24 February 1951, it logged its 10,000th combat sortie. Soon after, it began operating from two bases: Chinhae (K-10) was the Wing and Group HQ, located in the southern tip of the peninsula, and a forward operating base at Hoengsong (K-46) allowed them to launch squadron-strength strikes closer to enemy targets. Six months later (30 August 1951), the 18th posted its 25,000th sortie.

The Group's momentum never wavered, and in July 1952 it racked up it 45,000th sortie. At the time, this was almost unheard of, especially from a Group flying an outdated aircraft type! The loss of the 39th Squadron in June 1952 slowed down the Group somewhat, but it still retained all the existing F-51 airframes from that squadron. By the end of December, when Mustang operations were all but shut down, the 18th had made more than 50,000 combat sorties – and the Group's transition to the new F-86F ensured it would add to its existing records. By February 1953, the F-51 was being flown in combat only by the RoKAF. The aircraft's career had been one of the most illustrious in aviation history: it was good enough to have fought in the final world war and also in the first war of the new Cold War era.

Warren E. Thompson

Aerial victories by Mustang pilots in Korea

Pilot	Unit	Date	Aircraft
1st Lieutenant Harry T. Sandlin*	80th Fighter-Bomber Sqn	29 June 1950	Lavochkin La-7
1st Lieutenant Richard J. Burns*	35th Fighter-Bomber Sqn	29 June 1950	Ilyushin Il-10
2nd Lieutenant Orrin R. Fox*	80th Fighter-Bomber Sqn	29 June 1950	Yakovlev Yak-9 (2)
Captain A.R. Flake	67th Fighter-Bomber Sqn	1 November 1950	Yakovlev Yak-3
Captain A.R. Flake	67th Fighter-Bomber Sqn	2 November 1950	Yakovlev Yak-9
Captain Robert D. Thresher	67th Fighter-Bomber Sqn	1 November 1950	Yakovlev Yak-3
1st Lieutenant James Glessner	12th Fighter-Bomber Sqn	2 November 1950	Yakovlev Yak-9
Captain Howard I. Price	67th Fighter-Bomber Sqn	6 November 1950	Yakovlev Yak-9 (1½)
1st Lieutenant Henry S. Reynolds	67th Fighter-Bomber Sqn	6 November 1950	Yakovlev Yak-9 (½)
Major Arnold Mullins	67th Fighter-Bomber Sqn	5 February 1951	Yakovlev Yak-9
1st Lieutenant James B. Harrison	67th Fighter-Bomber Sqn	20 June 1951	Yakovlev Yak-9

** Pilots flying RoKAF aircraft during June 1950, before arrival of ANG aircraft aboard USS* Boxer

By the end of the war in Korea the last F-51s in the region were operated by the RoKAF which, by then, had an entire wing equipped with Mustangs. This example is seen at the RoKAF base at Kangnung (K-18) in late 1951. During the war 133 were operated by the Koreans; 39 were lost in combat and a further 29 in training and other accidents.

'Brief encounter'

Bell Airacobra in RAF service

The first American day fighter to enter squadron service with RAF Fighter Command during World War II, the Bell Airacobra had only a brief and unremarkable period of service in Britain as it proved unsuitable for combat in Europe's turbulent skies.

Bell's unconventional fighter was powered by a 12-cylinder, vee-type, in-line Allison V-1710-E4 (V-1710-35) embedded in the centre of the fuselage, aft of the pilot, around the aircraft's centre of gravity. This unusual feature had been incorporated to allow the fitting of a nose-mounted T-9 37-mm cannon firing through the airscrew hub, the propeller being driven by a 10-ft (3.05-m) long extension shaft from the engine which ran through the cockpit. It also necessitated the use of tricycle undercarriage – another novel feature in a fighter of the period. However, with its streamlined appearance and a loaded weight of just over 7,000 lb (3175 kg) driven by the 1,150-hp (858-kW) engine, the Airacobra showed a promising performance. It thus came to the attention of the British Purchasing Commission in the USA which eventually placed contracts for 505 of the Bell Model 14 fighters, as well as taking over an earlier French order for 170 examples. The type had entered service with the USAAC as the P-39 in early 1941 and the British aircraft were generally similar to the P-39D. However, to ensure commonality, the RAF had specified a 20-mm cannon in place of the larger weapon as well as six 0.303-in machine-guns, two above the nose firing through the airscrew and two more in each

***Skylark XIII** (seen here being rearmed with 0.303-in and 20-mm ammunition) was the personal aircraft of Sqn Ldr E.J. 'Jumbo' Gracie DFC, No. 601 Sqn's commanding officer, though he rarely flew it. The aircraft (selected for its appropriate serial number – AH601) carried the squadron's winged sword badge on the forward fuselage (in place of a code letter), and the aircraft's name and a squadron leader's pennant on the cockpit door.*

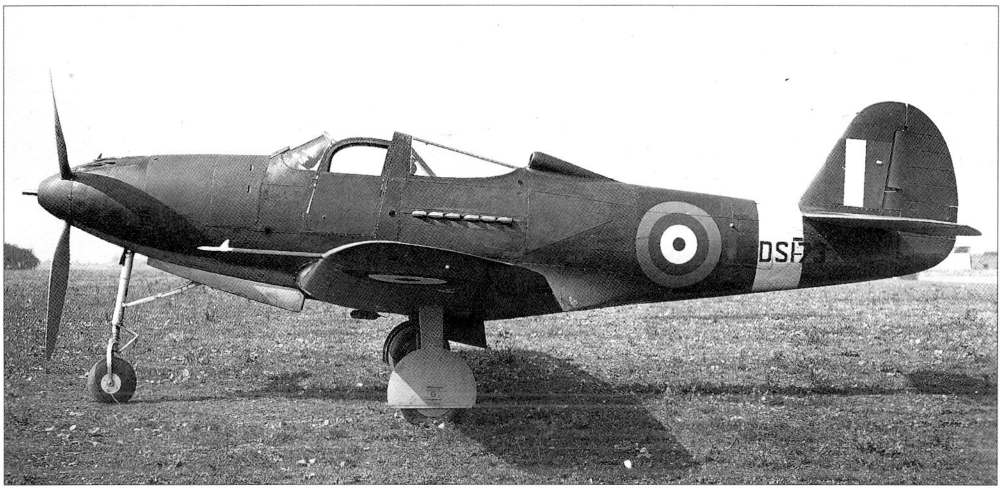

Left: Before Airacobra Mk I deliveries commenced, three ex-USAAF P-39Cs (40-2981, -2983 and -2984) were acquired to speed type tyesting. DS173 (40-2981) was one of two that reached Boscombe Down in July 1941, though testing was hindered by a lack of spares for these unique aircraft. A test pilot damaged DS173 in November when he raised the undercarriage instead of the flaps on landing; he continued testing in DS174, but suffered an engine failure four weeks later – an experience that is reputed to have convinced the hapless flyer to remuster as a Technical Officer.

Below: AH573, the fourth aircraft from the first batch of 170 aircraft, was photographed at Boscombe Down, in July 1941. Early examples, such as this machine, were delivered to the RAF in dark green/ dark earth camouflage with sky undersides in the unusual pattern shown here; by the time they reached No. 601 Sqn most had been resprayed in the later dark sea grey/dark green scheme, with medium sea grey undersides.

Left: A press day at RAF Duxford in late 1941 produced a number of well known photographs of No. 601's new charges lined up and seemingly ready for action. Thirteen aircraft are shown, including AH583 'UF-J' (nearest the camera), which took part in the type's first operational sortie – a 'Rhubarb' on 9 October 1941.

wing. Thus fitted, the RAF's Airacobras became known as the P-400 in US parlance and were initially given the name Caribou by the RAF, though this was changed to Airacobra by the time deliveries commenced.

In preparation for the arrival of the Airacobra Mk I the Auxiliary No. 601 (County of London) Sqn was designated to introduce it to service. On 1 July 1941 it moved with its Hurricanes from Manston, where it had been conducting cross-Channel operations, to Matlaske in Norfolk. It was under the command of Sqn Ldr E.J. 'Jumbo' Gracie DFC, a 30-year-old pilot who already had seven victories to his name, as well as three 'probables' and two damaged. Meanwhile, on 3 July the first Airacobra arrived at the A&AEE for trials and began flying three days later. It was, however, not one of the RAF order but one of three P-39Cs (serialled DS173, DS174 and DS175) shipped for trials and which retained their USAAC armament of a 37-mm cannon and two 0.50-in machine-guns and two 0.30-in machine-guns grouped in the nose. Later in the month, trials at the Air Fighting Development Unit at Duxford began when it took delivery of an Airacobra on the 30th. Comparative trials were conducted against both the Spitfire Mk V and a captured Messerschmitt Bf 109E, which showed the American type to match both in terms of speed and manoeuvrability at levels below 15,000 ft (4572 m). However, because of the lack of a turbo-supercharger (as fitted to the XP-39 but deleted in the optimistic belief that other methods could be used to improve performance and that the type's overall weight would not increase in service), performance fell off rapidly and above 20,000 ft (6096 m) the type was described as "utterly useless" – an ominous portent for the RAF's latest fighter.

Aircraft from the RAF order began soon afterwards and the first two Airacobras, AH576 and AH577 were delivered to Matlaske on 6 August, much to the excitement of No. 601 whose pilots were keen to receive them. So too was the Air Ministry, which saw the importance of the first American day fighters in service, so a press day was held at Duxford, to where No. 601 moved on the 16th, soon afterwards. There the sleek looking Airacobras were photographed in an impressive line-up, while AH577 showed off its paces in the air. Unlike the three P-39Ds delivered the previous month, which wore dark green dark earth and sky camouflage, the new arrivals wore the latest dark green, ocean grey and medium sea grey colours. No. 601 pronounced its ownership with the unit code letters 'UF' but also added (against regulations) its red winged sword badge on the fin flash. From the very beginning, No. 601 was disappointed with its new mount and soon came to regard being the pioneer squadron as a somewhat dubious honour, not least because of suspicion about the unorthodox layout. The engine mounting was of particular concern, partly over the effects on its longitudinal stability and partly over fears that it would tear off and crush the pilot in a crash landing, though this was largely groundless. For their part, the ground crew also considered the many novel features would lead to trouble. This was soon borne out when, while undertaking aerobatics in AH576 on 29 August, Sgt Briggs blacked out in a tight turn and so diverted to Mildenhall. On his approach an electrical fault jammed the undercarriage, necessitating a wheels up landing. The same day Sgt Bell force-landed AH577 at Oakington having suffered a glycol leak.

Conversion, operations – and finale

Conversion flying continued in earnest through September though there was a continuing spate of forced landings, Flt Lt Himr, a Czech and one of the flight commanders, landing one at Marham due to a drive shaft failure while Sgt Land took another into Langham because of a fuel feed snag, though he was concussed in the subsequent forced landing. Worse followed on the 29th, when AH596 crashed in another forced landing near Colchester and was written off. Nonetheless, in spite of manifest teething problems, Gracie considered it essential for the morale of his squadron that they should conduct some operations and his request was granted in early

AH573 was the first of seven Airacobras test flown by A&AEE Boscombe Down and was initially engaged in handling trials, beginning in August 1941, after the type had entered squadron service. Rudder and elevator movement was described as light, quick and effective and, although the fabric-covered ailerons were heavy above 300 mph (483 km/h), overall stability was judged satisfactory. However, the cockpit was strongly criticised as cramped and difficult to escape in an emergency, largely due to its small entrance door. AH573 crashed fatally in February 1942 after an engine failure shortly after take-off.

Above: The Air Ministry was keen to trumpet the arrival of the first American fighters to see service in Fighter Command. AH595 was one of the few RAF Airacobras to fire its guns in anger, when P/O Manak strafed barges near Dunkirk in 'UF-M' on 10 October 1941.

Left: Czech Flt Lt Jaroslav Himr led the No. 601 Sqn detachment that flew a total of eight operational sorties from RAF Manston during early October 1941.

Flt Lt Himr flew Airacobra 'UF-L' from Manston on the second of his two operational sorties over France. As built the Airacobras were fitted with unique 12-stack exhausts on either side of the engine. These were replaced in the UK with the six-stack 'fishtail' exhausts seen here, which offered improved flame-damping.

October for a week of trial operations.

Thus, on 6 October a section of Airacobras left Duxford for the squadron's old haunt at Manston on the Kent coast, the intention being to participate on some of Fighter Command's cross-Channel missions. With its cannon armament, it was felt that the Airacobra might be well suited to anti-shipping sorties and ground strafing. Led by Flt Lt Jaroslav Himr in AH589/'UF-L', AH583 flown by Sgt Briggs, Plt Off Manak in AH595/'UF-M', Sgt Scott flying AH581 and AH591 with Sgt Reynolds lifted off from Duxford at 1440 for the 25-minute flight to Manston, where they joined their fellow Auxiliary unit, the Hurricane-equipped No. 615 (County of Surrey) Sqn. Sadly, on this first day bad weather prevented a commencement of operations and so Reynolds returned to Duxford later in the afternoon. The remaining four pilots planned an operation for the following day but fog at Manston again prevented any operational flying, though F/O Chivers flew down in AH591 to join the detachment. Manston remained fog-bound on 8 October too, and there was no opportunity to fly, but the following morning Chivers in AH591 and Manak flying AH595/'UF-M' flew up to Duxford.

However, that afternoon, the Airacobra went on to operations when, at 1745 hours, Flt Lt Himr in AH583 and Sgt Briggs flying AH881 left Manston on a 'Rhubarb' (the Fighter Command code name for small-scale fighter missions against ground targets of opportunity). The pair flew low over the Channel to Dunkirk where they shot up a number of enemy personnel on the pier there and, as the squadron records stated, "severely hurt the feelings of a trawler"! They arrived back at 1820 hours. As the squadron also recorded, it was not much, but it was a start. The next morning, 10 October, Briggs flew AH581 to Duxford but 'Jumbo' Gracie, the CO, arrived at lunchtime in AH583. Shortly before he had arrived, Plt Off Manak had lifted off in AH595/'UF-M' and flew to France, where he shot up several barges in the canals behind Dunkirk before arriving back at 1300, having been up for just 45 minutes. At 1325 the CO also flew a Rhubarb but found nothing of interest, landing back at Manston after half an hour, while 10 minutes after him Sgt Scott flew his first operational Airacobra sortie in AH581, which was also uneventful. There was no further operational activity in the day due to the weather but in the late afternoon Manak flew back to Duxford, later returning in the company of F/O Chivers, who flew down in AH584.

Early on the 11th Jaroslav Himr flew a weather check over the Channel in AH583, returning at 0720 with the news that things looked suitable for operations. Accordingly, at 0800 Himr, again in AH583, Manak (AH595/'UF-M') and Chivers (AH584) took off as escort five minutes after eight of No. 615's Hurricanes, led by the five-victory ace Flt Lt 'Dutch' Hugo, who was flying Z3028 on a shipping reconnaissance off Boulogne. Unfortunately the formation found nothing and so returned to base, though the Airacobras did experience some accurate light flak, though were undamaged. No. 615 recorded the only ever Airacobra escort thus: "Proceedings commenced with a before-breakfast sweep up the 'single man's side' of the coast to Ostend with three Airacobras of No. 601 Sqn. Nothing of interest was seen except flak." That final statement was in the event a requiem for the Airacobra's operational service in the RAF, for at 1435 'Jumbo' Gracie in AH581 led Manak, Briggs and Himr back to Duxford, where they landed at 1515. So ended the first, and as it transpired, final operational detachment for the squadron's American fighters, which had flown just eight sorties before an accumulation of snags on the aircraft led to their withdrawal back to base.

Withdrawal

The previous day the appropriately serialled Airacobra AH601 was delivered and the coincidence of the number decreed that it became 'Jumbo' Gracie's 'own' aircraft. It carried the squadron's winged sword badge on the nose in place of an aircraft letter, though in the event Gracie seldom flew it, preferring AH577/'UF-M'.

After the return of the detachment No. 601 continued to try to iron out the bugs in the Airacobra, but continued to suffer a series of technical incidents. On 19 October the squadron suffered its first fatal accident when P/O Peter Hewitt was practising aerobatics in AH582 and spun, crashing near Bedford, though there was no obvious cause. In spite of its non-effective state, No. 601 continued to be the focus of attention, and in early November Lord Sherwood, the Under Secretary for Air, accompanied the US Ambassador, J. J. Winnant to Duxford, where they inspected the squadron and they left well pleased. However, the Airacobra had been tainted as far as the RAF was concerned and that same afternoon,

The RAF's disappointment with the Airacobra stemmed from its lacklustre performance at altitude. Bell had claimed a top speed of 400 mph, but only managed to attain 371 mph (597 km/h) in AH579 during testing. Anxious to make amends and under pressure from the British to do so, Bell conducted high-speed testing with AH571 (pictured) during April 1941. Highly polished and fitted with faired exhaust stacks, a smaller rudder and a number of other refinements, this aircraft was coaxed to 391 mph (629 km/h). However, few of the modifications made to the aircraft were suitable for incorporation on the production line and the Airacobra the RAF received was a sadly inferior product.

8 November, a delegation of Soviet officers visited the squadron. Their purpose was to assess the Airacobra for use in the USSR, as the Britain had offered to divert the bulk of the RAF order. As most combat over the Eastern Front tended to be at low level, it was felt that the Airacobra might be well suited. The Russians were clearly pleased, but No. 601 less so as incidents still abounded. On 21 November Sgt Land had another accident when the engine of AH603 cut as he lifted off and he ploughed through the airfield boundary hedge, while on 12 December, the same happened to the engine of AH601 as P/O Sewell took off from Duxford and it, too, ended on its belly. Worse followed the next day when AH581's engine cut near Debden; Sgt Sawyer died in the resulting crash.

Low morale and a move north

On Christmas Eve 'Jumbo' Gracie left for Malta where he was to take over No. 126 Sqn, his place at the helm of No. 601 being taken by Sqn Ldr E.J. Jones. Morale was low and not improved by a move north to Acaster Malbis near Tadcaster in Yorkshire on 6 January 1942. It is an irony that No. 601 was the only operational squadron ever based at this grass airfield – described as 'that frozen meadow' – which was covered in snow and slush when the Airacobras arrived, making flying virtually impossible. There was no flying at all between the 20th and 25th and only two weather checks were flown between then and the 31st, taking the total flying for January to just under 23 hours! Nothing daunted, however, the faithful groundcrew had got 13 aircraft serviceable for a visit by the AOC, but thick snow prevented any flying. Another aircraft suffered a problem on take-off on 7 February but on the 12th the getting of a four-ship section airborne became a cause for celebration! The following day, however, 601 had another fatality. Flying AH602/'UF-W', P/O McDonnell entered a roll at 6,000 ft (1829 m) over Acaster Malbis and, while inverted, lost control. The tail was seen to drop and a simultaneous increase in exhaust smoke indicated a dramatic increase in power, as if trying to correct the longitudinal trim. It was to no avail as the aircraft then entered an inverted vertical spin from which there was no escape, and it crashed on the banks of the River Ouse.

On 10 March Sqn Ldr John 'Bish' Bisdee DFC returned to operations after a period of instructing. He had six and two shared victories to his credit. His squadron was by now in a poor state, with serviceability low and poor operating conditions on the ground, a general thaw turning the airfield surface to mud which caused the nosewheels to dig in. Everything seemed to be conspiring against what had been a crack unit and later in the month it moved to Digby, re-equipped with Spitfire Mk Vs and in early April embarked for Egypt, where it soon re-established its reputation.

Andrew Thomas

P-400 in USAAF and Soviet service

No. 601 had taken a total of 24 Airacobras of the batch of 170 on strength – AH576-8, AH580-9, AH591-3, AH595-7 and AH601-3. Two of the P-39Ds, DS173 and DS174, were also used. Six others served with test establishments and three more were used by ferry units while, interestingly, one other, AH651, was used by No. 225 Sqn, presumably for trials in the army co-operation role, though few details are available. Most of No. 601's surviving aircraft were transferred to the USAAF, as were a few others of the batch before delivery, but the bulk were transferred direct to Russia. A further batch of 121 was delivered in early 1942 but only one, AP309, was used by the RAF, being flown at No. 218 MU at Colerne until being damaged beyond repair on 7 May. Twelve of the others went to the USAAF and the rest to the Soviet Union. Allocated batches of other aircraft, 84 serialed BW100-183 were also divided between the US and Soviet Air Forces, as were the 300 Airacobras of the batch BX135-434.

The USAAF made use of them in the southwest Pacific, as P-400s, initially as interceptors (alongside P-39Ds) until it was realised that they were no match for high-flying Japanese fighters. Strafing/dive-bombing then became the type's main role. Smaller numbers of P-400s also served with three groups in North Africa.

For the Soviets the RAF Airacobras were the first of many hundreds of P-39s (and later P-63 Kingcobras) delivered from US production. The first of the ex-RAF Airacobras delivered was AH628, which was despatched with the first batch to Archangel in November 1941. It was evaluated at Kol'tsovo and the type entered operational service in May 1942. Most retained their RAF serials and camouflage but with the Soviet red stars in place of the overpainted roundels. They were flown by a number of successful Soviet pilots, including Capt Ivan Bochkov of the 19 Guards Fighter Regt (eight single and 32 shared victories) who used AH962 '12', AH726 '36' and BX168 '15'; Capt Ivan Gaydayenko of the 19th, (29 personal and shared victories) who flew AH636 '33'; Lt Yefim Krivosheyev (29 victories) who died flying the Airacobra and regularly flew BX320 '16' and 1st Lt Nikolai Pasko of the 28th Guards Fighter Regt flew BX254 (15 victories).

Thus, the RAF's Airacobras had found their niche at low level over the frozen wastes of the Russian Front. Their primary mission was the protection of the Red Army from aerial attack, most P-39 operations involving combat with Luftwaffe bombers and their fighter escorts, with occasional forays into ground attack.

Airacobra BW114, pictured at RAAF Laverton in early 1942, is believed to have been used by the RAAF for familiarisation purposes prior to the receipt of 22 P-39D/Fs. The aircraft is known to have been among a batch of 178 P-400s modified by Bell to USAAF specifications during 1941.

Capt I.V. Bochkov, 19 GvIAP, East Karelia, 1942

P-400 'Yellow 16' (with the misquoted serial BX728) is believed to have been one of the last batch of 300 Airacobras ordered by Britain, but delivered instead to the Soviet Union. Bochkov was one of the early VVS aces and scored 39 kills, including 32 shared victories. Killed in action on 4 April 1943, he was awarded a posthumous Hero of the Soviet Union decoration on 1 May. Note that 'Yellow 16' is still finished in the dark green/dark earth camouflage in which it would have left the factory; a red star covers the RAF roundel.

Republic's legendary 'Lead Sled'

F-105 Thunderchief

The final member of Republic's 'Thunder' family, the F-105 Thunderchief was conceived to deliver a nuclear weapon against a point target while flying at supersonic speed at low level. Powered by the largest jet engine then available, the 'Thud', as it was mockingly known, was plagued with early problems, but as the 'bugs' were ironed out and the largest single-engined fighter yet built became embroiled in an entirely different sort of conflict, its reputation was transformed. Flying more missions than any other type over the jungles of Indo-China, the F-105 formed the backbone of the Rolling Thunder bombing campaign against North Vietnam between 1965 and 1968 and later helped pioneer USAF defence suppression techniques in the ground-breaking Wild Weasel III programme. To those that flew it the 'Thud' nickname had become a term of endearment.

F-105F 63-8326 of the 49th Tactical Fighter Wing at Spangdahlem AB was photographed over the Atlantic from a KC-135 tanker during December 1965. Like the vast majority of the Thunderchief fleet, 'FH-326' was destined to see service in Vietnam and eventually joined the F-105G Wild Weasel III programme.

"A decade of Thundercraft" reads the caption to this company publicity photograph. Republic's pedigree as a builder of fighter aircraft was beyond question. The F-105, represented here by F-105B 54-0102, was preceded by the widely-produced F-84 family (from right) – F-84E Thunderjet, RF-84F Thunderflash and F-84F Thunderstreak. The aircraft are pictured outside the company's Farmingdale, Long Island, factory on what would appear to have been a chilly winter's day.

On 22 October 1955 the YF-105A prototype 54-0098 (below) made its first flight, from Edwards AFB, California. The two prototypes had RF-84 type wing root intakes, also seen on 54-0099 (below right).

The aircraft was known by various names, most of which were derogatory in nature. Its official name was Thunderchief, but Air Force air and ground crews referred to it as 'Hog', 'Ultra Hog', 'Squash Bomber' and 'Lead Sled', because of its size and the amount of runway it consumed on take-off. It was said that if you built a runway that ran completely around the earth, Republic would build an aircraft that would use all of it. 'Thud' was the preferred name. It was said to be the sound an F-105 made when crashing into the jungle! However, 'Thud' was later to become a term of reverence; if you flew the F-105, you didn't want to fly anything else. Down 'in the weeds', where the Thunderchief operated, no other aircraft could catch it.

Initially, it was known simply as Advanced Project 63 Fighter Bomber Experimental, or AP-63FBX, a company-funded project begun in 1951 that had as its aim the construction of a Mach 1.5 fighter bomber with a nuclear strike capability. A formal proposal was made to the US Air Force in April 1952, resulting in a verbal 'go-ahead' commitment in May of that year. A formal contract was issued to Republic in September 1952, calling for initial production of 199 aircraft that would be operational in 1955.

However, in March 1953, the Air Force cut that contract down to 37 fighter-bomber and nine reconnaissance aircraft. This was followed in December by a complete shut-down of F-105 project development; it was reinstated in February 1954, but only for 15 development aircraft. By September 1954 even that contract had been cut, from 15 down to three, a figure quickly amended one month later to six and then back to 15 in February 1955.

GOR amendments

The original General Operational Requirement, GOR-49, issued by the Air Force in December 1954, called for three amendments: the F-105 was to have in-flight refuelling capabilities, a new, highly complex fire control system and enhanced performance compared to the aircraft envisaged in the original proposal. The latter would be accomplished by replacing the proposed Allison J71 engine with the new Pratt & Whitney J75 engine, then still under development.

The initial AP-63FBX proposal issued in December 1953 was intended to be powered by a pair of Allison J71s, later changed to a single J71-A-7 axial-flow turbojet engine rated at 14,500 lb (64.49 kN) of thrust with afterburner. The AP-63FBX was a single-seat aircraft with a 36.7-ft (11.19-m) wing span and a length of 52.3 ft (15.94 m). Combat weight was set at 27,500 lb (12474 kg), with a maximum take-off weight estimated to be 37,000 lb (16783 kg) with 6,000 lb (2722 kg) of ordnance. Defensive armament was to be four T-130 0.60-in machine-guns with 300 rounds per gun.

The second YF-105A demonstrates Republic's buddy refuelling system with the third F-105B. There were few outward differences between the first two prototypes, though 54-0099 had a one-piece, unpetalled exhaust, unlike the first aircraft. The differences between the YF-105As and the redesigned F-105Bs were more apparent, including the 'Coke bottle' area-ruled fuselage, taller fin and distinctive 'nostril'-type intakes of the latter.

The bomb load was estimated to be 2,000 lb (907 kg) of conventional weapons or a single 'special store' – i.e., nuclear weapon – weighing up to 3,400 lb (1542 kg), in an internal rotary weapons bay. Externally, the AP-63FBX would be capable of carrying up to six 1,000-lb (454-kg) bombs on underwing pylons, or another 'special store'. The weapons were to be delivered using the new MA-8 fire control system, using a 'toss-bombing' technique in the case of nuclear weapons. Initially, the AP-63FBX design looked very similar to the RF-84F Thunderflash reconnaissance aircraft – though much bigger.

By 1 January 1954, the design had evolved into the basic shape of the F-105 production aircraft. The nose was much more pointed and the horizontal stabilisers had been moved from the middle of the vertical stabiliser to low on the fuselage. Wing sweep remained at 45º at 25 per cent of the wing chord, but the span had been reduced from 36.7 to 35 ft (16.65 to 15.88 m). The aircraft's length was drastically increased from 52.3 to 62 ft (15.94 to 18.90 m).

The proposed engine remained the Allison J71-A-7, but it was modified with a variable-area tail pipe that increased maximum thrust from 14,500 lb (64.49 kN) to 15,500 lb (68.93 kN) in afterburner. The maximum weapons load was set at 6,000 lb (2722 kg), more external hardpoints allowing loading variations. The four T-130 0.60-in guns were replaced by a single General Electric T-171D six-barrelled, 20-mm rotary cannon that had a rate of fire of up to 6,000 rounds per minute and an ammunition drum that held up to 1,028 20-mm shells.

Designated the F-105A, an official mock-up was inspected and approved in October 1953. The MA-8 fire control system comprised an AN/APG-31 ranging radar, K-19 sight, a toss bomb and time-of-flight computer, and the T-145 special stores release system. Combat weight was increased to

Back to the drawing board - F-105B

It became apparent after the type's first flight that, as originally designed, the Thunderchief was not going to achieve supersonic flight because it was underpowered and was held back by transonic drag produced by its fuselage. The assistance of aerodynamicists from NACA (NASA after October 1958) was requested and their solution was Richard T. Whitcomb's area-ruled, so-called 'Coke bottle' or 'wasp-waisted' fuselage, as already applied to Convair's similarly-handicapped F-102. Other changes included the new Pratt & Whitney J75 engine, revised air intakes to assist the increased airflow and shock waves produced during supersonic flight and a taller tail fin to improve directional stability at high angles of attack.

Far left: Republic's chief engineer, Alexander Kartveli took some convincing as to the necessity of the area-ruled fuselage and, in the event, the bulged rear fuselage of the F-105B was, at his insistence, only 80 per cent of the size recommended by NACA.

Below: Four F-105Bs were constructed for service test, the first taking to the air on 26 May 1956. The following month Republic applied to the Air Force for approval of the name 'Thunderchief'; this was granted in July, an official naming ceremony taking place in August.

The third of the F-105B-1-RE aircraft, 54-0102 (top), was used extensively for publicity photographs, usually with the designation painted on its nose in place of the more usual 'FH-102' buzz number (above). In the latter view the aircraft carries a dummy bomb load, including 3,000-lb weapons on the inboard pylons.

28,530 lb (12941 kg) and maximum take-off weight with a single 3,400-lb (1542-kg) 'special store' rose more than 4,000 lb (1814 kg), to 41,248 lb (18710 kg). However, a normal nuclear strike would have been accomplished with one of the smaller 1,700-lb (771-kg) 'special stores'.

Maximum speed of the F-105A was estimated to be 765 kt (1417 km/h) at 35,000 ft (10668 m) with a single 1,700-lb (771-kg) 'special store' and external drop tanks. Combat radius was estimated to be 959 miles (1543 km) on internal fuel but, with its in-flight refuelling capability, the F-105A's combat radius was virtually unlimited.

Right: Officers inspect 54-0105, the first of the former RF-105B prototypes. The flat nose panels for camera ports are evident; five cameras were envisaged for the aircraft.

By April 1955, the GOR-49 had been amended to include installation of the new Pratt & Whitney YJ75-P-3 engine in place of the Allison J71. The YJ75 engine was a twin-spool, axial-flow turbojet with a two-position convergent tail pipe with a Republic-designed ram air ejector. The YJ75-P-3 was rated at 23,500 lb (104.51 kN) of thrust in afterburner, an 8,000-lb (35.58-kN) increase over the J71. However, development of the powerplant at Pratt & Whitney lagged behind schedule, which in turn caused several setbacks to the F-105A programme. The end result of these setbacks was that both YF-105As were built with P&W J57-P-25 powerplants.

Through the sound barrier

The first prototype, YF-105A 54-0098, was rolled out from the Republic factory in early October 1955. Ground testing and taxi tests began shortly after it had been shipped to Edwards AFB, California. On 22 October 1955, Republic's chief test pilot, 'Rusty' Roth, lifted the YF-105A off the dry lake runway at Edwards for its first flight. During this initial flight, which lasted some 45 minutes, Roth took the new aircraft through the sound barrier for the first time. However, ensuing high-speed testing revealed that the design would not come close to the requirements laid out in GOR-49. The problem was not simply the lack of power available with the substitute J57 powerplant; drag along the fuselage contours during transonic flight also played a part.

The second YF-105A made its first flight on 28 January 1956. The programme suffered a variety of problems that ultimately led to the crash of both YF-105A prototypes. In March, Roth had almost completed a series of flight tests when the aircraft suddenly lost power: the right main land-

Reconnaissance variant – RF-105B

Known to Republic as AP-71, the specification for the RF-105B was issued in August 1955 and described a day photo-reconnaissance fighter based on the F-105B. Two months prior to the maiden flight of the first F-105B $10 million was set aside from FY1957 funds for 65 F-105Bs and 17 RF-105Bs; in the event only three aircraft were completed with camera noses – 54-0105, -0108 and -0112. When the RF programme was cancelled in July 1956 all were completed, less camera equipment, as JF-105B test airframes and joined the other F-105B aircraft in the ongoing test programme.

The RF-105 was revisited in late 1960, when an F-105D variant was proposed with a pod-mounted SLAR, IR sensors, various cameras and in-flight film development and ejection capability. The project was finally abandoned in December 1961, in favour of the RF-110A (later RF-4C) Phantom.

The last of the RF-105Bs and the only RF-105B-2-RE, 54-0112 sits on the ramp at Farmingdale. The 'JF-3' legend on the tail indicates the aircraft's new role as the third of the JF-105Bs.

By early 1958 there were 11 F-105Bs involved in various test programmes, namely the two YF-105As, the four pre-production F-105Bs, three F-105B-5-REs and a pair of JF-105B-1-REs (the former RF-105B prototypes). Here F-105B-5-RE 54-0106 is seen in typically garish test colours at Farmingdale.

ing gear had extended at over 500 kt (926 km/h) in a 5.5-*g* turn, shearing off the landing gear and wheel. Roth was forced to make a belly landing on the lakebed, breaking the back of the first prototype.

Republic test pilot Hank Beaird also suffered an in-flight emergency, in the second aircraft. After a normal take-off the nose gear had retracted exceptionally quickly, breaking the 'unlock' that held it in place in the nose gear bay. The gear would neither stay retracted nor extend and lock in the down position. Beaird had no choice but to crash land on the lakebed, also breaking the YF-105A's back. Both pilots were able to walk away with only injured pride.

F-105B debuts new engine

In mid-May 1956, the third prototype arrived at Edwards AFB. It was the first F-105B service test aircraft, 54-0100, and was quite different from both YF-105A prototypes. In addition to installation of the new YJ75-P-3 engine, the design of the air intakes and fuselage was completely changed. The fuselage was redesigned to incorporate Richard Whitcomb's area-ruled 'Coke bottle' or 'wasp waist' design, developed by the National Advisory Committee for Aeronautics (NACA). Area ruling the fuselage reduced drag caused by shock waves that occurred along the straight fuselage contours of conventional designs. Republic engineers bulged the fuselage at a point 1 ft (0.3 m) in front of the spot where wind tunnel tests indicated that increased drag first took effect. On the F-105, this was near the trailing edge of the wing-fuselage joint.

Concurrently, Antonio Ferri completely redesigned the air intake system. His new intake design was swept forward and had a variable air inlet (VAI) control system that used a set of intake 'ramps' to increase or decrease airflow. These were controlled by a Bendix central air data computer (CADC) that measured the amount of air needed for maximum power at any speed, and moved the ramps in or out accordingly. For low-speed operation, the ramps retracted, allowing maximum air ingestion; at high Mach numbers, the ramps were extended to decrease air flow. These two changes – area ruling and the controllable air intakes – were all that was needed to put the F-105 design into the Mach 2 speed range.

However, this additional speed exacerbated another problem with the original design – vertical tail flutter. It had been discovered on both the earlier F-100A Super Sabre and the YF-105A that their vertical tails were much too short, causing flutter and a loss of lateral control. Republic engineers thus increased the height and chord of the vertical tail and rudder. In addition, on production aircraft rear vision windows behind the canopy were deleted; these had been carried over from the F/RF-84F.

The first production F-105B-10-REs were accepted by the Air Force on 27 May 1958, some three years late. This view of the crowded Farmingdale production line show F-105B-15-REs in final assembly with fuselages for later aircraft in the foreground. A number of the aircraft in this view would later equip the USAF's 'Thunderbirds' display team.

The final pair of F-105B test airframes (B-5-REs 54-0109 and -0110), plus the sole B-6-RE, 54-0111 (pictured) were employed to test the type in simulated combat conditions in a programme run jointly by Republic, TAC, Air Research and Development Command and Air Proving Ground Command. 54-0111 was later among the first F-105Bs accepted for service with the Air Force and was issued to the 4th Tactical Fighter Wing.

Six newly delivered F-105B-10-REs, complete with 4th TFW markings, make an impressive sight on the ramp at Eglin AFB. The 335th TFS was soon moved to Seymour-Johnson AFB, North Carolina, where it would conduct operational testing of its new mount.

Above right: This echelon formation of unmarked F-105Bs belonged to the 334th TFS, the second unit to receive Thunderchiefs. All are examples of the F-105B-20-RE, which introduced the uprated J75-P-19W engine.

It was on the F-105B that the four-petal, clover-leaf air brakes were first installed. Designed to fit around the exhaust jet orifice, the four segments could be deployed in several different configurations, depending on requirements. During landing, only the two side segments were deployed (as the bottom segment had insufficient ground clearance and brake 'chute deployment negated the use of the top segment), while all four segments could be deployed simultaneously in flight.

First F-105B's first flight mishap

On 26 May 1956, the first F-105B, 54-0100, took to the air for the first time. Republic test pilot Lin Hendrix put the aircraft through the required flight tests, including operation of the new air brakes, and came around for a landing. Checking his instrument display, Hendrix found he did not have a green 'down and locked' light on the main landing gear. Chase pilots confirmed that the main landing gear had not extended.

Hendrix had forgotten to retract the air brakes, causing him to apply more power to keep the aircraft in the air. With the VAI intakes fully opened, the central air data computer had determined that still more air was needed, and opened the auxiliary air intake doors (normally used only during ground operations), which were located inside the main wheel bays. Operation of the auxiliary air intakes, combined with a set of brand new seals on the main landing gear doors, caused a suction effect in the wheel bay that kept the doors closed.

Nothing Hendrix tried seemed to work and the main gear remained firmly in the 'up' position. Hendrix followed his F-105 predecessors and put the F-105B prototype down on its belly on the dry lakebed at Edwards. Following a controlled crash, Hendrix started to walk away from the aircraft when he heard a noise – a 'whoosh'. Turning to look at the aircraft, he watched as the main landing gear slowly lowered on to the lakebed. The suction seal on the main gear doors had broken after engine shutdown, allowing the gear to cycle into the down position. Unlike the previous crashes, the F-105B was back on flight status six weeks later.

F-105B testing continued into 1957. On 28 July the public got its first look at the huge new fighter, which had been named Thunderchief in 1956, at the Andrews AFB Open House and Air Show celebrating the 50th anniversary of military aviation. Two sub-types were also under development – the RF-105B and the F-105C. The production contract called for development of three reconnaissance aircraft using the F-105 airframe, to be designated RF-105B. Five cameras in its nose would replace the T171E-3 Vulcan

Project Fast Wind

As the first unit in the world equipped with a Mach-2 strike capability, the 4th TFW was keen to demonstrate the performance of its new aircraft, especially in view of the problems plaguing the F-105 programme. Project Fast Wind saw a new world closed-circuit speed record set by the wing's commanding officer, Brigadier General Joseph H. Moore, on 11 December 1959. Moore flew F-105B 57-5812 over a 100-km (62.14-mile) course at Edwards AFB at 36,000 ft (10973 m) and at an average speed of 1216.48 mph (1957.73 km/h), breaking an earlier record set by the French Nord Griffon II by a substantial margin.

Production delays and modifications meant that it would be mid-1959 before the 335th TFS was fully equipped with F-105Bs and another year before the other two squadrons in the 4th TFW had received their aircraft. Despite this the Thunderchief had been declared operational in January 1959. Among the modifications were the retrofitting of J75-P-19W engines and a Goodyear anti-skid braking system to earlier B-10 and B-15 aircraft. Additionally, toward the end of 1959, problems with the air data computer, GE MA-8 fire control system and the F-105's autopilot prompted Project Optimise, a 26-point modification programme. It was little wonder that at this time fewer than one quarter of the F-105 fleet was flyable.

'Thunderbirds'

Nine F-105Bs (seven B-15s, plus single examples from the B-10 and B-20 blocks) were allocated to the 4520th Air Demonstration Squadron 'Thunderbirds' in May 1963, as follows: 57-5782, 57-5787, 57-5790, 57-5793, 57-5797, 57-5798, 57-5801, 57-5802 and 57-5814. Despite misgivings expressed by some that the aircraft was too big and heavy to be flown in close formation, the team had few problems adapting to the F-105.

***Left:** Ground crews look on during a photo shoot involving 'Thunderbirds' pilots and an immaculately presented Thunderchief.*

***Below:** A pair of the 'Thunderbirds' F-105B-15s demonstrates the team's 'Calypso' routine over Nellis during early 1964. The sooty marks on the tail of the trailing aircraft suggest that it normally flew the 'slot' position below and behind the lead aircraft. Four of the 'Thunderbirds' F-105s were fitted with stainless steel tail fins so that they could fly the slot without risking damage from the exhaust of the lead aircraft.*

cannon and E-34 ranging radar. Defensive armament would consist of a pair of M39A1 high-speed 20-mm cannon, mounted inside bulges on the fuselage sides under the air intakes, while a 340-US gal (1287-litre) fuel tank would be installed in the area originally designed as the bomb bay. However, the USAF cancelled the RF-105B project in July 1956, and these three airframes were completed as JF-105B service test aircraft.

An amendment to the contract also called for development of five airframes as F-105C two-seat fighter-bomber or trainer aircraft. The F-105C would utilise the same fuselage as the F-105B, with the second cockpit in a tandem position behind the original cockpit and taking the place of one of the fuselage fuel cells. Otherwise, the F-105C would be identical to the B model, with all its combat capabilities, weapons, electronics, and ordnance carriage. However, the Air Force cancelled the F-105C in 1957 as the aircraft entered the mock-up stage. Republic completed 15 pre-production F-105s: two YF-105As, four F-105Bs, three JF-105Bs, and five F-105B-5s.

***Left:** Most of the modifications made to the F-105 for the 'Thunderbirds' and carried out by Republic between October 1962 and April 1964, were prompted by the need to reduce its weight. The M61A1 cannon, Doppler equipment and other items were removed and ballast added to maintain the aircraft's centre-of-gravity. The twin pipes directing smoke oil into the aircraft's exhaust are visible in this view of F-105B-10-RE 57-5782.*

The 141st TFS, 108th TFG had been operating F-105Bs for just over a year when a smartly turned out 57-5784 was photographed at the New Jersey ANG unit's station – McGuire AFB – in 1965. The 141st acquired most of the ex-'Thunderbird' aircraft when the team abandoned the Thunderchief in 1964.

The first true production version of the F-105B came off the Republic assembly line in mid-May 1958 and was accepted by the Air Force on 27 May. It was designated F-105B-6, 54-0111, and was the only B-6 built. Delivered to Eglin AFB test facility in August 1958, the F-105B-6 was handed over to pilots and crews of the 335th Tactical Fighter Squadron, who conducted the Category I and II flight tests in addition to bringing the type into operational service. The 335th Squadron was a unit within the veteran 4th Tactical Fighter Wing, which was slated to be the first combat unit to operate the F-105B.

However, the many problems within the F-105B development programme, including all the modifications needed to meet the often-revised GOR-49 requirements, set back the production schedule by several months. The 335th Squadron would not be at full strength until summer 1959. By late spring 1960, three of the four squadrons within the 4th TFW had been equipped with F-105Bs – or should have been; in early 1960, all 56 F-105B aircraft in the 4th TFW were grounded for various reasons, including spare parts availability problems. The fourth squadron, the 333rd TFS, would not be equipped until 1961 when the first F-105Ds became available.

In December 1959, the last six F-105B-20 aircraft, featuring the J75-P-19W engine with 24,500 lb (108.95 kN) of thrust in afterburner, came off the Republic assembly line. This brought total production of the B model to 75 examples, including the 15 pre-production service test aircraft.

Four units operated the F-105B in first-line service with the US Air Force. The 4th TFW had three squadrons equipped with the type – the 334th, 335th, and 336th Squadrons – and the 4520th CCTW had one squadron of F-105Bs at Nellis AFB, the 4537th CCTS. The 23rd TFW at McConnell AFB, Kansas, converted to F-105D Thunderchiefs in 1964, and flew many ex-4th TFW F-105Bs during its initial transition to the Thunderchief. The other Thunderchief unit was the 4520th Air Demonstration Squadron, otherwise known as the 'Thunderbirds'.

'Thuds' for the 'Thunderbirds'

In May 1963, the Air Force decided to re-equip the 'Thunderbirds' with the F-105 Thunderchief. The first of nine F-105B aircraft were withdrawn from service with the 4th TFW, then flown to Republic for modification to 'Thunderbird' demonstration aircraft standard. The weapons and fire control system were removed and replaced by ballast. Two 50-US gal (189-litre) smoke oil tanks (one for red and the other for blue smoke) were installed in the ammunition bay in lieu of one of the fuel cells, operated by buttons on the control stick. The flaps were modified for employment at speeds up to 500 kt (926 km/h) and the fuel system was changed to allow for extended periods of time in inverted flight. To the onlooker, however, the most obvious external change was the application of the team's spectacular red, white and blue paint scheme.

All nine aircraft had been delivered to Nellis AFB, home of the 'Thunderbirds', by 16 April 1964. Less than a month later, the team flew its first air show with the 'Chief. However, it was at one of these early public shows that a disastrous crash occurred: on 9 May 1964 at Hamilton AFB, California, F-105B 57-5801 disintegrated during landing and

Reactivated in 1964 to operate the F-105D, the 23rd TFW at McConnell AFB, Kansas, operated a number of ex-4th TFW F-105Bs during work-up. From early 1966 until late 1970 the 23rd was responsible for F-105 replacement training (taking over from the 4th TFW) and then trained crews for ANG 'Thud' units until April 1971. The wing's constituent training unit, the 4519th CCTS (later TFTS) flew F-105Bs (as well as Ds and Fs) until the wing was disestablished later that year. Coded 'MG' for the 4519th, F-105B-20-RE 57-5826 is seen at McConnell in April 1969 (right). It would appear that at least one of the 23rd TFW's front-line units was still operating F-105Bs as late as August 1968. This F-105B-20-RE (above right), taking off from NAS Miramar, carries the 'MD' tail code of the 561st TFS.

Cancelled two-seaters - F-105C and F-105E

Republic's first attempt at generating Air Force interest in a fully operational two-seat Thunderchief came in early 1956, when the F-105C was proposed. Based on the F-105B, the new variant was designed around a specification issued by Training Command. A mock-up was completed and Air Force approval gained in April; five examples were to be built for evaluation. However, before any were built the Air Force had a change of heart and the two-seat Thunderchief was cancelled the following year.

The F-105E – similar to the F-105C, but based on the F-105D – fared better than had its predecessor; no fewer than eight forward fuselages were 45 per cent complete when it, too, was cancelled, on 18 March 1959. The USAF felt that Republic should concentrate on the F-105D in order to make up lost time and reduce costs.

The F-105C mock-up (right) had tandem seats under a large double canopy with bulged sides to improve visibility. The F-105E (below) was essentially similar, but based on the F-105D.

pilot Captain Gene Devlin was killed. The F-105B had suffered a structural break along its upper fuselage spine. The 'Thunderbirds' transitioned back to F-100 Super Sabres for the remainder of the 1964 season, with the intention that the Thunderchief would be re-introduced in 1965. The Air Force grounded all F-105s and introduced Project Backbone, under which the structural brace in the upper fuselage was replaced by a much stronger design in all F-105Bs and a handful of early F-105Ds.

The 'Thud' never rejoined the 'Thunderbirds'. All remaining front-line F-105Bs, including the display aircraft, were transferred to reserve status with the 141st Tactical Fighter Squadron, New Jersey Air National Guard, beginning on 16 April 1964. The 141st TFS operated the F-105B until 1981. F-105Bs were also later operated by the 119th TFS, New Jersey ANG, and the 466th TFS, Air Force Reserve, at Hill AFB, Utah until 1981.

F-105D – the ultimate Thunderchief

With the B-model service test aircraft in the initial stages of development, the Air Force and Republic began a design proposal for the follow-on aircraft, the F-105D. It was drastically different from the F-105B and showed so much more promise that the Air Force verbally committed to no fewer than 1,500 aircraft to equip 14 tactical fighter wings, though subsequently production of the Air Force version of the renowned McDonnell F-4 Phantom led to reductions in the F-105D contract of almost 50 per cent.

The revision to the original General Operating Requirement on the F-105, i.e., GOR 49-1, called for three major differences between the F-105B and the F-105D: a far more sophisticated fire control system, a more powerful engine, and a completely revised instrument display in the cockpit. These, in turn, required several minor changes throughout the design. All the changes mandated in GOR 49-1, plus those called for when newer advanced offensive and defensive weapons systems became available, made the F-105D into the ultimate Thunderchief.

A completely new fire control system was installed in the D-model – the AN/ASG-19 Thunderstick system. The ASG-19 could be used in both a visual or limited blind-bombing mode; with both air-to-air and air-to-ground capability; and was capable of handling the latest in conventional or nuclear weapons.

The new NASAAR R-14A (North American Search and Range Radar) had a ground and contour mapping feature. It also featured the APN-131 Doppler radar for terrain avoidance, plus an X-band beacon. The R-14A ranging radar had a much larger antenna dish, which required a complete redesign of the forward fuselage contours of the F-105D. Installation of the R-14A also required a 15-in (38.1-cm) increase in the overall length of the fuselage through an extension of the nose. The pitot tube was moved from the left wingtip on the B-model to the tip of the nose on the F-105D.

Power for the D-model would come from the J75-P-19W engine that had been introduced on the F-105B-20. With water injection, the -19W engine offered a maximum thrust of 26,500 lb (117.85 kN) in afterburner. The additional thrust required more air, so Republic engineers redesigned the interior contours of the air intakes and ramps, and

Enter the 'D'

Demonstrated at Fort Bragg, North Carolina in 1961, F-105D 58-1173 set a weight-lifting record for a single-engined aircraft with this load of 16 750-lb M117 bombs. Given that an M117 actually weighed around 820 lb (372 kg), the total load was 13,120 lb (5951 kg). In combat bomb loads like this were rarely realistic; certainly in Southeast Asia loads of 4,000-6,000 lb (1814-2722 kg) were more typical.

Left: In time-honoured fashion the first F-105D (58-1146) poses for the company photographer on the ramp at Farmingdale around the time of its first flight on 9 June 1959. There was no prototype, the first production examples going to the 4th TFW and the 4520th CCTW at Nellis AFB.

Apart from an uprated engine, the major new feature in the F-105D was its AN/ASG-19 Thunderstick fire control system (which could, more accurately, be described as a bombing and navigation system). This included the NASAAR R-14A radar set, housed in the D-model's distinctive black radome. The latter represented a major improvement over the E-34 ranging radar fitted to the F-105B as part of its MA-8 fire control system. While Thunderstick was an advanced system it was primarily a visual system intended to aid the delivery of a nuclear weapon. In this view the fourth production F-105D is pictured, without radar fitted, carrying a pair of B43-0 parachute-retarded nuclear weapons.

reconfigured the central air data computer that controlled the amount of air going to the engine. The installation of the water injection plumbing inside the rear fuselage also necessitated some redesign work.

In the cockpit, the entire instrument panel was changed. Gone were many of the old circular design dials, replaced by new vertical 'tape' instruments, which were far more accurate and easier to read in a combat situation. Four major instruments were changed – the attitude direction indicator, horizontal situation indicator, Mach meter, and altitude vertical velocity indicator, which measured rate of climb.

Other new systems included an ASQ-37 radio, ARA-48 direction-finder, ARN-61 instrument landing system, ARN-62 TACAN system, APX-27 IFF beacon, and an APN-131 Doppler radar. With all this new equipment, plus the additional length of the redesigned fuselage, the D-model weighed over 1,000 lb (454 kg) more than the B, requiring that its landing gear be strengthened accordingly.

The first F-105D (58-1146) came off the Republic production line in May 1959. Lin Hendrix made the maiden flight on 9 June 1959, and this time the first flight went off without a hitch. Less than a year later, in May 1960, the first F-105Ds were delivered to the 335th TFS, which again was to bring the type into operational service as well as conduct the Category II flight test programme at Eglin AFB, Florida.

Above: F-105D training was carried out by the 4520th CCTW (the 4525th FWW from September 1966), pilot conversion courses beginning in March 1961. This quartet of newly delivered F-105D-6-REs from the wing demonstrates the type's air braking system.

Right: As the most experienced Thunderchief unit, the 335th TFS/4th TFW was once again called upon to carry out operational type testing of the F-105D.

Engine problems delayed the Category II testing until late December 1960. Once again, the F-105 programme was in jeopardy, and once again, the Air Force needed 'something' to save the F-105. This time it was a head-to-head fly-off competition between the two best fighter-bomber types in the US Air Force – the McDonnell F-101C Voodoo and the F-105D. The Thunderchief won the competition easily and the Air Force gave Republic a commitment that it would take 1,500 Thunderchiefs. However, Secretary of Defense Robert McNamara's so-called 'all-service' fighter, the F-4 Phantom, was coming off assembly lines. McNamara ordered that no more than seven Air Force wings would be equipped with the F-105D, accounting for roughly half of the production order.

Pilot training for the F-105D began when the 4520th Combat Crew Training Wing at Nellis AFB received its first aircraft on 28 September 1960. Both the aircraft delivered to the 335th TFS at Eglin and those delivered to the 4520th at Nellis were basically flight test machines and were not equipped with many of the weapons systems required under GOR49-1. The first true combat-operational aircraft were delivered to the 4th TFW at Seymour Johnson AFB in early 1961. The second unit to convert to the F-105 was the 36th TFW at Bitburg AB, Germany, in May 1961, followed by the 49th TFW at Spangdahlem AB in October 1961. Both of these units had a nuclear strike mission and were based within minutes of the 'Iron Curtain'.

Project Look Alike

Problems encountered by these units brought about the next major external change. The Europe-based Thunderchiefs were subject to different weather conditions on the Continent. Although the F-105D was considered an 'all-weather' aircraft, that reference was to its combat capabilities. USAFE discovered that moisture was getting through seals and penetrating avionics and electronics compartments.

The Air Force countered this with Project Look Alike, which corrected the moisture problems by applying silver acrylic paint to the entire surface of the aircraft. The 'down time' was also used to correct and enhance the Thunderchief's effectiveness with the newest weapons. All the aircraft were brought up to the latest factory standard, F-105D-25, to correct flight control problems and hydraulic line chafing. The necessary wiring to launch and guide the Martin AGM-12 Bullpup air-to-ground tactical missile was installed. A tail hook was installed under the fuselage for short or rough field operations. The Look Alike modifications, which included remaining F-105Bs, were carried out at Brookly AFB, Alabama.

First combat and new problems

In August 1964, North Vietnamese gunboats attacked US Navy destroyers in the international waters of the Gulf of Tonkin. It marked the beginning of 11 long years of war in Southeast Asia (SEA), nine of which saw heavy participation by US forces. Several of the first squadrons involved were Air Force F-105D units. Combat in SEA soon created new problems for the F-105D. The first was excessive heat

Left: The first F-105Ds went abroad in May 1961 as the 36th TFW at Bitburg AB, West Germany, re-equipped with F-105Ds. Here a pair of 23rd TFS F-105D-10-REs takes fuel from a KB-50J tanker during 1965.

Below left: Operational flying in Europe included not only regular exercises with other NATO air forces, but also armament training courses. This 49th TFW is pictured during a practice bombing run during a mission from Wheelus AB, near Tripoli, Libya. Both of the USAFE F-105 wings had a small detachment permanently based at Wheelus and crews undertook 50 per cent of their flying from the North African base. Weapons courses were run by the 7272nd Flying Training Wing and included live firing at the nearby El Uota range.

Worldwide deployment

The USAF had hoped to equip a total of 16 wings with Thunderchiefs – 14 in Tactical Air Command and two in US Air Forces Europe. However, the changes brought by the Kennedy Administration in 1961, under the guiding hand of Secretary of Defense Robert S. McNamara, saw to it that only eight wings would operate F-105s; the remaining six would receive F-4 Phantoms. The eight original Thunderchief wings were: 36th and 49th TFWs (Seventeenth Air Force, USAFE); 23rd, 355th and 388th TFWs (Twelfth Air Force, TAC); 4th TFW (Ninth Air Force, TAC) and the 8th and 18th TFWs (Fifth Air Force, PACAF).

build-up in the aft fuselage around the afterburner, for which Republic designed a pair of small air scoops that attached to the rear fuselage over the auxiliary air intakes. The scoops pulled in cooling air for the affected area around the afterburner.

When the first flights began flying combat over North Vietnam, they encountered a very formidable air defence system, which only increased in intensity and effectiveness as the war dragged on. The flight crews discovered a potentially deadly problem with the stabiliser unit on the Thunderchief. It was controlled by a single hydraulic system, and one hit in a vital area could take out the entire flight control system, locking the horizontal stabiliser in full 'up' position, which forced the Thunderchief into a nose-down attitude. Several aircraft were hit and ended up in the jungle.

Initially, Republic came up with a 'quick fix' before modifying the aircraft permanently to eliminate the problem. A mechanical lock was installed on the stabiliser, which the pilot could engage manually from the cockpit as soon as he realised the hydraulic system had been damaged. Later, Republic engineers designed and installed a third, redundant hydraulic system in a small tunnel on top of the rear fuselage.

Another change was made to the appearance of the aircraft when, from 1965, the Air Force ordered all tactical aircraft to be camouflaged to hide them from high-flying interceptors. The entire F-105 fleet and all other tactical aircraft in the Air Force, both in the combat zone and at bases in Europe and the United States, was painted in what was dubbed the 'Southeast Asia scheme': dark green (FS 34079), olive green (FS 34102), and tan (FS 30219) over the upper surfaces, with light grey (FS 36622) on the underside. Not only were the aircraft camouflaged against the jungle, but the paint added still another moisture barrier over the Look Alike silver. Finally, the internal weapons bay would not be used during combat in SEA, so an addi-

PACAF Thunderchief assets were assigned to the 8th and 18th Tactical Fighter Wings, which had received F-105Ds in 1962 and 1963, respectively. Pictured are aircraft from the 12th TFS/18th TFW (left) and the 80th TFS/8th TFW (above).

Project Look Alike

Prompted by the conclusions of investigations into the cause of two crashes at Nellis AFB during June 1962, the entire F-105B/D fleet was grounded on 20 June and Project Look Alike implemented. Expected to take a matter of weeks to complete, Look Alike ran for two years and cost $50 million, the work being carried out by the Mobile Air Material Area (MOAMA) from October 1962. As well as involving the re-rigging of flight controls and replacement of chaffed hydraulic lines and wiring looms, Look Alike saw (in the case of the earliest production aircraft) over 300 technical orders issued to bring all existing F-105Ds up to the latest F-105D-25-RE standard, allowing all aircraft to, among other things, carry maximum 12,000-lb (5443-kg) bomb loads and fire AGM-12 Bullpup missiles. Among other changes was the fitting of arrester hooks to all aircraft and introduction of better sealing of avionics and instrument bays and cockpits to reduce the effects of damp and condensation. The latter changes had been suggested by the Europe-based wings and resulted in aircraft being finished in a new silver acrylic lacquer to seal them against the elements. Many of these modifications were also applied to the remaining F-105Bs.

This F-105D-6-RE is finished in the overall silver lacquer applied during the Look Alike modification programme. The air scoop above and to the rear of the national insignia on the fuselage was a later modification made in 1964 after a spate of crashes attributed to in-flight fires and explosions. Found on both sides of the rear fuselage, the scoops reduced temperatures in the engine bay. Fuel system piping was replaced at around the same time.

tional 390-US gal (1476-litre) bomb bay fuel tank was installed.

One of the biggest changes was one of the smallest to the naked eye. It was a modification directly linked to the North Vietnamese radar-directed anti-aircraft and surface-to-air missile (SAM) defence net. Applied Technologies Inc., a small computer company in California, had studied the Soviet air defence radars and created a 'black box' that monitored all Soviet radar signals. The box could advise the pilot of when he was being tracked and of the location of the radar emitter and its type. They even discovered a characteristic in the signal that told the pilot when a SAM had been launched toward him.

This box was the APR-25/26 radar homing and warning system, or RHAWS. Externally, the APR-25(V) had four antennas – two on the trailing edge of the vertical fin (with small plates covering the pre-amp located forward of the actual antenna) and two more in a small fairing under the nose just behind the radome – plus three small 'tit' antennas for the ER-142 OMNI radar under the nose. A small strike camera was installed immediately aft of the APR-25 fairing on most aircraft. The APR-26 launch warning receiver antenna was a short blade type under the fuselage. In the cockpit, a small strobe scope was mounted atop the right side of the instrument shroud. A threat display unit, responsible for identifying the threat and indicating whether a SAM had been launched, was attached to the underside of the upper part of the windscreen. The APR-25/26 RHAWS equipment was installed on all aircraft flying into North Vietnamese airspace, both Air Force and Navy/Marine.

All of these modifications took a while to install, and by the time the Thunderchiefs were ready to take the war to the heart of North Vietnam, that country's defences had already downed over 100 of the big fighters. North Vietnamese anti-aircraft gunners exacted a terrible toll, shooting down over 90 F-105s. The North Vietnamese SAM threat, which had been greatly feared by the Western press, had not materialised, due in great part to a bunch of daring F-105 crews known as the 'Wild Weasels'. They flew the two-seat Thunderchief, the F-105F/G. However, the F-105F did not begin its combat career as the devastating specialised weapons platform it became.

A two-seat Thunderchief, at last

The F-105F was the third-generation two-seat design based on the F-105 airframe. The first was the previously discussed F-105C based on the B-model Thunderchief. Next was a similar effort based on the F-105D, designated the F-105E: it used a D-model airframe with the rear cockpit in tandem behind the front one, taking the place of one of the fuselage fuel cells. As with the F-105C, the E was proposed with a single, rear-hinged bubble canopy covering both cockpits. The F-105E would have had all the combat capability of the D, with a slight reduction in unrefuelled range. Also like the F-105C, the F-105E was cancelled (in 1959) before a prototype could be built.

However, as the Thunderchief went into full operation, the Air Force quickly discovered that it needed a trainer version of the Thunderchief, and needed it badly. The 'Thud' was a handful of an aircraft to fly, and available trainers such as the T-33A and the F-100F simply could not duplicate its flight characteristics. Initially, thought was

A handful of F-105Fs were issued to each F-105D-equipped squadron; this 22nd TFS, 36th TFW example is seen at a foggy Bitburg Air Base during 1964.

F-105F

The primary mission of the F-105F was that of tactical bomber; after all, the last 143 F-105D-31-RE single-seaters had been sacrificed to make way for the F-105F and the two-seat variant was fully equipped to D standard. However, the primary reason for the F-105F's development lay in the fact that the USAF urgently needed a two-seat derivative to smooth the hitherto protracted type conversion process. The first of 143 F-105Fs (62-4412, right) flew on 11 June 1963.

'Thud' at war, 1965-68

War service in Southeast Asia began for the F-105 in August 1964, as the last F-105Ds rolled off the Farmingdale production line. Among the USAF aircraft despatched to the region after the Gulf of Tonkin Incident were eight 36th TFS F-105Ds sent to Korat RTAFB on TDY from Yokota AB, Japan. Initially these Thunderchiefs were confined to operations in support of the CIA's secret war in Laos and their presence in Thailand was officially denied, though from 1965 until the bombing halt in 1968 the F-105 was to bear the brunt of operations against North Vietnam. Thunderchiefs flew more missions than any other type over the three years of Rolling Thunder raids 'up north' – and more were lost than any other type, with the exception of the F-4.

Below: 2 March 1965 saw the commencement of Rolling Thunder strikes against targets in North Vietnam, 25 18th TFW Thunderchiefs joined Martin B-57s in bombing an ammunition depot at Xom Bong; for the next three years the campaign was to become the almost exclusive province of the F-105s. Here an F-105D of the 354th TFS, 355th TFW hauls a load of eight 750-lb M117s on an early Rolling Thunder mission during April 1965.

Above: Pictured carrying the red markings of the 36th TFS, F-105D 62-4371 was one of the quartet of aircraft that undertook the type's first combat mission on 14 August 1964 – a ResCAP for an Air America T-28 over the Plain of Jars, Laos. Flown by Lt David Graben the aircraft was making a strafing pass over an anti-aircraft gun position when it was hit by 37-mm shells and caught fire. Graben nursed the aircraft back to Korat and made a successful emergency landing, though the F-105 was deemed beyond repair and scrapped – the first of close to 400 Thunderchiefs lost in the conflict.

given to equipping Thunderchief squadrons with F-100Fs to fill the need for a transition aircraft between the T-33A and the F-105D. However, a lack of F-100Fs, combined with the great difference in performance between the Super Sabre and the Thunderchief, created a need for a two-seat version of the F-105. In May 1962, Secretary of Defense Robert McNamara authorised the Air Force to contract with Republic to build a number of two-seat Thunderchiefs. These aircraft would not be just a trainer: they would be the first dual-mission aircraft with the Air Force – a transitional trainer and an attack aircraft, complete with a nuclear strike capability.

The two-seat Thunderchief, designated the F-105F, was a straightforward conversion based on the latest D airframe. An F-105D-31 fuselage was used for the conversion, but rather than try to put a second cockpit inside the already crammed fuselage contours of the F-105D at the expense of internal fuel, Republic engineers added 31 in (79 cm) to the forward fuselage just behind the rear bulkhead of the front cockpit. The rear cockpit was a virtual duplicate of the front cockpit, with flight controls, throttles, weapons control panels, and a second R-14A radar scope. Thus, the rear pilot could fly the mission even if the front pilot became incapacitated in some way.

Another change from previous two-seat proposals was in the canopy design. Rather than use a single canopy over both cockpits, as proposed on the C and E, the F model had two smaller individual canopies and ejection systems. The aircraft's overall length increased from 64 ft 4 in (19.58 m) in the F-105D to 66 ft 11 in (20.38 m), and Republic felt the aircraft would need more lateral stability, so increased the height and chord of the vertical tail assembly. The vertical tail was 6 in (15 cm) taller, standing 20 ft 1 in (6.12 m). Neither the wings nor the stabilisers were increased in size. The engine, air intakes, and CADC were the same as their counterparts on the D-31 model.

The weapons capabilities of the F-105F were identical to those of the F-105D-31, including the ability to carry out a nuclear strike mission. Defensive armament was the General Electric M61A1 Vulcan rotary cannon, with an ammunition drum that held a total of 1,000 rounds. The internal weapons bay was the same size as in the D, and could hold up to 2,000 lb (907 kg) of conventional weapons, up to 3,400 lb (1542 kg) of 'special stores', or a 390-US gal (1476-litre) bomb bay fuel tank.

F-105D-25-RE 61-0169* Hazel *was flown by Captain Bill Sparks of the 563rd TFS 'Ace of Spades', 23rd TFW from Da Nang during May 1965. The squadron was about a third of the way through a three-month TDY deployment during which it lost 11 aircraft.

Thunderchiefs were among the first Air Force aircraft to be repainted in what became known as 'Southeast Asia' camouflage, as set out in Technical Order 1-1-4 and introduced during 1964. Here F-105Ds of the 334th TFS are pictured during 1965. Both photographs feature F-105D 61-0105 which fell to 37-mm AAA on 8 May 1967, at the height of the Rolling Thunder campaign.

During 1966 a typical F-105 strike mission included a lead element of F-4Cs and Iron Hand Thunderchiefs, the F-105 bombers in flights of four aircraft, 'College Eye' EC-121s to warn of MiG launches and 'Brown Cradle' EB-66s providing high-altitude radar jamming support. Equipped with radar, the latter were also employed as pathfinders for F-105 formations bombing obscured targets. Often attached to Thunderchief wings, Brown Cradle aircraft remained an important part of F-105 operations, even after the 'Thuds' were equipped with their own ECM equipment.

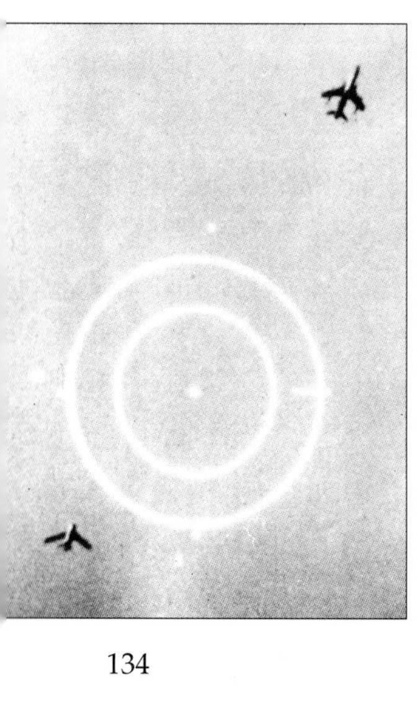

Below: This well-known view of an F-105 being pursued by a VNAF MiG-17 was captured by the gun camera of second Thunderchief. The incident took place north of Hanoi on 19 December 1967, after an attack on the Paul Doumer Bridge.

Externally, the F model had the same five hardpoints under the wings and fuselage to carry up to an additional 6,000 lb (2722 kg) of ordnance or additional droppable fuel tanks. Electronically, the F-model had the complete AN/ASG-19 Thunderstick fire control system. It was wired at the factory to carry and launch the latest in both air-to-air and air-to-ground missiles, such as the AIM-9 Sidewinder and the AGM-12 Bullpup.

The first F-105F, 62-4412, came off the Republic assembly line on 23 May 1963. There was no prototype or service test model. The aircraft were delivered to the Air Force with all the Look Alike modifications, including the silver acrylic paint finish, and the installation of a tail hook. Republic test pilot Carlton Ardery made the first flight of the F on 11 June 1963, and pushed the aircraft through the speed of sound to Mach 1.15 without any problems.

Less than six months later, on 7 December 1963, the Air Force took delivery of the first F-105Fs for operational duties. They were assigned to the 4520th Combat Crew Training Wing at Nellis AFB, home of the Air Force's Fighter School. On 23 December, the first F-105Fs were delivered to the first combat unit, the 4th TFW at Seymour Johnson AFB. All F-105 units within the Air Force, and later the Air Guard and Air Force Reserve, had at least one F-105F in the inventory, serving as both a transitional trainer and an additional attack aircraft.

F-105 versus MiG

April 1966 saw an upsurge in the activities of the VNAF's fighters in the point defence role and, although F-4 Phantoms were tasked with MiGCAP duties, F-105s inevitably came into contact with MiG-17s, -19s and -21s. Between June 1966 and December 1967 F-105s were officially credited with 27½ MiG-17 kills (25 by F-105Ds; 2½ by F-105Fs); in a handful of incidents AIM-9B AAMs were employed, though most kills were scored using the 'Thud's M61 cannon, demonstrating to the USAF the continuing value of gun armament. On the other side of the coin, the first air-to-air encounter of the war saw a pair of F-105Ds fall to MiG-17s on 4 April 1965 – the USAF's first air-to-air losses of the conflict. Seventeen F-105 kills by MiG-17s and -21s have been matched with known USAF air-to-air losses. In addition, the Vietnamese claim that another 10 F-105s acknowledged as lost by the USAF to other causes (usually ground fire) were actually shot down by MiG crews, plus a further 13 where no corresponding USAF loss is acknowledged. Conversely, seven F-105 losses believed by the USAF to have been caused by air-to-air action cannot be traced in VNAF records.

Right: The loss of an F-105D on 14 December 1966 was the first kill acknowledged by the USAF as caused by an AAM (an RS-3/AA-2 'Atoll' fired by a MiG-21), though the VNAF claim that there had been a number of other such victories scored by this stage of the conflict. Here North Vietnamese personnel inspect the wreckage of the F-105, flown by Captain Robert Cooley, who ejected safely.

Republic built a total of 143 F-105Fs, funded with monies originally set aside for additional F-105Ds. The final F-105F, 63-8566, was the last Thunderchief to come off the Republic assembly line; it was finished and accepted by the Air Force in January 1965, bringing total Thunderchief production to 833 aircraft of all variants, including prototypes.

The F-model was not the final Thunderchief variant, however. There were two more, with very specialised missions brought about by the Vietnam War.

Wild Weasel 'Thuds'

As the war in Southeast Asia heated up, President Johnson attempted to halt North Vietnamese aggression by bombing targets in North Vietnam, using F-105s and Navy fighter-bomber aircraft. With every incursion into North Vietnamese airspace, American pilots encountered more and better anti-aircraft defences, culminating with the introduction of the Soviet SA-2 surface-to-air missile.

Soviet engineers began building the launch sites for the SA-2 'Guideline' missile in and around Hanoi in spring 1965. For unknown reasons, Secretary McNamara put all such sites 'off limits' to attack until after they became operational. The first site went on the air in late June 1965. The first launch took place on 24 July 1965 and was a complete success, destroying one 44th TFS F-4C and damaging two of the other three F-4Cs in the flight.

It was an entirely new air war. The Air Force responded to the threat with various types of new equipment and tactics, such as underwing jamming pods. The answer, though, was the development of an entirely new mission and aircraft type. It was to be known as the 'Wild Weasel'.

Applied Technologies Inc. had developed the VECTOR IV radar homing and warning system soon after encountering Soviet air defences over Cuba. The RHAWS consisted of the IR-133 panoramic scan receiver, which had four antennas – two under the nose and two more on the trailing edge of the vertical fin. The IR-133C monitored Soviet radar signals in the S-, C-, and X-band frequencies, displaying their presence to the pilot on a small cathode ray scope in the cockpit. The IR-133C was standardised as the AN/APR-25(V).

Another 'black box' developed by ATI was the WR-300 launch warning receiver. ATI engineers had noted that the SA-2's Fan Song radar signal increased in intensity immediately prior to launch of the missile. The WR-300 monitored this specific frequency and noted any changes on a threat display unit (TDU) in the cockpit. The TDU had a small field of lights that indicated the threat – AAA or SAM – and lit up according to what it identified. It was also able to detect a launch. The WR-300 launch warning receiver was standardised as the AN/APR-26.

Now that the Air Force had the equipment, it launched a crash programme to develop an aircraft to counter the SAM

***Main picture:* These 354th TFS, 355th TFW F-105Ds are pictured topping up their tanks from a KC-135 en route to a target during 1966. Each carries a pair of the largest bombs able to be carried by the 'Thud' – the M118 3,000-pounder.**

***Inset above:* Sortie rates in the early months of the Rolling Thunder campaign were so high that a bomb shortage resulted in mid-1966. This pair of F-105Ds (F-105D-10-RE 60-0469 and, nearest the camera, F-105D-31-RE 62-4287), led by an F-100D flak suppression aircraft, appear to carry incomplete loads of Mk 83 750-lb bombs.**

threat. The aircraft chosen was the two-seat F-100F Super Sabre. The USAF dubbed the programme Project Wild Weasel and its mission was to ferret out North Vietnamese radar signals, locate them, and put them out of action by using ordnance such as bombs and HVAR rockets, or by simply scaring the radar operator into shutting down the radar. The first Wild Weasel F-100Fs arrived at Korat AB, Thailand, in late November 1965, less than five months after the SAM threat had materialised. 7th Air Force code-named the defence suppression mission Operation Iron Hand. The aircraft were an instant success, finding both SAM and AAA radars, then directing F-105 fighter-bombers to destroy the site.

EF-105F makes its debut

However, the F-100F was not the right aircraft for the job. It was too slow and the strike force it was protecting was forced to slow down to its speed, making it even more vulnerable. Moreover, the numbers of F-100Fs were dwindling. The strike force was made up primarily of F-105 Thunderchiefs, so the natural solution was to build a follow-on 'Wild Weasel' aircraft using the F-105, specifically the F-105F. On 8 January 1966, the Air Force accepted the recommendation of Brigadier General K.C. Dempster's defence suppression task group that the next generation Wild Weasel aircraft be based on the F-105F. Eight days later the first Wild Weasel III prototype, F-105F 62-4412, made its first flight. It was known within the Air Force as the EF-105F, for 'electronic fighter', although the official designation remained F-105F and it carried the same electronics package as the F-100F, comprising the APR-25(V) and the APR-26 as an integrated radar homing and warning system.

The first four EF-105Fs deployed to Thailand in late May 1966, arriving at Korat on 28 May. They flew their first mission on 3 June, working with veteran F-100F aircrews. The EF-105Fs led their first mission on 6 June, scoring their first 'kill' on the next day. The aircraft carried a new weapon for the defence suppression mission – the Texas Instruments AGM-45 Shrike anti-radar missile (ARM). With a top speed of Mach 1.5, Shrike homed on enemy radar signals, destroying the radar with a 145-lb (66-kg) warhead. If the enemy radar was destroyed, the SAM missile was blind.

A 'Weasel' mission was considered a success if the enemy radar was either destroyed or forced off the air through fear of destruction. The EF-105Fs had a large variety of weapons other than the Shrike, including gravity

***Below:* This busy scene on the ramp at Takhli AB, Thailand, during 1968 shows F-105s being prepared for another mission. During October F-105D 62-4364, nearest the camera, was among the first four 'Thuds' to complete 3,000 combat flying hours.**

As the North Vietnamese strengthened their defences with additioned radar-controlled AAA and, from April 1965, SA-2 'Guideline' SAMs, the carriage of ECM pods became a necessity. In this October 1968 view, all three 355th TFW F-105Ds carry at least one AN/ALQ-71 or -87 jamming pod on their outer wing pylons. Also evident is the comparatively low altitude at which refuelling often took place, due to the 'Thud's limited ceiling when loaded.

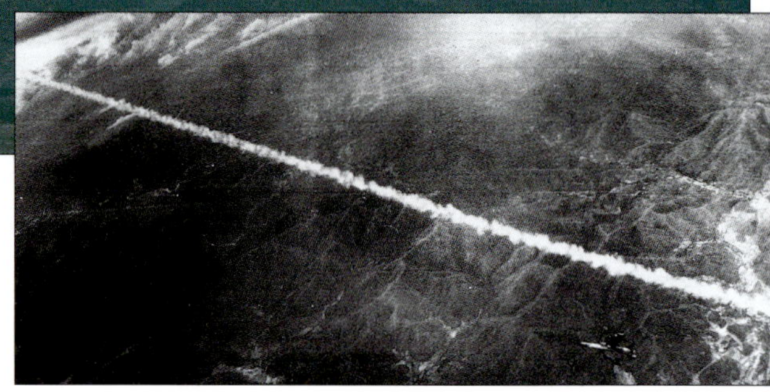

Inset: An F-105 banks sharply to port to avoid a SA-2 launched at its strike package.

bombs, napalm, cluster bombs, and rocket pods. The 'fear factor' was one of the best weapons. The North Vietnamese gunners knew if a Weasel aircraft turned in their direction, it was probably going to launch a missile at them; they would turn off the radar, blinding the SAM and leaving the strike force relatively free to perform its task.

The 'Weasel' mission was easily the most dangerous of the entire war. The EF-105Fs preceded the strike force to the target area and 'trolled for SAMs', i.e. they flew around the hostile North Vietnamese skies making a target of themselves in the hope an enemy radar would try to pinpoint the aircraft, thereby becoming a target itself for the Shrike-equipped Weasel. It was this description of the Wild Weasel mission that led to their motto – "You Gotta Be Shittin' Me!", or "YGBSM".

On 11 July 1966, the last F-100F Wild Weasel aircraft departed Korat. The Iron Hand mission was now entirely in the hands of the Wild Weasel III aircrews. Slowly but surely, the number of Wild Weasel EF-105Fs and crews increased until both F-105 combat wings had the equivalent of a squadron of EF-105Fs available for the defence suppression mission. The two wings utilised the Weasel assets in different ways. The 388th Wing at Korat had an entire squadron, the 13th TFS, dedicated to the suppression mission; at the 355th Wing at Takhli, though, Weasels were split into flights of four within the three squadrons.

As the missions mounted, engineers in the US began to look at ways to improve the capabilities of the 'Weasel' aircraft. The IR-133C launch warning receiver was replaced by the improved ER-142 unit. Many aircraft were equipped with the AZ-EL (azimuth-elevation) antennas to improve the capabilities of the APR-25(V). A new weapon was introduced, the General Dynamics AGM-78 Standard ARM. The AGM-78A was the combination of a US Navy Standard missile, originally developed as a shipboard surface-to-air missile, with the seeker head from an AGM-45 Shrike. The Standard ARM weighed over 1,350 lb (612 kg) and an F-105 needed a special heavy-duty underwing launch pylon, the LAU-80A/80B, to carry the big missile. Later variants of the Standard ARM had internal memories that 'remembered' where a radar site was located even if the operator turned off his equipment. The first EF-105Fs with Standard ARM capability arrived at Takhli in late summer 1968.

SEESAMS and the F-105G

Many people equate the F-105G with the large ECM fairings on the sides of the fuselage, but it was the installation of something called SEESAMS that created the G-model. Air Proving Ground Command and North American had developed the SEESAMS (Search, Exploit, and Evade Surface to Air Missile Systems) as an improvement to the AN/APR-25. SEESAMS scanned the frequencies used by the 'Fan Song' radar that was searching for the strike aircraft, and monitored the tracking frequency of the SA-2 guidance radar, indicating to the EF-105F electronic warfare officer what the threat was, where it was, and how imminent the threat might be, i.e. 'search activity' or 'launch'.

It was developed under Quick Reaction Contract 317. Loral Industries improved the SEESAMS system as part of QRC-317A, which was designated AN/ALR-31. Installation of the ALR-31 required a redesign of the wingtip configuration, in which a pair of antennas took the place of the navigational lights, which were themselves moved to the leading edge of the wingtip. At least 14 aircraft were modified with the ALR-31 SEESAMS equipment, and were known as the first-generation F-105G.

Installation of the fuselage ECM fairings was prompted by a 7th Air Force requirement that all F-105 aircraft flying into North Vietnamese airspace be equipped with underwing ECM or jamming pods, including the 'Wild Weasel' EF-105Fs. Installation of such a pod on one of the limited

Below: Although Bullpup missile performance against hardened targets had proved disappointing, Thunderchief units continued to use the weapon until the F-105Ds were withdrawn. This 44th TFS aircraft carries an AGM-12C missile during operations in February 1970. The naval AGM-12C Bullpup B, with its 1,000-lb (454-kg) warhead, was preferred over the AGM-12B which, though it had been developed for USAF service, was equipped only with a 250-lb (113-kg) warhead and clearly lacked sufficient punch. Bullpup other disadvantage was its radio command guidance system, which required the pilot of the launch aircraft to maintain line-of-sight with the missile until impact, placing him and his aircraft at considerable risk from ground fire.

Right: Radar installations and bridges located in the DMZ between the 17th and 19th Parallels were targets during the first months of Rolling Thunder. This 1965 photograph shows a Bullpup launch against a rail bridge 110 miles (177 km) northwest of Hanoi during 1965.

This 354th TFS F-105D is pictured over typical North Vietnamese terrain of the type associated with the infamous 'Thud Ridge'. Stretching between the Red River, northwest of Hanoi and paralleled by a major railway link with China, and another rail line to the northeast, 'Thud Ridge' became the final resting place for numerous F-105s.

number of underwing stores pylons meant the loss of an offensive weapon to the Wild Weasel crews. Republic and Westinghouse, prime contractor for the underwing ECM pods, worked together to develop an internal ECM system to free up the vital weapons pylon on the EF-105F. Under QRC-380, the engineers took an AN/ALQ-101 ECM pod, split it in two, and installed the pods on the sides of the fuselage above the bomb bay doors. Inside each pod was a modified version of the ALQ-101 ECM system. Standardised as ALQ-105, these ECM fairings were the second part of the complete F-105G modifications.

An upgrade of the APR-25/26 RHAW systems to APR-35/36, which improved the ER-142 panoramic scan receiver, became the third aspect of the F-105G. Finally, the next-generation Standard ARM, the AGM-78B – which had a dual seeker head that monitored the E- and G-bands with one seeker, and the I-band with the second seeker – was made available to the Wild Weasel III crews. The AGM-78B-2 had a single all-band head.

The first ALR-31-equipped F-105Gs began arriving in Thailand in April 1968, and in late 1970 the first F-105Gs with the QRC-380 internal ECM systems began to appear in Southeast Asia. With the exception of constant upgrades in the RHAWS and further development of the AGM-78, the F-105G was now complete. Republic and Sacramento Air Material Area modified a total of 61 F-105F aircraft, including veteran EF-105Fs and some F-105F trainer/attack aircraft, to F-105G specification.

Refining the F-105D – Thunderstick II

Information received from the crews flying the Project Northscope F-105F missions in the night skies over North Vietnam was the basis for an upgrade of the capabilities of the F-105D strike aircraft that offered much greater accuracy during any weather condition, including night bombing.

The programme, called Thunderstick II, applied to both F-105Ds and F-105Fs and involved installation of the AN/ARN-92 Long Range Navigation (LORAN) radar. The NASARR radar system was also improved by installation of solid-state components, now designated R-14K. Basically, the North American Aviation engineers who had created the NASAAR R-14 radar performed the 'fine tuning' that had been done to the Northscope F-105Fs operating in SEA – only they did it better, and permanently.

With the Thunderstick II, the F-105 pilot now had a circular error probability of +/- 50 ft (15 m) from a bombing altitude of 15,000 ft (4572 m). Republic Aviation modified at least 30 F-105Ds with the Thunderstick II between early 1969 and July 1971. All could easily be identified, as the installation of the new Doppler and ARN-92 LORAN systems required a large spinal fairing between the cockpit and the base of the vertical tail.

Although designed to help the F-105 crews who were flying combat over North Vietnam, none of the Thunderstick II aircraft left the US. All were initially assigned to the 563rd TFS, 23rd TFW at McConnell AFB, Kansas. When the F-105 was phased into Air Guard/Air Force Reserve squadrons, the Thunderstick II F-105Ds were all transferred to the 457th TFS, Air Force Reserve, at Carswell AFB, Texas, where they remained until January 1980 when the 457th TFS transitioned to F-4Ds.

Thunderchief at war

The first Thunderchief units into combat were those based in the Far East. The 36th TFS at Yokota had been on a rotational Zulu Alert deployment to Osan AB, Korea, when President Johnson called for retaliatory strikes in response to the Gulf of Tonkin incidents. It was easy for the squadron to deploy in early August 1964 to Korat AB, Thailand, where – officially – it was to supply ResCAP support for any airman downed over Laos or Cambodia. Its actual mission was combat air support of CIA and other friendly units fighting in Laos, Cambodia, and South Vietnam. On 9 August, the 36th TFS set up at Korat, putting eight F-105Ds on strip alert.

The Department of Defense began rotating squadrons in and out of Korat on a regular basis, initially from the 6441st TFW at Yokota and the 18th TFW at Kadena. In February 1965, the 36th Squadron redeployed back to the war. This

After the bombing halt

During 1968, with the Tet Offensive at an end, missions against targets in North Vietnam were scaled back. From 1 April all bombing north of the 20th Parallel ceased; from 3 April this was extended to all offensive missions north of the 19th Parallel. Operations south of this latitude continued however, especially around the DMZ and the border with southern Laos. Ceasefire negotiations and mounting pressure from public opinion at home forced President Johnson to end all offensive missions against targets in North Vietnam on 1 November 1968, though reconnaissance missions (permitted by the Paris ceasefire accord) continued, often with an F-4 or F-105 escort. If hostile intent on the part of North Vietnamese AAA or SAM defences was shown, so-called 'protective reaction' retaliatory strikes were launched. By early 1970 10 F-105Ds had reached 3,500 flying hours, nine of them with the 355th TFW; in February the first F-105D (61-0169 of the 354th TFS) had reached 4,000 hours – the type's theoretical maximum fatigue life. By the end of 1970 the last of the surviving F-105Ds had been withdrawn to the US, replaced in the 355th and 388th TFWs by F-4s, including new F-4Es, the first of which had entered service in November 1968.

Left: Increasingly the F-105F was adapted for specialised roles other than tactical bombing and conversion training. During 1967 RHAWS-equipped EF-105Fs were modified, under Project Northscope for precision night attacks. Initially operated by the 13th TFS, the so-called 'Ryan's Raiders' were distinguished by their wrap-around camouflage.

Lower left: Combat Martin was an attempt to jam North Vietnamese air defence radar. Around 12 EF-105Fs were fitted with a large QRC-128 communications jamming set (variously known as 'Colonel Computer' and 'Fartin' Martin') in the rear cockpit, replacing both the EWO and his ejection seat, and a large blade aerial behind the rear cockpit canopy. This example is pictured in 561st TFS markings at McConnell AFB during 1971.

Below: Det. 1, 561st TFS F-105Gs refuel during a mission from Korat during 1972. Heavily utilised during the Linebacker raids, the F-105Gs of the 561st were among the last to operate over North Vietnam.

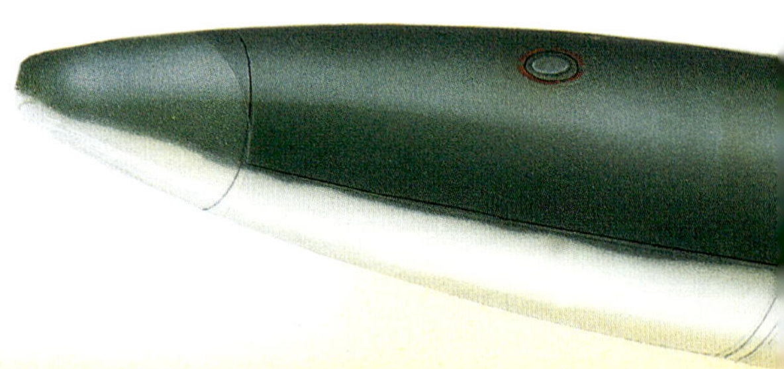

What made an F-105G an F-105G?
The 'EF-105F' became an F-105G once it had been fitted with the AN/ALR-31 RHAW set, the first part of a four-phase modification exercise. An aircraft equipped with ALR-31 could be distinguished by changes to its wingtips; the ALR-31's antennas displaced the aircraft's navigation lights; these, in turn, were moved to a position further forward to the leading edge of the wingtip. The most distinctive external features of the F-105Gs were their external AN/ALQ-105 ECM blisters. Though these are often quoted as being the definitive identifying feature of the variant, around 14 early F-105Gs operated without them until such a time as modifications could be completed. The upgrading of AN/APR-26 RHAW and -27 LWR systems to Standard-compatible APR-36/-37 and the introduction of AGM-78B Mod 1 and Mod 2 missiles completed the package (though RHAW sets were constantly subject to modification as the nature of the SAM threat changed; later versions of the Standard ARM were also introduced).

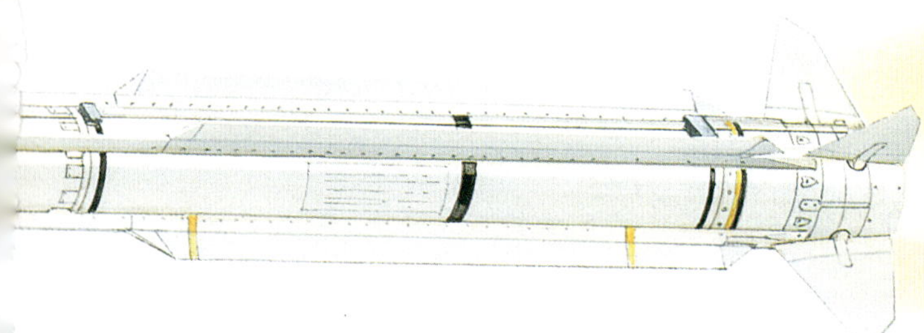

AGM-78 Standard ARM
An attempt to address the performance weaknesses of Shrike, the General Dynamics AGM-78 Standard ARM was developed from an existing US Navy shipborne SAM, by simply combining the RIM-66 Standard missile with the seeker head of the AGM-45A-3A Shrike. The AGM-78A Standard ARM Mod 0 was introduced in 1968 and equipped a handful of US Navy A-6B and E Intruder aircraft, as well as being produced in larger numbers for the USAF's Wild Weasel force. AGM-78A was soon superseded by the AGM-78B Mod 1 variant, with a new Maxson broad-band seeker. This allowed the missile to be used against targets with differing guidance frequencies and incorporated a memory capability so that the missile would remain locked-on to a target if the latter's emitter was switched off.

A considerably larger weapon than Shrike, Standard ARM was 15 ft (4.57 m) in length, weighed 1,370 lb (621 kg) and carried a 215-lb (97-kg) blast-fragmentation warhead. Travelling at speeds of up to Mach 2.5, the missile had range of around 35 miles (56 km).

Because of its expense, Standard ARM was not procured in sufficient numbers to completely replace Shrike; both missiles remained in use into the 1980s, latterly equipping the F-4G Phantom Wild Weasel V.

F-105G-RE Wild Weasel III production
Outlined in USAF technical order (TO) 1F-105-1133, dated 23 February 1970, the F-105G modification programme initially called for 51 conversions, though two of the selected airframes were deleted in an amended TO in May and a further 12 aircraft added in July 1971. 63-8266 *White Lightning* was one of the original batch of 51 originally identified machines, the others being 62-4415, 62-4416, -4422/4425, -4427/4428, -4434, -4436, -4438/4439, -4442, -4446; 63-8265, -8281, -8284/8285, -8292, -8296, -8300/8307, -8311, -8313, -8316, -8319/8321, -8326, -8327/8328, -8332/8334, -8339/8440, -8342, -8345, -8347, -8350/8351, -8355 and -8359/8360. 62-4415 and 63-8281 were later deleted. The dozen additional aircraft added to the programme in July 1971 were 62-4432, -4440, -4443/4444, 63-8274/8276, -8278, -8291, -8318, -8336 and -8363.

While most of these aircraft were selected from among the 86 'EF-105Fs' a number were converted from standard F-105Fs; modification work was carried out by Republic at Farmingdale and at the USAF's Sacramento Air Material Area at McClellan AFB, California.

Republic F-105G Thunderchief

17th Wild Weasel Squadron
388th Tactical Fighter Wing
Korat RTAFB, Thailand
United States Air Forces Pacific
c. late 1972

The sole remaining SAM-suppression unit in Southeast Asia, the 17th WWS was activated on 1 December 1971 from the assets of the 6010th WWS, itself a redesignation of Detachment 1, 12th TFS (comprising the surviving Wild Weasel assets from the deactivating Takhli RTAFB), equipped with 'ZB'-coded F-105Gs. Reinforced by Detachment 1, 561st TFS (deployed from the F-105 training base at McConnell AFB, Kansas) between April 1972 and September 1973, the 17th was operational until 29 October 1974 (when the last F-105 left the region), the squadron inactivating on 15 November, having turned its assets over to the 562nd TFS, 35th TFW at George AFB.

F-105 Wild Weasel development
The Wild Weasel I programme, which had seen the modification of a limited number of the F-100F Super Sabres for the defence suppression role, was limited in its scope to the installation of radar-homing and warning (RHAW) and launch warning receiver (LWR) equipment (installed in a wide range of USAF and USN aircraft of all types from late 1965) and the addition of an AGM-45A Shrike ARM launch capability. Having proved too slow to keep up with strike packages equipped with more powerful aircraft and unable to carry a sufficient bomb load for an effective lone attack, the F-100Fs were sidelined, early the following year, in favour of a modified F-105F variant. In all, 86 F-105Fs were already equipped with the aforementioned RHAW/LWR gear, comprising the Itek AN/APR-25 (VECTOR IV) radar-homing and warning set (detecting S-band emissions from SA-2 'Guideline' SAM, EW and GCI radars, C-band emissions from improved SA-2 radars and X-band radiation from AI radar) and the Itek AN/APR-26 (WR-300) SAM launch warning set (which sensed a power-level change in the SA-2's L-band command guidance radar). As part of what was dubbed Wild Weasel III, the F-105Fs (thereafter known in some quarters, unofficially, as 'EF-105F's) also received an IR-133 receiver (for the detection of S-band emissions at long range), a KA-60 panoramic strike camera and a two-track combat event tape recorder, plus the ability to carry a pair of Shrike ARMs.

By February 1968 14 of the 86 F-105F(WW)s had been further modified to carry AGM-78 Standard Mod 0 ARMs. To locate, identify and acquire targets for the Standard guidance system, a Bendix AN/APS-107 RHAW system was also installed, along with a new AN/APR-35 RHAW set, and two sets of active ECM equipment – a Hallicrafters AN/ALQ-59 VHF communications jammer and a GE AN/ALQ-71 noise jammer. Towards the end of 1968 a further batch of 16 F-105F(WW)s was reworked so that they could fire the latest version of the Standard ARM – AGM-78B Mod 1. Concurrently, a package of equipment changes, many of them destined for the later F-105G, was also introduced.

The F-105G was an attempt to standardise the Wild Weasel III fit as well as introducing a number of new capabilities. Enshrined in a USAF technical order (TO) dated 23 February 1970, the F-105G modification programme eventually covered 61 aircraft. The variant's expanded array of systems comprised AN/APR-35, -36 and -37 RHAW systems (the latter to operate with Standard ARM), a Loral AN/ALR-31 (SEESAM) RHAW set (which scanned frequencies for 'Fan Song' radar activity, providing the location and status of any threat detected), an Itek ER-142 high-sensitivity video receiver (giving automated direction-finding and homing on a threat), a 14-channel combat event tape recorder, a QRC-373 noise jammer and fuselage-mounted Westinghouse AN/ALQ-105 ECM blisters. The latter system was effectively a repackaged ALQ-101 ECM pod, half of which was mounted on either side of the aircraft's lower fuselage in order to keep another wing station free for weapons carriage.

Republic F-105D-5-RE Thunderchief

334th TFS, 4th TFW Takhli RTAFB, 1966

One of the first camouflaged F-105s to see combat in Vietnam, 59-1745 carries the fin-top 'polka dot' marking of the 334th Tactical Fighter Squadron, which spent just over a year on TDY at Takhli during 1965/66.

31 March 1967
Flown on this day by Captain Harry Hennigar of the 421st TFS, 388th TFW at Korat, 59-1745 was undertaking an armed reconnaissance mission along the Song Ca River in North Vietnam when it was hit in the tail by automatic weapons fire. Hennigar flew on for 15 miles (24 km) before ejecting and was rescued by a USAF HH-3. It was his 25th mission; he went on to complete 110 in all.

Survivability modifications
By December 1967 $33 million had been spent on changes to the F-105's airframe and systems to maintain combat efficiency. Among new features were an independent duplicate pitch control system, a new zero-zero ejection seat, an improved braking parachute and speed brake, duplicated hydraulic and fuel systems and improved fuel cell fire suppression.

Shouldering the burden
By the end of 1967 the F-105 had flown three-quarters of the 18,583 missions mounted by the USAF over North Vietnam. Over 250 F-105 pilots had completed 100-mission tours.

Armament and fuel load
This aircraft carries a typical load comprising six 1,000-lb Mk 83 LDGP bombs on a centreline MER plus a pair of 500-lb Mk 82 LDGP bombs with 'daisy-cutter' fuse extenders and the obligatory pair of 450-US gal (1703-litre) drop tanks. An additional 390 US gal (1476 litres) of fuel was held in a tank housed in the redundant internal weapons bay.

Inside the Thunderchief

The F-105F and F-105G (right) shared essentially the same cockpit instrumentation as the F-105D, simplified tape instrumentation for the Airspeed Mach Indicator (AMI) and the Altitude-Vertical Velocity Indicator (AVVI) replacing the conventional dials of the F-105B (above). Between the tape instruments were an enlarged Attitude Director Indicator (ADI) and Horizontal Situation Indicator (HSI) and, below these, a radar scope – a feature new to the F-105D and subsequent variants.

Above: During conversion from F-105F to F-105G standard (pictured), the aircraft's rear cockpit instrumentation was extensively remodelled to make room for mission equipment, most of which was installed above the radar scope. A full set of flight controls and basic flight instrumentation was retained, though tape-type instruments were deleted.

F-105D Thunderchief cutaway

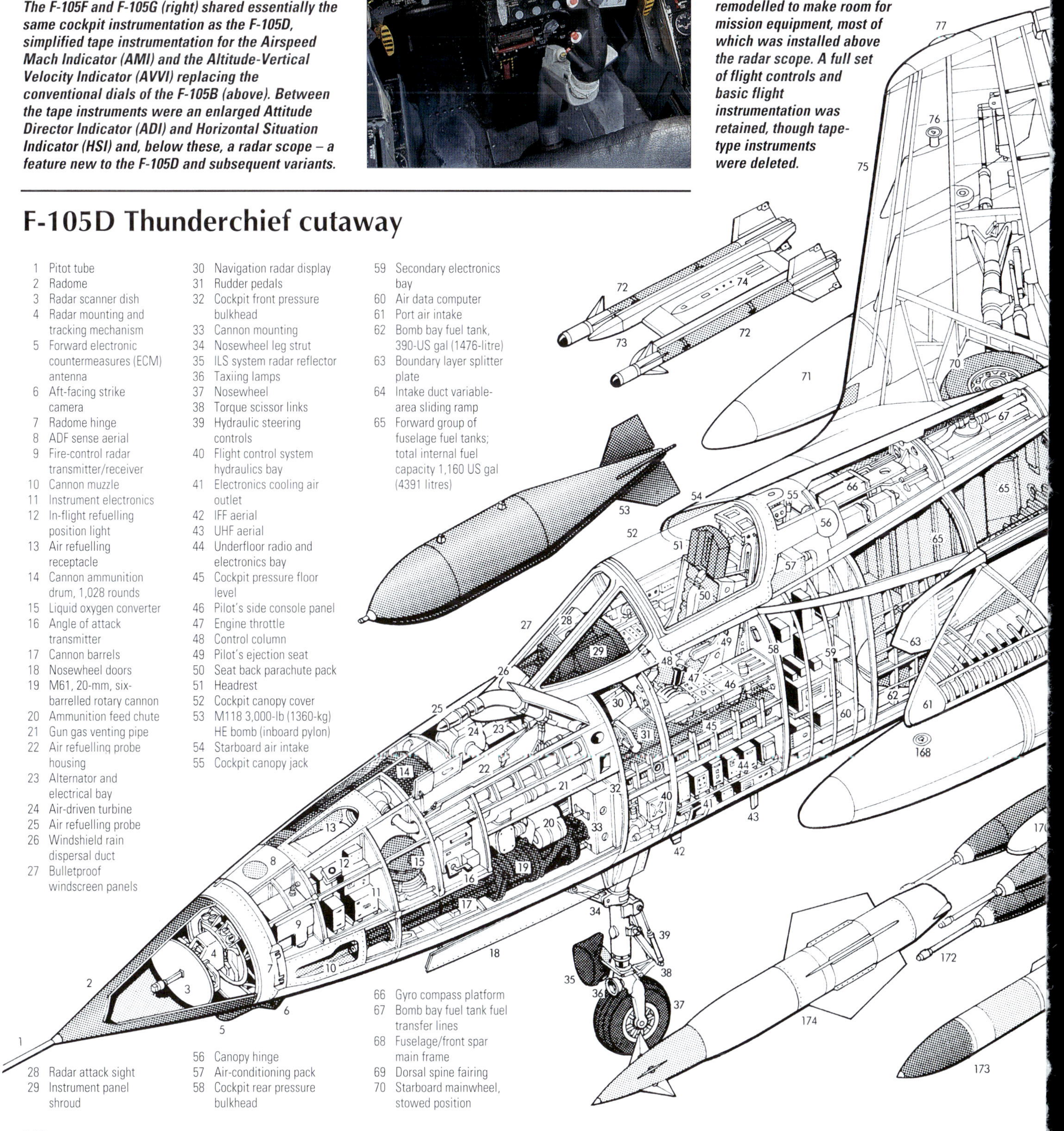

1 Pitot tube
2 Radome
3 Radar scanner dish
4 Radar mounting and tracking mechanism
5 Forward electronic countermeasures (ECM) antenna
6 Aft-facing strike camera
7 Radome hinge
8 ADF sense aerial
9 Fire-control radar transmitter/receiver
10 Cannon muzzle
11 Instrument electronics
12 In-flight refuelling position light
13 Air refuelling receptacle
14 Cannon ammunition drum, 1,028 rounds
15 Liquid oxygen converter
16 Angle of attack transmitter
17 Cannon barrels
18 Nosewheel doors
19 M61, 20-mm, six-barrelled rotary cannon
20 Ammunition feed chute
21 Gun gas venting pipe
22 Air refuelling probe housing
23 Alternator and electrical bay
24 Air-driven turbine
25 Air refuelling probe
26 Windshield rain dispersal duct
27 Bulletproof windscreen panels
28 Radar attack sight
29 Instrument panel shroud
30 Navigation radar display
31 Rudder pedals
32 Cockpit front pressure bulkhead
33 Cannon mounting
34 Nosewheel leg strut
35 ILS system radar reflector
36 Taxiing lamps
37 Nosewheel
38 Torque scissor links
39 Hydraulic steering controls
40 Flight control system hydraulics bay
41 Electronics cooling air outlet
42 IFF aerial
43 UHF aerial
44 Underfloor radio and electronics bay
45 Cockpit pressure floor level
46 Pilot's side console panel
47 Engine throttle
48 Control column
49 Pilot's ejection seat
50 Seat back parachute pack
51 Headrest
52 Cockpit canopy cover
53 M118 3,000-lb (1360-kg) HE bomb (inboard pylon)
54 Starboard air intake
55 Cockpit canopy jack
56 Canopy hinge
57 Air-conditioning pack
58 Cockpit rear pressure bulkhead
59 Secondary electronics bay
60 Air data computer
61 Port air intake
62 Bomb bay fuel tank, 390-US gal (1476-litre)
63 Boundary layer splitter plate
64 Intake duct variable-area sliding ramp
65 Forward group of fuselage fuel tanks; total internal fuel capacity 1,160 US gal (4391 litres)
66 Gyro compass platform
67 Bomb bay fuel tank fuel transfer lines
68 Fuselage/front spar main frame
69 Dorsal spine fairing
70 Starboard mainwheel, stowed position

Vulnerability

Designed as a fast, stable, low-level platform for the delivery of nuclear weapons, the F-105 was a sturdy aircraft (in typical Republic fashion, thanks to the use of large forgings at points of high stress and machine-milled skins both of which saved weight without sacrificing strength) but had not been designed with an adequate degree of survivability in mind. There was very little in-built redundancy for its flight controls and fuel and hydraulic systems, and the fact that it was single-engined put it at a disadvantage compared to other types like the F-4 Phantom. Though there were numerous examples of aircraft returning to base with major damage to control surfaces and the like, the Thunderchief proved particularly vulnerable to fires caused by AAA and small arms damage.

AGM-45A Shrike

Developed directly in response to the threat posed by the Soviet S-75 (SA-2 'Guideline') SAM and its 'Fan Song' guidance radar, Shrike (normally carried on the outer wing stations) was an ARM derivative of the AIM-7C air-to-air missile with a larger warhead, a smaller motor and smaller tailfins. Entering service with the US Navy in 1965 and adopted for use aboard USAF F-100F Wild Weasel I aircraft in April 1966, Shrike was used with some success, though it suffered from a number of weaknesses that limited its effectiveness: its seeker had only a fixed frequency range (necessitating the time-consuming development of a raft of sub-variants); the seeker head was not gimballed and had only a limited field-of-view (and therefore needed to be aligned with a target in order to detect it) and it lacked a target 'memory', which meant that if an SA-2 site switched off its 'Fan Song' radar the missile lost lock and went ballistic. Measuring 10 ft (3.05 m) in length and weighing 400 lb (181 kg), Shrike had a range of 19-25 miles (30-40 km), had a Mach 2 top speed and carried a 145-lb (66-kg) proximity-fused, fragmentation warhead.

Though the missile is most often associated with the F-105F/G Wild Weasel aircraft, all Thunderchiefs from F-105D-5-RE onwards (as well as earlier aircraft modified to Look Alike standard) were able to carry Shrike, on the same pylon adapter developed to carry AGM-12 Bullpup.

Fuel and weapons loads

Though used simply to accommodate an extra fuel cell for much of its service, the Thunderchief's internal weapons bay was among its key features. Measuring 15 ft 10 in (4.83 m) in length, with a 32- x 32-in (81- x 81-cm) cross-section, the weapons bay was positioned directly below the wing centre-section and enclosed by a pair of doors, which retracted into the fuselage to minimise drag. The bay's dimensions were determined by the size and shape of the nuclear weapons of the period; those identified with the F-105 include Mk 28, Mk 43, Mk 57 and Mk 61 devices, single examples of which could be accommodated internally. Other than an MN-1A practice bomb dispenser, the only other store available for internal carriage by the Thunderchief was a 390-US gal (1476-litre) fuel tank; this became a routine fitting for most of the type's career.

The F-105 had five external stores stations for nuclear and conventional weapons. Using multiple ejector racks (MERs) on all five pylons allowed a maximum of 16 750-lb M117 iron bombs to be carried; maximum loads of the order of six tons were routinely flown in Vietnam by F-105Ds and Fs. External fuel could be carried on the inner wing pylons and on the centreline station. The latter, as illustrated by 63-8266, was usually used for a 650-US gal (2460-litre) tank, while 450-US gal (1703-litre) tanks of Republic or Fletcher manufacture were carried beneath the wings.

Defensive armament

The F-105's primary defensive weapon was its General Electric T-171E3 (later M61) Vulcan 20-mm rotary cannon, carried by all F-105s on the left side of the nose. Hydraulically driven, the M61 weighed 275 lb (125 kg) and was fed with 1,029 rounds of 20-mm electrically-primed ammunition, which it was able to deliver at a rate of 6,000 rpm, at a muzzle velocity of 3,380 ft/sec (1030 m/sec). The entire ammunition supply could be exhausted by an 11-second burst.

In order to maintain the aircraft's centre-of-gravity during firing (during which up to 500 lb/227 kg of ammunition would be expended) Republic's engineers devised a dual-feed ammunition system for the F-105B, which collected links and empty shells, storing them in separate compartments. In the F-105D, which required a redesign to accommodate radar, the feed system was altered so that the separate shell and link compartments could be eliminated; linkless ammunition was used and empty shells stored in the ammunition drum, saving 16 cu ft (0.45 m^3) of space.

Though seldom carried, up to four AIM-9B/E Sidewinder AAMs could be mounted on a double launcher hung from the aircraft's outer wing pylon.

Access to the F-105's massive Pratt & Whitney J75 was gained by removing the entire rear fuselage on a purpose-built trolley. The open flap at the base of the rudder concealed the braking parachute.

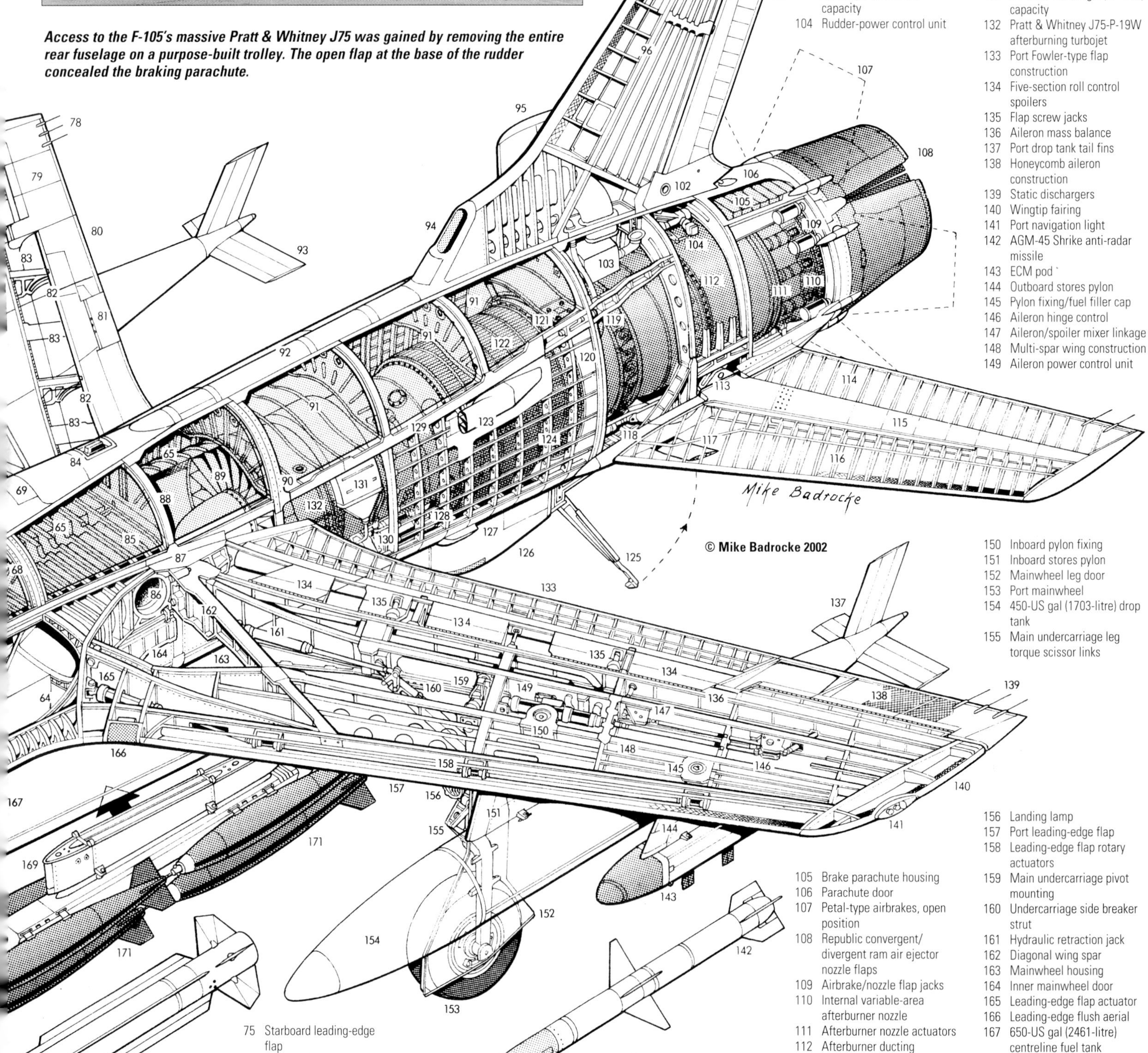

© Mike Badrocke 2002

71 450-US gal (1703-litre) external fuel tank
72 AIM-9 Sidewinder air-to-air missile
73 Missile launcher rail
74 Twin missile carrier (outboard pylon)
75 Starboard leading-edge flap
76 Outboard pylon fixing/drop tank filler cap
77 Starboard navigation light
78 Static dischargers
79 Starboard aileron
80 Starboard fowler flap
81 Trim tab, starboard only
82 Flap guide rails
83 Roll control spoilers
84 Anti-collision light
85 Air intake ducting
86 Ground running secondary air intake
87 Wing spar attachment joint
88 Fuselage/rear spar main frame
89 Engine compressor face
90 Forward engine mounting frame
91 Rear fuselage group of fuel tanks
92 Fuel pipe ducting
93 Drop tank tail fins
94 Afterburner duct cooling ram air intake
95 Starboard all-moving tailplane
96 Tailfin construction
97 Fin tip ECM aerials
98 Tail position light
99 Static dischargers
100 Rudder mass balance
101 Rudder
102 Formation light
103 Water injection tank 36-US gal (136-litre) capacity
104 Rudder-power control unit
105 Brake parachute housing
106 Parachute door
107 Petal-type airbrakes, open position
108 Republic convergent/divergent ram air ejector nozzle flaps
109 Airbrake/nozzle flap jacks
110 Internal variable-area afterburner nozzle
111 Afterburner nozzle actuators
112 Afterburner ducting
113 Tailplane pivot fixing
114 Port all-moving tailplane construction
115 Tailplane titanium box spar
116 Leading-edge nose ribs
117 Ventral fuel vent
118 All-moving tailplane control jack
119 Rear fuselage break point
120 Engine firewall
121 Rear engine mounting
122 Engine turbine section heat shroud
123 Engine bay venting ram air intake
124 Rear fuselage frame and stringer construction
125 Runway arrester hook
126 Ventral fin
127 Accessory cooling air duct
128 Cartridge starter
129 Fuselage top longeron
130 Engine-driven accessory gearbox
131 Oil tank, 4.5-US gal (17-litre) capacity
132 Pratt & Whitney J75-P-19W afterburning turbojet
133 Port Fowler-type flap construction
134 Five-section roll control spoilers
135 Flap screw jacks
136 Aileron mass balance
137 Port drop tank tail fins
138 Honeycomb aileron construction
139 Static dischargers
140 Wingtip fairing
141 Port navigation light
142 AGM-45 Shrike anti-radar missile
143 ECM pod
144 Outboard stores pylon
145 Pylon fixing/fuel filler cap
146 Aileron hinge control
147 Aileron/spoiler mixer linkage
148 Multi-spar wing construction
149 Aileron power control unit
150 Inboard pylon fixing
151 Inboard stores pylon
152 Mainwheel leg door
153 Port mainwheel
154 450-US gal (1703-litre) drop tank
155 Main undercarriage leg torque scissor links
156 Landing lamp
157 Port leading-edge flap
158 Leading-edge flap rotary actuators
159 Main undercarriage pivot mounting
160 Undercarriage side breaker strut
161 Hydraulic retraction jack
162 Diagonal wing spar
163 Mainwheel housing
164 Inner mainwheel door
165 Leading-edge flap actuator
166 Leading-edge flush aerial
167 650-US gal (2461-litre) centreline fuel tank
168 Fuel tank filler cap
169 Centreline stores pylon
170 Triple ejection rack
171 Six Mk 83, 1,000-lb (454-kg) HE bombs
172 Anti-personnel extended bomb fuse
173 AGM-78 Standard anti-radar missile
174 AGM-12C Bullpup air-to-surface missile

Above: The F-105D carried its General Electric AN/ASG-19 Thunderstick fire control system (which could be more accurately described as a bombing and navigation computer) in a modified nose that added 15 in (38.1 cm) to the length of the aircraft. Thunderstick did not have a truly all-weather capability until modified to T-Stick II standard – a change only applied to 30 F-105Ds.

Production

Thunderchief production ran to 833 examples (of 926 ordered), the bulk of which were F-105Ds and F-105Fs (610 and 143, respectively). All were built at Republic's Farmingdale, New York, factory.

variant	serial range	number
YF-105A-1-RE	54-0098/0099	2
F-105B-1-RE	54-0100/0103	4
F-105B-5-RE	54-0104	
	54-0106, 0107	
	54-0109, 0110	5
JF-105B-1-RE	54-0105, 0108	2
F-105B-6-RE	54-0111	1
JF-105B-2-RE	54-0112	1
F-105B-10-RE	57-5776/5784	9
F-105B-15-RE	57-5785/5802	18
F-105B-20-RE	57-5803/5840	38
F-105D-1-RE	58-1146/1148	3
F-105D-5-RE	58-1149/1165	
	58-1166/1173	
	59-1717/1757	
	(59-1775/1816, cancelled)	66
F-105D-6-RE	59-1758/1774	
	58-1817/1826	
	60-0409/0426	45
F-105D-10-RE	60-0427/0535	
	(60-0536/0546, cancelled)	
	60-5374/5385	121
F-105D-15-RE	61-0041/0106	66
F-105D-20-RE	61-0107/0161	55
F-105D-25-RE	61-0162/0220	
	(61-0221/0260, cancelled)	
	62-4217/4237	80
F-105D-30-RE	62-4238/4276	39
F-105D-31-RE	62-4277/4411	135
F-105F-1-RE	62-4412/4447	36
	63-8260/8366	107

F-105 weapons

Right: Among the weapons arrayed before this F-105B are two types of 'secret' nuclear weapon, iron bombs of various types and weights, napalm tanks, an M61 rotary cannon (with 1,029 rounds of 20-mm ammunition), AIM-9B Sidewinder AAMs, weapons bay, centreline and underwing drop tanks, 2.75-in FFAR pods and the Republic 'buddy' refuelling tank. The 'Bravo Chief' had a maximum ordnance load of 7,500 lb (3402 kg).

Left: The third JF-105B was used to trial the F-105's GAR-8 (AIM-9 from 1962) Sidewinder armament; a pair of AIM-9Bs could be carried on each outboard pylon.

Above: A new weapon introduced with the F-105D was the Martin Marietta AGM-12 Bullpup air-to-surface missile, two of which could be carried. Under the Look Alike modification programme, surviving F-105Bs were also given a Bullpup capability. Bullpup development was initiated by the US Navy in 1953, under the designation ASM-N-7. Two years later the USAF contracted Martin to develop an advanced derivative of Bullpup, known as GAM-79 White Lance, which introduced a nuclear warhead option and a much improved radio command guidance system, the latter allowing guidance from an offset position. In the interim the USAF procured ASM-N-7 (as GAM-83) and later elected to cancel GAM-79 in favour of a Bullpup variant fitted with the new guidance system – GAM-83A (later AGM-12B) Bullpup A. F-105s carried both AGM-12B and the larger AGM-12C Bullpup B.

Specifications and performance

	YF-105A	F-105D	F-105F	F-105G
powerplant:	one P&W J57-P-25	one P&W J75-P-19W	one P&W J75-P-19W	one P&W J75-P-19W
rating, dry	10,200 lb st (45.37 kN)	17,200 lb st (76.51 kN)	17,200 lb st (76.51 kN)	17,200 lb st (76.51 kN)
rating, afterburner	15,000 lb st (66.73 kN)	26,500 lb st (117.88 kN)	26,500 lb st (117.88 kN)	26,500 lb st (117.88 kN)
span:	34 ft 11 in (10.64 m)	34 ft 11 in (10.64 m)	34 ft 11¼ in (10.65 m)	34 ft 11¼ in (10.65 m)
length:	61 ft 5 in (18.72 m)	64 ft 3 in (19.58 m)	69 ft 7½ in (21.22 m)	69 ft 7½ in (21.22 m)
height:	17 ft 6 in (5.33 m)	19 ft 8 in (5.99 m)	20 ft 2 in (6.15 m)	20 ft 2 in (6.15 m)
wing area:	385 sq ft (35.77 m²)	385 sq ft (35.77 m²)	385 sq ft (35.77 m²)	385 sq ft (35.77 m²)
weight, empty:	21,010 lb (9530 kg)	27,500 lb (12473 kg)	–	–
weight, combat:	28,966 lb (13139 kg)	35,637 lb (16165 m)	–	41,091 lb (18638 kg)
weight, mto:	40,561 lb (18398 kg)	52,546 lb (23834 kg)	54,027 lb (24506 kg)	54,580 lb (24757 kg)
maximum load:	8,000 lb (3629 kg)	14,000 lb (6350 kg)	14,000 lb (6350 kg)	–
internal fuel:	850 US gal (3217 litres)	1,550 US gal (5867 litres)	1,160 US gal (4391 litres)	1,051 US gal (3978 litres)
maximum fuel:	2,500 US gal (9463 litres)	3,100 US gal (11734 litres)	3,100 US gal (11734 litres)	2,991 US gal (11321 litres)
maximum speed:	857 mph (1379 km/h) at 36,000 ft (10973 m); 778 mph (1252 km/h) at sea level	1,420 mph (2285 km/h) at 38,000 ft (11582 m); 836 mph (1345 km/h) at sea level	1,386 mph (2231 km/h) at 38,000 ft (11582 m); 876 mph (1410 km/h) at sea level	– 784 mph (1262 km/h) at sea level
stalling speed:	185 mph (298 km/h)	208 mph (335 km/h)	–	216 mph (348 km/h)
time-to-height:	30,000 ft (9144 m) in 17.6 mins	–	–	30,000 ft (9144 m) in 28 mins (combat speed)
climb rate, clean:	–	34,500 ft (10516 m)/min	34,500 ft (10516 m)/min	–
service ceiling	–	32,100 ft (9784 m)	–	–
combat ceiling:	49,950 ft (15225 m)	48,500 ft (14783 m)	–	43,900 ft (13381 m)
combat radius:	–	778 miles (1252 km)	740 miles (1191 km)	450 miles (724 km)
normal range:	1,100 miles (1770 km)	–	–	–
maximum range:	2,720 miles	2,208 miles (3553 km)	–	1,866 miles (3004 km)

T-Stick II

The high level of maintenance skill required to keep the more complex T-Stick II aircraft serviceable ultimately precluded their deployment to Southeast Asia. All 30 were issued to the 563rd TFS, 23rd TFW at Carswell instead and eventually finished up with the 457th TFS, Texas ANG before retirement in 1982. Known T-Stick II conversions, which are believed to have numbered 30, include 60-0458, -0464, -0465, -0471, -0485, -0490, -0493, -0500, -0513, -0527, -0533, -5375, 61-0044, -0047, -0063, -0075, -0080 (prototype), -0100 and -0110. The T-Stick II's AN/ARN-92 LORAN receiver was housed in a 'saddleback' fairing fitted along the spine of the aircraft, as seen in this view of 60-0464 of the 563rd TFS.

time, it was sent to Takhli AB, Thailand, about 100 miles (161 km) northwest of Korat. These were the two bases from which most of the F-105 sorties were launched.

Initially, all the squadrons operated under control of two provisional fighter wings – the 6234th TFW (Provisional) at Korat, and the 6235th TFW (Provisional) at Takhli. The first missions and losses took place within a week after arrival at Korat. On 14 August, the 36th TFS was assigned a ResCAP mission regarding a downed CIA T-28 pilot in Laos. Lieutenant Dave Graben's F-105D, 62-4371, was struck by 37-mm anti-aircraft fire. Graben brought the badly damaged 'Thud' back to Korat, but it never flew again.

On 10 December 1964, 13th AF began Operation Barrel Roll, an open air support campaign of Royal Laotian troops who were fighting both North Vietnamese and Communist Pathet Lao troops.

Flaming Dart

The first mission into North Vietnam was conducted under Operation Flaming Dart. F-105s with F-100 MiGCAPs attacked the Donghoi military barracks and Vinh Linh communications centre about 100 miles (161 km) north of the DMZ that separated North and South Vietnam. The Viet Cong answered with an attack against US personnel at bases in South Vietnam. Flaming Dart II was the 13th AF's answer to those attacks. And so it went, punch and counter-punch. More and more F-105 squadrons were deployed to the combat zone, with units from Japan and the US arriving in Thailand in early 1965.

By late February 1965, enough Thunderchief squadrons were in place to begin a systematic destruction of all the military targets throughout North Vietnam. This was known as Operation Rolling Thunder. The first Rolling Thunder missions included an attack by F-105s against the Xom Biang military depot, just north of the DMZ. Although the target was destroyed, so were three F-105s from the 67th Squadron, with one pilot killed.

Thirteenth Air Force had divided North Vietnam into six military 'regions' for targeting purposes and called them Route Packages, Route Packs or simply RPs. The higher the RP number, the heavier the concentration of its defences. The first Rolling Thunder missions against Xom Biang were in RP-1, just north of the DMZ. The most dangerous was any mission to Route Pack 6, and later to -6A, the immediate area around the North Vietnamese capital city of Hanoi.

The target list was chosen by military and political leaders in Washington, D.C. Just prior to the start of Rolling Thunder, the Joint Chiefs of Staff (JCS) drew up an initial list of 94 targets inside the borders of North Vietnam. Pilots were not free to deviate from the assigned target for any reason other than being attacked themselves; in other words, if the first F-105 pilot completely destroyed the target on his bomb run, the other flights had to bomb the rubble! If a Thunderchief flight passed over a column of North Vietnamese tanks and troops heading directly into South Vietnam, they could not attack them. They could report them to 13th AF and possibly attack them the following day – by which time, the column would be long gone. The same rules applied to anti-aircraft artillery (AAA) sites, SAM sites, and MiGs.

The SAM and MiG threats were heaviest in and around Hanoi. The North Vietnamese Air Force was made up of Soviet-supplied MiG-17, MiG-19, and MiG-21 interceptors. Some of their pilots were very good, especially in a MiG-17, which was much more manoeuvrable than the

Above: A number of Thunderchiefs were assigned to test units throughout the type's career, including this unusual silver USAF Systems Command F-105G, pictured at Eglin AFB during August 1970.

Right: The second production F-105F, 62-4413, was with the 3246th Test Wing, Armament Development Test Center at Eglin AFB during 1977.

F-105. The MiGs were armed with 'Atoll' air-to-air missiles in addition to the 23-mm and 37-mm cannon. Although the MiGs did relatively minor damage to the F-105 forces striking North Vietnam, they could force the 'Thud' pilots to jettison their bomb loads in order to evade the MiGs. F-105 pilots shot down 27½ MiG interceptors during the war.

The North Vietnamese SAM threat was not as effective as the Western press would have led people to believe. The SA-2 could be defeated a number of ways, such as forcing it out of its flight envelope or forcing the guidance radar off the air. If the pilot saw it coming, he could perform a 'split S' manoeuvre and fly directly at the missile, something its limited flight controls could not follow. Or he could close the gap before the SAM could arm itself. The addition of Wild Weasel flights was very effective at shutting down the 'Fan Song' radars that controlled the SAMs. Losses to SAMs totalled 32 by the end of the war.

By far the deadliest North Vietnamese defensive weapons were the radar-directed anti-aircraft artillery that ranged from the mobile ZSU-23/4 23-mm automatic cannon to fixed 37-mm, 57-mm, and 85-mm artillery guns. The 'Weasels' could monitor and shut down the AAA radars in the same manner as with the SAM, but there were simply too many radar-directed guns for the 'Weasels' to be effective. Anti-aircraft artillery accounted for 280 F-105 losses during the war.

Despite the fierce defences – often referred to as "the heaviest air defences in the history of air warfare" – the Rolling Thunder missions continued successfully. Every major target on the JCS list was struck, some several times. The Thai Nguyen Steel Works, Thanh Hoa Bridge, the Northeast Railway that brought war supplies in from China; all were struck. The only viable target that steadfastly remained standing was the Paul Doumer Bridge in Hanoi, which refused to go down despite repeated attacks by the F-105s using bombs that weighed up to 3,000 lb (1360 kg). (F-4s with 2,000-lb/907-kg laser-guided bombs eventually dropped the Doumer Bridge.)

Of course, facing this formidable array of defences on a daily basis took a deadly toll on the Thunderchief aircraft and crews. In 1965, F-105 forces in Thailand lost 60 aircraft. In 1966, the 355th and 388th Wings lost an incredible 111 aircraft, 103 to AAA; another 97 were shot down in 1967. On 31 October 1968, President Johnson ordered a complete cessation of offensive air strikes against targets inside North Vietnam – an action that probably prevented the eventual loss of the entire F-105D/F inventory. The following month, the 388th Wing at Korat traded in its veteran F-105 Thunderchiefs for new F-4E Phantoms.

Project Northscope and Combat Martin

In January 1967, General John Ryan perceived a need for an all-weather, night capability over North Vietnam. By March of that year, training had begun at Yokota AB, Japan. All the aircraft modified for Northscope were EF-105Fs. The main modification was to adjust the R-14A radar to allow an expanded radar picture with a faster sector sweep. This gave the rear pilot a better target definition and much greater accuracy during night attacks.

Northscope crews, known as 'Ryan's Raiders', arrived at Korat on 24 April 1967 and were assigned to the 13th TFS, which also had the Wild Weasel mission in the 388th Wing. Its first night mission came on 26 April, striking the Ron Ferry. By 22 May, there were 12 Northscope crews at Korat. Between 26 April and 4 October 1967, 'Ryan's Raiders' flew 415 'one-plane, one target' sorties to RP-1, -5 and -6A. In July, the 'Raider' aircraft were used as pathfinders on daylight missions. Three 'Raider' Thunderchiefs were shot down before the programme came to a close in late 1967.

With the success of the Wild Weasel programme, Air Force engineers reasoned that if they could develop a system to counter the ground air defences of North Vietnam, they could perhaps find one to restrict the MiG threat. It was known as Project Combat Martin. At least 12 Wild Weasel EF-105Fs were modified with the Hallicrafters QRC-128 airborne jamming system to jam communications between North Vietnamese ground controllers and the MiGs.

The QRC-128, which pilots called 'Colonel Computer', was installed in the rear cockpit in place of the rear ejection seat. The large blade antenna on top of the fuselage easily distinguished the Martin aircraft, which were assigned to both combat wings of F-105s. In December 1970, all remaining Martin aircraft were demodified, brought up to F-105G 'Weasel' standard and returned to combat in Southeast Asia.

Post-war 'Gs'

While F-105Ds returning from Southeast Asia were passed to ANG and Air Force Reserve units, the F-105Gs joined the 35th TFW at George AFB. Eventually including three squadrons from the 23rd TFW, the 35th was the USAF's last front-line F-105 unit, training crews in the Wild Weasel mission in preparation for the arrival of F-4Gs in the late 1970s. Some systems upgrading took place post-war, AN/ALR-46 RHAWS equipment, with improved digital displays for the EWO, replacing the earlier ALR-36/37 set. The 'Thud' finally bowed out of front-line service beginning in April 1978, the 25 best airframes going to the 128th TFS, Georgia ANG. The last 'G' left George on 12 July 1980, bring the 22 years of eventful front-line service to an end.

***Top:** A trio of 'GA'-coded 561st TFS F-105Gs, each carrying a practice bomb dispenser, returns from a training mission. Towards the end of its time operating F-105s, the 561st was averaging 10 1½-hour sorties per day, these including a high proportion of simulated and live weapons firing at a number of ranges.*

***Above left:** 63-8300 of the 563rd TFS undergoes maintenance at George AFB in the late 1970s. 'WW' tail codes replaced 'GA' during 1978/79.*

The US bicentennial celebrations were marked by the 141st TFS, New Jersey ANG, which applied appropriate markings to one of its veteran F-105Bs.

The first F-105Ds to join an ANG unit were those handed to Virginia's 149th TFS in 1971. F-105D-5-RE 59-1743* Hanoi Express *sported artistic reminders of its service in Southeast Asia when pictured in 1978. Visible beneath the aircraft is the weapons bay fuel tank, lowered for inspection.

Below: Upon its conversion to the F-105 shortly after the 149th, the 127th TFTS, Kansas ANG, took over the RTU role from the 23rd TFW. This F-105D-10-RE, returning after a training sortie, is pictured at McConnell AFB during 1976.

After 1968, only the 355th Wing at Takhli continued to fly missions with the F-105. These included 'protective reaction' strikes against air defences inside North Vietnam that were attempting to restrict RF-4 reconnaissance flights, which were trying to monitor North Vietnamese activity. The 355th was also very active against the North Vietnamese supplies flowing down the Ho Chi Minh Trail through Laos and Cambodia into South Vietnam.

Last F-105D missions flown

The days of the Thunderchief were numbered through both attrition and advances in aircraft design. One of the first things accomplished by President Nixon's 'Vietnamisation' plans was to end operations with the 355th Wing at Takhli. On 9 October 1970, the last F-105D mission was flown, and the 355th Wing was shut down and rotated back to the United States on 9 December. Most of the F-105D assets were transferred to the 18th Wing at Kadena, or returned to the US and the 23rd Wing at McConnell AFB, Kansas. The remaining Wild Weasel F-105F/G assets, about a dozen aircraft, were combined with Det. 1 F-105Gs from the 12th TFS and transferred back to Korat. Det. 1, which evolved into the 6010th Wild Weasel Squadron and later the 17th WWS, flew 'protective reaction' support of RF-4C reconnaissance flights until Easter 1972.

When North Vietnam launched its Easter Offensive on 30 March 1972, 13th AF had insufficient in-theatre assets to counter the initial attack. The bombing halt imposed by President Johnson on 31 October 1968 remained in effect until President Nixon cancelled it in early April 1972. Of the F-105 units that could have been redeployed to SEA, only the F-105F/G 'Weasels' returned to full combat, supporting 13th AF F-4 and B-52 strikes throughout North and South Vietnam. The small force of F-105Gs, i.e., the under-strength 17th WWS at Korat, was quickly overwhelmed with the mission requirements of Operation Linebacker.

Initially, PACAF decided that the best way to increase the effectiveness of the defence suppression units was to have 'hunter-killer' teams of F-105Gs paired with F-4E Phantoms. This virtually doubled the size of the 'Weasel' force overnight, but more aircraft were needed to stem the North Vietnamese offensive. Operation Constant Guard sent many different squadrons to the combat arena of Vietnam, one being the F-105G-equipped 561st TFS from McConnell AFB. The 561st combined with the 17th WWS at Korat and acted as 'hunter-killer' teams with 388th Wing F-4Es, becoming a very credible presence in the skies over North Vietnam.

The crews of the 17th and 561st Squadrons flew up to four sorties per day, supporting the Linebacker and Linebacker II missions that inflicted heavy losses on North Vietnamese military capabilities. The decisive raids during the 18-day Christmas air offensive proved to be enough to bring the North Vietnamese back to the peace talks in earnest. The last F-105G mission into North Vietnam returned to Korat in the early morning hours of 29 December 1972.

However, this was not the final mission to be flown by 'Thuds' in the region. For the next two years, F-105G

Georgia's 128th TFS transitioned from the F-100D/F to the F-105F/G over a 270-day period during the summer of 1979. Aircraft serviceability was a problem from the outset, the 'Thuds' suffering from the rigours of war service, several years with a training unit and plain old age. The situation came to a head in 1980/81 when fuel cell and engine combustion chamber problems required attention; the latter problem reduced the squadron to just three flyable aircraft in July 1981 as 31 of the unit's 34 J75 engines awaited repairs. Landing gear and wing cracks were also discovered, but perseverance on the part of the 128th eventually saw 12 of the unit's establishment of 24 aircraft in service. Despite its problems the 128th managed to take part in 17 exercises during the year ending 30 June 1982, including eight that required deployments. Pictured is F-105F 63-8299, callsign PEACH 91, which made the last official flight by an ANG Thunderchief, on 25 May 1983.

Like the Air National Guard, the Air Force Reserve picked up Thunderchiefs returning from the war in Vietnam. The first examples were issued to the 305th TFS at Tinker AFB, Oklahoma, in May 1972.

Weasels escorted B-52 Arc Light strikes into Laos and Cambodia, or RF-4C flights into North Vietnam that were monitoring the effectiveness of the 1973 Paris Peace Accords which supposedly ended North Vietnamese aggression against South Vietnam.

On 15 August 1973, all US involvement in the wars in Southeast Asia, apart from reconnaissance flights, was halted. In September, the 561st TFS returned to the US, assigned to the 35th TFW at George AFB. On 29 October 1974, the last 17th WWS F-105Gs left Korat and also flew to George AFB, where the unit was redesignated the 562nd TFS. The war in Southeast Asia had decimated F-105 forces. A total of 395 F-105D/F/G Thunderchiefs was lost during the war – 296 F-105Ds and 38 F-105F/Gs to enemy action, and 61 more to operational causes such as engine problems, mid-air collisions or hydraulic failures.

Post-war Thunderchiefs

In 1974, the only F-105s remaining on operational status with the US Air Force were the Wild Weasel squadrons, the 561st and 562nd Squadrons at George AFB, California. They continued to perform the defence suppression mission for Air Force units throughout the world until next-generation 'Weasel' aircraft became operational. A third squadron, the 563rd TFS, was added to the wing at George AFB in 1975. Along with carrying out the defence suppression mission, the F-105G crews taught the new F-4G Wild Weasel IV crews the tactics that had been learned the hard way and had proved effective in the skies over Vietnam.

On 28 April 1978 the first F-4Gs became available and the phase-out of the F-105Gs began, though it was 12 July 1980 before the last F-105G left the 35th, that day being marked with a ceremony at George which brought together a 'reunion' of F-105Ds from AFRES and ANG units. Twenty five of the remaining F-105Gs were transferred to the 128th TFS, Georgia Air National Guard at Dobbins AFB; the remainder either went directly to the scrapyard at Davis-Monthan AFB or were used as targets on the gunnery ranges in Nevada. In October 1982, the F-105Gs of the Georgia ANG were replaced by the F-4D Phantom. The last F-105G flight took place on 25 May 1983.

When PACAF closed down the 355th Wing at Takhli in December 1970, the remaining D-models were distributed to units that still had F-105Ds on the inventory, including the 347th Wing at Yokota, the 23rd Wing at McConnell AFB, Kansas, and the 57th Fighter Weapons Wing at Nellis AFB. Their time in active Air Force service was brief, and the D-models were rapidly transferred to Air Guard squadrons, including the 121st TFS, District of Columbia ANG at Andrews AFB; the 127th TFS, Kansas ANG at McConnell AFB; and the 149th TFS, Virginia ANG at Byrd Field ANGB.

Below: The unique F-105D T-Stick II aircraft, which had seen service with the 563rd TFS, were passed to the Carswell AFB, Texas-based 457th TFS. During May 1981 60-0471 was the commanding officer's aircraft.

In addition, three Air Force Reserve squadrons flew F-105Ds – the 466th TFS at Hill AFB, Utah; the 457th TFS at Carswell AFB, Texas; and the 465th TFS at Tinker AFB, Oklahoma. On 25 February 1984, the last official F-105 flight took place when a 466th TFS Thunderchief landed at Hill AFB, where the type was replaced by F-4D Phantoms.

It was the end of an era. F-105 Thunderchiefs remain on static display as gate guards or museum aircraft at many Air Force installations in the United States. One of the best places to view the huge fighter that symbolised the air war in Vietnam is the US Air Force Museum in Dayton, Ohio, where F-105D 60-0504 *Memphis Belle II* and F-105G 63-8320 *Wham Bam*, both veterans of the conflict in Southeast Asia, sit proudly as visitors still marvel at the immense size of the 'Chief.

Larry Davis

The 'Thud' finally bowed out of service on 25 February 1984 when an 466th TFS F-105D made the final flight (the last of the unit's F-105Bs had left during 1981). During the latter years of their service the unit's 'HI'-coded aircraft became well known for their non-standard camouflage schemes, as illustrated in this 1983 view of three of the squadron's aircraft over Salt Lake City with an example of the squadron's new equipment – a recently acquired F-16A.

Thunderchiefs were first assigned to the 335th TFS, 4th TFW in May 1958 and remained in front-line USAF service until 1979, when the 35th TFW finally relinquished its last F-105Gs to the Georgia Air National Guard. Here active-duty wings are listed in the order in which they were equipped with F-105s.

UNITED STATES AIR FORCE

4TH TACTICAL FIGHTER WING

333rd Tactical Fighter Squadron
334th Tactical Fighter Squadron
335th Tactical Fighter Squadron
336th Tactical Fighter Squadron

The 4th Fighter-Day Wing returned from its Far East deployment (which had commenced in 1950 at the outbreak of the Korean War) on 8 December 1957, taking over the few F-86H and F-100C assets of the 83rd FDW at Seymour Johnson AFB, North Carolina. On 1 July 1958 the unit was redesignated the 4th Tactical Fighter Wing, with four squadrons assigned.

On 1 May 1958, the first F-105Bs arrived and were assigned to the 335th Squadron which was TDY at Eglin AFB, Florida, and would conduct the Category II Operational Flight Testing on the new Thunderchief. Operational tests continued as the F-105B was brought into active service with the 4th Wing. By mid-1960, three squadrons – the 334th, 335th, and 336th – were operational with F-105Bs.

On 18 October 1962, at the height of the Cuban Missile Crisis, the 4th TFW deployed to McCoy AFB, Florida, for possible combat over Cuba.

In the spring of 1964, the 4th Wing began the transition from F-105Bs to F-105D/Fs, a process that was completed on 23 June.

As the war in Southeast Asia heated up in the late summer of 1964, the 4th Wing was alerted for deployment to the Far East. On 3 July 1965, as part of Operation Two Buck 13, the 335th TFS deployed to Yokota AB to take the place of units assigned there but which were on TDY to Thailand, standing the normal VICTOR nuclear alert rotations at Osan AB, Korea.

On 8 November 1965, the 335th TFS moved TDY to Takhli RTAFB, Thailand, for combat operations against North Vietnam. At Takhli, the squadron was under the control of the newly-activated 355th TFW. The 335th TFS returned to Seymour Johnson AFB on 15 December 1965.

On 28 August 1965, also under Operation Two Buck 13, the 334th TFS deployed TDY direct to Takhli RTAFB for combat operations against North Vietnamese targets, coming under the control of the 6235th TFW(P) at Takhli. The 334th TFS returned to Seymour Johnson AFB on 10 October 1966, leaving its aircraft at Takhli.

On 4 December 1965, the 333rd TFS deployed to Korat RTAFB for combat operations. However, the 333rd then transferred to Takhli to relieve the 335th TFS. On 8 December 1965, the 333rd TFS deployment was changed from TDY to PCS (Permanent Change of Station), and the squadron was permanently assigned to the 355th TFW at Takhli. Aircraft and pilots of the 336th TFS rotated between Seymour Johnson and Takhli in 1965/66 but the squadron did not deploy to the war in Southeast Asia.

Aircraft markings on the F-105B included a natural metal finish; silver lacquer after Operation Look Alike, with a green band with white trim around the nose and across the vertical fin. A 'native American' logo was painted either on the side of the fuselage under the cockpit, or on the tail band. These markings were retained on F-105D/F aircraft as delivered, though in SEA coloured tail bands were used for squadron identification, as follows: 333rd TFS – red tail stripe with white borders; 334th TFS – blue band with white 'polka dots'; 335th TFS – green band with white borders or green band with a white 'V' superimposed. Camouflaged aircraft retained coloured tail bands.

Above: The 4th TFW 's colourful markings are clearly displayed on this pair of F-105Bs. The 'native American' logo is situated next to the aircraft's buzz number below the cockpit; a TAC badge and lightning bolt were applied over the green tail band.

The first Air National Guard unit to operate the F-105D variant, Virginia's 149th TFS, converted to the Thunderchief in early 1971. Its F-105D/F complement was replaced by Vought A-7D/Ks in 1982.

4520TH COMBAT CREW TRAINING WING (4525TH FIGHTER WEAPONS WING)

4537th Combat Crew Training Squadron
(4537th Fighter Weapons Squadron)
4520th Air Demonstration Squadron

The 4520th Combat Crew Training Wing was established at Nellis AFB, Nevada, on 1 July 1958, taking over the role of the Air Force Fighter School. Two squadrons with F-105B assets were assigned to the 4520th CCTW – the 4537th CCTS and the 4520th Air Demonstration Squadron, 'Thunderbirds'.

The 4520th ADS had nine F-105B aircraft assigned between May 1963 and May 1964. Following a fatal crash, the 'Thunderbirds' converted to F-100D Super Sabres for the remainder of the 1964 season. The remaining F-105Bs were transferred to the 141st TFS, New Jersey ANG, following the implementation of Operation Backbone.

On 1 September 1966, the 4520th CCTW was redesignated the 4525th Fighter Weapons Wing, with one F-105D/F squadron assigned – the 4537th FWS.

Unit markings for the 4537th included a black/yellow checkerboard band across the vertical fin on silver aircraft. Aircraft were camouflaged beginning in 1965, the tail code 'WC' being assigned from 1967.

Below: The 4537th FWS operated a number of F-105Ds and Fs in the training role. This F-105D-10-RE was pictured at George AFB in June 1962.

57TH FIGHTER WEAPONS WING

4537th Fighter Weapons Squadron
66th Fighter Weapons Squadron

Activated at Nellis AFB, Nevada on 15 October 1969, the 57th FWW took over the assets of the 4525th FWW. F-105D/F/G assets within the 57th FWW were assigned to the 66th FWS and the 4537th FWS which conducted combat crew training, test and evaluation, and flew weapons systems demonstrations until 1975.

Aircraft of the 57th FWW continued to carry the black/yellow tail band; when tail codes were adopted 66th FWS aircraft carried 'WA' and 4537th FWS aircraft 'WC'. The left side of 66th FWS aircraft carried a unit badge, with the badge of the 57th FWW appearing on right side of all aircraft.

Right: From late 1969 the F-105 Wild Weasel community learnt its trade with the 66th FWS at Nellis AFB. F-105G 63-8318 was typical of aircraft assigned to the unit; its markings include a 57th FWW badge and black/yellow checkerboard tail band.

36th Tactical Fighter Wing

22nd Tactical Fighter Squadron
23rd Tactical Fighter Squadron
53rd Tactical Fighter Squadron

Based at Bitburg AB, West Germany, the 36th Wing had three squadrons attached – the 22nd, 23rd, and 53rd TFSs. The 36th Wing was the second unit to re-equip with the F-105D Thunderchief, beginning in May 1961.

In 1966, the 36th Wing began to transition to the F-4D Phantom, its F-105D/F assets passing to combat units in SEA.

Aircraft markings on natural metal/silver aircraft included diagonal stripes on the vertical tail in squadron colours, as follows: 22nd TFS – red bands with white trim; 23rd – blue bands with white trim; 53rd – yellow bands with black trim. In late 1964, all aircraft had tri-colour tail bands in red/blue/yellow. In 1965, F-105D/Fs of the 36th Wing were among the first aircraft to receive the new dark green/olive green/tan tactical camouflage scheme, with light grey undersides.

These views of 36th TFW F-105Ds show the initial use of individual squadron tail striping (above, red representing the 22nd TFS) and the later adoption of a generic wing marking with all three squadrons represented (below).

18th Tactical Fighter Wing

12th Tactical Fighter Squadron
44th Tactical Fighter Squadron
67th Tactical Fighter Squadron

The 18th TFW began operations with F-105D/F aircraft at Kadena AB, Okinawa, in 1962. Three squadrons were attached – the 12th, 44th, and 67th TFS – deployments to Southeast Asia beginning in 1964, as follows: 67th TFS deployed TDY to Korat RTAFB, Thailand, in August 1964, returning to Kadona in May 1965. Det. 2, 18th TFW deployed from Korat to Da Nang AB, RVN, on 25 December 1964 with aircraft and crews from the 67th and 44th Squadrons.

The 44th TFS deployed TDY to Korat RTAFB with the 67th TFS in August 1964, returning to Kadena in May 1965. Det. Two, 18th TFW, deployed to Da Nang AB, RVN, on 25 December 1964 with aircraft and crews from the 67th and 44th TFS. Aircraft of Det. 2, 18th TFW, returned to Korat in January 1965. F-105D/F assets of the 44th TFS were absorbed by the 13th TFS at Korat when the 13th was activated in May 1966.

The 12th TFS deployed six F-105D aircraft to Da Nang as part of Det. Two, 18th TFW, which began operation on 25 December 1964. With deactivation of Det. Two, 18th TFW, the aircraft and crews of the 12th TFS deployed TDY to Korat beginning on 1 February 1965, remaining at Korat until 15 March 1965. The 12th re-deployed to Korat on 15 June 1965, returning to Kadena on 25 August 1965.

During the TDY deployments, the 12th TFS lost four aircraft, the 44th TFS lost one F-105D, and the 67th TFS lost nine aircraft, including three on the first day of the Rolling Thunder operations.

F-105F/G Wild Weasel aircraft from the 12th TFS deployed to Korat AB as Det. 1, 12th TFS, beginning on 24 September 1970. Det. 1 aircraft were integrated with the remaining F-105D/F/G assets transferred from the 355th Wing at Takhli. Under the control of the 388th TFW at Korat, Det. 1 was operational for one month before being redesignated the 6010th WWS on 1 November 1970.

Aircraft markings on natural metal/silver F-105D/F aircraft included a PACAF badge on both sides of the vertical fin, and a coloured band around the nose directly behind the radome, as follows: 12th TFS – yellow; 44th TFS – medium blue; 67th TFS – red. On camouflaged aircraft, the radar reflector, canopy sills or fin cap was painted in the appropriate squadron colour. 12th TFS aircraft are known to have carried 'ZA' tail codes; Det. 1, 12th TFS machines are known to have had 'ZB' codes during 1970.

The medium blue canopy of this F-105D (above) indicates service with the 44th TFS; the 44th was absorbed by the 13th TFS before tail codes were introduced in 1967; the 12th TFS was allocated 'ZA' (below).

49th Tactical Fighter Wing

7th Tactical Fighter Squadron
8th Tactical Fighter Squadron
9th Tactical Fighter Squadron

The 49th Tactical Fighter Wing was the third unit to transition into the F-105D/F, beginning in October 1961. The 49th Wing was based at Spangdahlem AB, West Germany. Three squadrons were attached – the 7th, 8th, and 9th Tactical Fighter Squadrons.

The 49th received two Air Force Outstanding Unit Award for F-105 operations during the periods 1 March 1964 to 28 February 1966, and 1 March 1966 to 30 June 1967.

In 1967, the 49th Wing transitioned from the F-105D/F to the F-4D Phantom, its remaining F-105 assets being transferred to other combat units.

Aircraft markings on natural metal/silver aircraft included the 49th Wing badge on both sides of the vertical tail; above this was a red/yellow/blue flash. Around the rear of the radome was a band in the squadron colour: 7th TFS – blue; 8th TFS – yellow; 9th TFS – red. Later, the three-colour lightning bolt was added behind the radome and nose gear doors were painted in squadron colours.

This view shows both the 49th Wing's tail marking and the three-colour lightning bolt applied to the nose. Camouflage was applied to Europe-based F-105s from 1965.

8th Tactical Fighter Wing

35th Tactical Fighter Squadron
36th Tactical Fighter Squadron
80th Tactical Fighter Squadron

The 8th Tactical Fighter Wing with three squadrons – the 35th, 36th, and 80th TFS – was based at Itazuke AB, Japan with F-105D/F aircraft from 1963 to 10 July 1964, at which time the Wing HQ moved to George AFB, California. All squadrons and F-105 assets remained in Japan, but were now attached to the 6441st TFW(P) based at Yokota AB, Japan.

Unit markings included coloured engine air intakes, canopy rails and ailerons. Squadron colours were: 35th TFS – blue; 36th TFS – red, and 80th TFS – yellow.

6441st Tactical Fighter Wing (P)

35th Tactical Fighter Squadron
36th Tactical Fighter Squadron
80th Tactical Fighter Squadron

The 6441st TFW (Provisional) took over the squadrons and F-105D/F assets of the 8th TFW in July 1964.

At regular intervals, the squadrons of the 6441st TFW(P) deployed first to Korat RTAFB, then to Takhli RTAFB. In May 1965, the entire 6441st TFW(P), moved TDY from Yokota AB, Japan to Takhli RTAFB, Thailand, returning to Yokota in July 1965 when the 6235th TFW(P) was established at Takhli.

On 9 August 1964, following the Gulf of Tonkin Incident, the 36th TFS deployed TDY to Korat RTAFB, Thailand, for combat operations against North Vietnam. TDY ended in November and was followed by further TDYs in February 1965 (to Takhli RTAFB, returning to Yokota in May 1965) and August 1965 (to Takhli, returning to Yokota in October).

The 80th TFS deployed TDY to Takhli from February 1965 to 26 August, when it returned to Yokota AB.

The 35th TFS deployed TDY to Takhli in May 1965, replacing the 36th TFS. The 35th TFS was relieved in June 1965, then returned to Takhli in October before returning to Yokota once again on 9 November 1965. The 35th TFS, although not officially assigned to the 355th TFW or 388th TFW, rotated aircraft and crews to augment both combat wings because of attrition.

Aircraft markings remained as they had been when the squadrons were under the control of the 8th TFW.

F-105D 59-1743* Arkansas Traveler *of the 34th TFS was the mount of Colonel Paul P. Douglas Jr, commander of the 388th TFW between January and July 1968. Douglas achieved ace status flying P-47s in the ETO during World War II (note the German kill markings below the cockpit); the same name also appeared on his P-47s as well as an F-4E Phantom he flew after the 388th re-equipped in 1969.

347th Tactical Fighter Wing

35th Tactical Fighter Squadron
36th Tactical Fighter Squadron
80th Tactical Fighter Squadron

The 347th Tactical Fighter Wing, under control of Fifth Air Force, took over the F-105D/F assets of the 6441st TFW at Yokota AB, Japan, on 15 January 1968. As F-105Ds were phased out with the 388th TFW at Korat, these Thunderchiefs were assigned to the 347th TFW at Yokota. The 347th's mission was the air defence of Japan, with a tactical and Zulu Alert mission in support of the Republic of Korea. The 347th TFW had F-105Ds for a matter of months during early 1968; when the wing re-equipped with F-4Cs, all were transferred to stateside units.

Aircraft were finished in standard T.O.1-1-4 camouflage; tail codes were as follows: 35th TFS – 'GG'; 36th TFS – 'GL'; 80th TFS – 'GR'

Thunderchiefs were seen with 'Gx' codes for only a brief period, during early 1968, flown by the Yokota-based 347th TFW.

388th Tactical Fighter Wing

560th Tactical Fighter Squadron
561st Tactical Fighter Squadron
562rd Tactical Fighter Squadron
563rd Tactical Fighter Squadron

On 1 October 1962, the 388th Tactical Fighter Wing was activated at McConnell AFB, Kansas, with four squadrons assigned – the 560th, 561st, 562nd and 563rd Tactical Fighter Squadrons, all equipped with F-105D/F aircraft. Once conversion to the Thunderchief had been completed, in the spring of 1963, the 388th TFW was inactivated and replaced at McConnell by the 23rd TFW, on 28 January 1964.

13th Tactical Fighter Squadron
34th Tactical Fighter Squadron
44th Tactical Fighter Squadron
421st Tactical Fighter Squadron
469th Tactical Fighter Squadron
Det. 1, 12th Tactical Fighter Sqn (TDY)
6010th Wild Weasel Squadron
17th Wild Weasel Squadron
561st Tactical Fighter Squadron (TDY)

The 388th Tactical Fighter Wing was reactivated PCS at Korat RTAFB, Thailand, on 8 April 1966, taking over the F-105 assets of the 6234th TFW(P), which included the 421st and 469th Tactical Fighter Squadrons.

The 469th TFS originally deployed to Kadena on 30 November 1964 from McConnell AFB, assuming the mission assignment of squadrons under the control of the 18th TFW. In January 1965, the 469th TFS moved from Kadena to Korat AB, Thailand, where it was under the control of 2nd AD. On 13 March 1965, the 469th TFS returned to McConnell AFB. The 469th returned to Korat PCS in early November 1965, and was assigned to the 6234th TFW(P). On 8 April 1966, the 469th TFS was assigned to the newly activated 388th TFW, remaining with the 388th Wing until the F-105D was replaced with F-4Es.

The 421st TFS originally deployed TDY from McConnell AFB, Kansas, to Korat RTAFB on 15 September 1964, the deployment lasting until 23 November 1964. The squadron returned to Korat TDY on 7 April 1965 (and back to McConnell AFB on 20 August 1965). The 421st TFS lost three F-105D aircraft during the two TDY deployments.

On 8 April 1966, the 421st TFS moved PCS to Korat RTAFB, where it was assigned to the 388th TFW. The 421st TFS returned to the US on 25 April 1967, and was assigned to the 15th TFW at McDill AFB, Florida. The 421st lost 21 F-105Ds during the PCS tour at Korat. Aircraft markings on camouflaged aircraft included red canopy sills and radar reflector; and red trim with white stars on the bottom of the main landing gear doors. No tail codes were applied to 469th TFS aircraft.

The 13th TFS was activated at Korat RTAFB, under the control of the 388th TFW, on 15 May 1966 as a PCS assignment, taking over the F-105 assets from the 44th TFS. On 11 July 1966, the 13th TFS absorbed the EF-105F Wild Weasel assets from the Project Wild Weasel Detachment, which had been at Korat under control of the 6234th TFW(P).

The mission of the 13th TFS became defence suppression – the entire squadron being equipped with EF-105F aircraft. Aircraft and crews from the 13th TFS also carried out the all-weather, night attack mission of Project Northscope. On 20 October, the EF-105F assets of the 13th TFS were given to the 44th TFS, and the 13th TFS was then assigned to the 432nd TFW at Udorn. The 13th TFS lost 13 F-105D/F aircraft to enemy action.

The 34th TFS was activated PCS at Korat on 15 May 1966 under the control of the 388th TFW, and was equipped with F-105D aircraft until November 1968 when F-4E Phantoms replaced the type in service at Korat. During the period of time that the 34th TFS was equipped with F-105Ds, a total of 28 aircraft was lost to enemy action. Pilots of the 34th TFS shot down a total of 5½ NVAF MiGs.

The 12th TFS at Kadena sent a detachment of 12 EF-105Fs TDY to Korat on 24 September 1970. Detachment 1, 12th TFS augmented the F-105F/G assets that were transferred from the 355th TFW at Takhli. Det. 1 was the only unit assigned to the defence suppression mission. On 1 November 1970, Det. 1 was redesignated the 6010th WWS.

The 6010th Wild Weasel Squadron was activated on 1 November 1970, taking over the F-105G assets of Det. 1, 12th TFS at Korat. The 6010th WWS remained at Korat until 1 December 1971, when the unit was redesignated the 17th WWS.

The 17th Wild Weasel Squadron was activated at Korat on 1 December 1971, taking over the F-105F/G assets of the 6010th WWS. The 17th WWS performed valiantly during the Linebacker I and II campaigns against North Vietnam between April 1972 and January 1973. Six Wild Weasel F-105Gs were shot down during the Linebacker operations. The 17th WWS remained at Korat until 15 November 1974 when the unit was transferred to George AFB, California. The 17th was then inactivated and the F-105G assets were absorbed into the 562nd TFS at George AFB.

Detachment 1, 561st TFS deployed TDY to southeast Asia as part of Operation Constant Guard I on 7 April 1972. Under the operational control of the 388th TFW at Korat, the F-105G assets of Det. 1 were integrated with those of the 17th WWS, and flew defence suppression mission in support of 13th AF aircraft during Operations Linebacker I and II. One Det 1 F-105G was shot down. On 5 September 1972, Det 1 was relieved and returned to George AFB, California, to begin the training of F-105G air and ground crews.

Markings on the 469th's silver-lacquered aircraft consisted of 4-in (10-cm) green/white/green stripes with the TAC badge and lightning bolt on both sides of the vertical tail. Camouflaged 469th TFS aircraft had a green tail band, canopy rails and radar reflector, with 'JV' tail codes from 1968. Camouflaged 13th TFS aircraft sported a yellow canopy rail and radar reflector. No tail codes were applied. Aircraft involved in Project Northscope had a distinctive 'wrap-around' camouflage of olive green/dark green/tan on the upper surfaces; with tan and blue-green on the undersides.

'ZB' codes were first assigned to Det. 1, 12th TFS – then the USAF's only dedicated defence suppression unit. On 1 November 1970 Det. 1 became the 6010th WWS, retaining 'ZB' codes. A year later it became the 17th WWS ('JB' codes).

Squadron colours (usually on canopy rails and radar reflectors) and tail codes applied to the aircraft of other units were: 34th TFS – black ('JJ'); Det. 1, 12th TFS – yellow ('ZA'); 6010th WWS – blue ('ZB'); 17th WWS – blue ('JB'); Det. 1, 561st TFS – yellow ('MD', later 'WW').

Having taken over the assets of the 6010th WWS, the 17th WWS recoded its aircraft 'JB'. The last Thunderchief unit in the region, the 17th finally withdrew in October 1974, its aircraft passing to the 562nd TFS, 35th TFW at George AFB. This F-105G is at Korat in September 1973.

23rd Tactical Fighter Wing

560th Tactical Fighter Squadron
561st Tactical Fighter Squadron
562rd Tactical Fighter Squadron
563rd Tactical Fighter Squadron
4519th Combat Crew Training Squadron
(4519th Tactical Fighter Training Squadron)

Activated at McConnell AFB, Kansas on 28 January 1964, the 23rd TFW took over the F-105D/F assets of the 388th TFW and its four squadrons – the 560th, 561st, 562nd and 563rd TFSs.

The mission of the 23rd TFW changed to that of the F-105 Replacement Training Unit in 1966, the 4519th Combat Crew Training Squadron being added on 1 August 1967. The 560th TFS was inactivated on 25 September 1968.

The 23rd Wing deployed three squadrons on a TDY basis to combat in Southeast Asia, beginning in February 1965. These units were initially under the control of 2nd AD until the 6441st TFW(P) was activated at Takhli. The TDY squadrons then came under the control of the 6235th TFW(P), beginning in July 1965.

The 561st Squadron deployed to Takhli on 6 February 1965, returning to McConnell on 10 July 1965. The 563rd TFS deployed to Takhli on 8 April 1965, returning to McConnell AFB on 10 July 1965. The 562nd TFS deployed to Takhli and relieved the 563rd Squadron, returning to McConnell on 8 December 1965. Detachments of the 561st TFS were also deployed to Da Nang AB, RVN, for operations within the borders of the Republic of Vietnam.

During combat operations in Southeast Asia, the 562nd TFS lost three aircraft; while the 563rd TFS lost 11 aircraft. On 1 July 1972, the 23rd TFW moved to England AFB, Louisiana, to begin operations with the LTV A-7D. The 561st, 562nd and 563rd Squadrons were assigned to the 35th TFW at George AFB, California, and equipped with F-105G Wild Weasel aircraft.

Squadron markings on the natural metal/silver lacquered aircraft included the following: 561st TFS – black/yellow checkerboard on rudder; 562nd TFS – a red, white and black 'sharkmouth' on the nose of the aircraft; 563rd TFS – red and white stripes on the rudder, wing tips, and stabilisers, with a white band on the top of the vertical fin, on which was painted a black 'Ace of Spades' decoration. All aircraft had the TAC badge on both sides of the vertical fin. Camouflaged aircraft carried the following tail codes: 561st TFS – MD'; 562nd TFS – 'ME'; 563rd TFS – MF'; 4519th TFTS – 'MG'.

Above: 562nd TFS F-105Ds pause on Guam, en route to Takhli during August 1965. Each of the 23rd TFW's squadrons adopted distinctive markings; the 562nd's aircraft were adorned with a 'sharkmouth'.

Below: Pictured a short time before the wing inactivated, this 563rd TFS F-105F carries the squadron's 'MF' code with a red canopy rail and tail band.

2nd Air Division

Korat-based units
36th Tactical Fighter Squadron
67th Tactical Fighter Squadron
44th Tactical Fighter Squadron
357th Tactical Fighter Squadron
35th Tactical Fighter Squadron
469th Tactical Fighter Squadron
12th Tactical Fighter Squadron
354th Tactical Fighter Squadron
421st Tactical Fighter Squadron

Takhli-based units
36th Tactical Fighter Squadron
80th Tactical Fighter Squadron
561st Tactical Fighter Squadron

The 2nd Air Division had control over all TDY squadrons deployed for combat in SEA from August 1964 until the activation of the 6234th TFW(P) at Korat on 5 April 1965, and the transfer of the 6441st TFW(P) from Yokota AB to Takhli in May 1965.

6234th Tactical Fighter Wing (P)

67th Tactical Fighter Squadron
44th Tactical Fighter Squadron
469th Tactical Fighter Squadron
12th Tactical Fighter Squadron
354th Tactical Fighter Squadron
421st Tactical Fighter Squadron

The 6234th Tactical Fighter Wing (Provisional) was activated at Korat RTAFB on 5 April 1965 to assume command of the TDY squadrons that were operating from Korat under the temporary control of 2nd Air Division. The status of the 6234th TFW(P) was changed from TDY to PCS on 8 July 1965, remaining in control of all TDY squadron assignments until activation of the 388th TFW on 8 April 1966.

6235th Tactical Fighter Wing (P)

35th Tactical Fighter Squadron
36th Tactical Fighter Squadron
80th Tactical Fighter Squadron
334th Tactical Fighter Squadron
335th Tactical Fighter Squadron
354th Tactical Fighter Squadron
357th Tactical Fighter Squadron
561st Tactical Fighter Squadron
562nd Tactical Fighter Squadron
563rd Tactical Fighter Squadron

The 6235th Tactical Fighter Wing (Provisional) was activated at Takhli RTAFB, Thailand, on 8 July 1965, taking over control of the TDY F-105 squadron assets of the 6441st TFW(P). The 6235th TFW(P) remained in control of TDY squadrons operating from Takhli until the 355th TFW was activated on 8 November 1965.

The 36th TFS was TDY to Takhli, and assigned initially to the 2nd AD, from February 1965 until the 6441st TFW(P) was activated at Takhli in May 1965, when it passed to the 6235th TFW(P), on 8 July 1965. In October 1965, the 36th TFS returned to Yokota AB, Japan. During the TDY assignment, the 36th TFS lost seven aircraft to enemy action.

The 80th TFS made a TDY deployment to Takhli in February 1965, initially under the control of 2nd AD, then the 6441st TFW(P), and finally the 6235th TFW(P) beginning July 1965. On 26 August 1965, the 80th TFS returned to Yokota AB and resumed the Zulu Alert mission. Two 80th TFS F-105Ds were shot down over North Vietnam.

The 561st TFS deployed TDY to Takhli from McConnell AFB on 6 February 1965, under control of the 2nd AD initially, then the 6441st TFW(P) and the 6235th TFW(P) in July 1965. The 561st TFS returned to McConnell AFB, Kansas on 10 July 1965.

355th Tactical Fighter Wing

44th Tactical Fighter Squadron
333rd Tactical Fighter Squadron
354th Tactical Fighter Squadron
357th Tactical Fighter Squadron

The 355th Tactical Fighter Wing was activated at George AFB, California, on 8 July 1962, with four squadrons assigned – the 354th, 357th, 421st, and 469th TFS. On 21 July 1964, the 355th TFW, now equipped with F-105D/F aircraft, made a PCS move from George AFB to McConnell AFB, Kansas. All four squadrons assigned to the 355th TFW deployed on a TDY basis to either Korat or Takhli following the 4 August 1964 Gulf of Tonkin Incident. On 8 November 1965, the 355th TFW moved PCS from McConnell AFB to Takhli RTAFB, Thailand, taking over the assets of the 6235th TFW(P). The 355th TFW was inactivated at Takhli on 10 December 1970, and all remaining F-105D/F assets were transferred to units in Southeast Asia, the Far East, and in the United States that were still operating the type.

When the 355th TFW was activated at Takhli, there were two squadrons assigned PCS – the 354th and 333rd TFS. The 354th TFS was transferred PCS on 27 November 1965, and remained assigned until deactivation of

The 355th TFW operated F-105Ds and Fs until December 1970. Here 44th TFS ('RE' tail code) and 357th TFS ('RU') F-105Ds are seen during combat operations from Thailand.

Most Thunderchiefs units flew a mixture of F-105Ds and Fs; this heavily-loaded F-105F is a 44th TFS machine.

the Wing on 10 December 1970. Aircraft losses totalled 46 aircraft to enemy action.

The 333rd TFS was transferred PCS from the 4th TFW at Seymour Johnson on 8 December 1965 to the 355th TFW at Takhli, remaining with the 355th Wing until inactivation on 10 December 1970. F-105D/F losses totalled 37, with 333rd TFS pilots accounting for 7½ NVAF MiG interceptors.

The 357th TFS was the third squadron assigned PCS to the 355th TFW at Takhli. After a TDY assignment between 12 June 1965 and 29 January 1966, the status of the 357th was changed to PCS and assigned to the 355th Wing. The squadron remained assigned to the 355th Wing until deactivation on 10 December 1970.

The 354th TFS lost 40 aircraft to enemy action, while shooting down four MiGs.

On 15 October 1969, the 44th TFS was transferred PCS from the 388th TFW at Korat to the 355th TFW at Takhli. The 44th TFS remained with the 355th TFW until deactivation on 10 December 1970. Following deactivation of the 355th TFW, all remaining F-105F/G Wild Weasel assets were transferred to the 388th TFW at Korat.

Aircraft markings included coloured canopy sills and radar reflectors, with an 'arrow device' in the intake, also in the squadron colour, with tail codes applied in 1967: 354th TFS – medium blue/'RM'; 333rd TFS – red/'RK'; 357th TFS – yellow/'RU'; 44th TFS – silver with black trim/'RE'.

35th Tactical Fighter Wing

561st Tactical Fighter Squadron
562nd Tactical Fighter Squadron
563rd Tactical Fighter Squadron

The 35th Tactical Fighter Wing was re-activated at George AFB, California on 1 October 1971, replacing the 479th TFW. Three F-105G squadrons were transferred from the 23rd TFW at McConnell AFB – 561st, 562nd, and 563rd TFS, beginning on 1 July 1973, when the 561st TFS returned from duty with the 388th TFW at Korat. The 562nd TFS was activated on 31 October 1974; with the 563rd TFTS being activated as a tactical training squadron on 31 July 1975.

Initially, the mission of the F-105G units in the 35th TFW was that of defence suppression; in 1975, the F-105G squadrons were given the mission of training new Wild Weasel pilots in missions and tactics for use with the newly developed F-4G Wild Weasel aircraft. F-105G operations phased out, beginning in 1978, and all F-105G assets were transferred to the 128th TFS, Georgia ANG.

Aircraft markings included 'GA' tail codes on 561st TFS F-105Gs returning from combat, and 'WW' on 17th WWS F-105Gs returning from combat. Tail codes for all 35th TFW F-105G, F-4E and F-4G aircraft were changed in 1978/79 to 'WW' to reflect the Wild Weasel mission of the 35th TFW.

Reflecting their Wild Weasel role, the 35th TFW soon adopted 'WW' tail codes, as previously used by the 561st TFS at Korat during 1973. F-105G 63-8332, pictured at Nellis AFB, retains T.O.1-1-4 camouflage, but has black codes as introduced in the late 1970s.

Air National Guard

Thunderchiefs joined the Air National Guard in 1964, when F-105Bs from the 4th TFW were transferred to the 141st TFS. This was the only Guard F-105 unit until 1970/71, when a further four squadrons received aircraft made available as front-line units re-equipped with F-4 Phantoms. Here the six squadrons that operated Thunderchiefs are listed in the order in which they were equipped.

141st TFS/108th TFG, New Jersey ANG

Based at McGuire AFB, New Jersey, the 141st Tactical Fighter Squadron, New Jersey ANG converted from F-86H Sabres to F-105Bs transferred direct from the 4th TFW in April 1964 when the 4th Wing had completed the transition to the F-105D. In May 1981, the 141st TFS converted from F-105Bs to F-4D Phantoms.

Markings on 141st TFS F-105Bs, whether in silver lacquer or SEA camouflage, was the Air Guard badge on the vertical fin with the logo 'New Jersey' in black or white. Most aircraft also had the Air Force OUA ribbon painted on the top of the vertical fin. On the fuselage below the intakes was a miniature 141st TFS emblem.

The sign on the hangar in the background, proclaiming McGuire AFB as the 'Home of the Air Guard Thunderchiefs', seems appropriate given that New Jersey's F-105Bs operated from the base for some 17 years before their retirement in 1981. Like their front-line counterparts, all received T.O.1-1-4 camouflage during the 1960s.

119th TFS/177th TFG, New Jersey ANG

In June 1970, the 119th Tactical Fighter Squadron, New Jersey ANG, converted from F-100C Super Sabres to F-105Bs at Atlantic City ANGB, New Jersey. The 119th TFS was the second and only other Air Guard squadron to operate the F-105B. In January 1973, the mission of the 119th Squadron was changed to air defence, and the unit, now designated the 119th FIS, converted from F-105Bs to Convair F-106A Delta Daggers.

Differing slightly from those of the 141st TFS, the markings applied to 119th TFS F-105Bs incorporated a thin red band on the top of the tail fin and the 119th TFS emblem on the fuselage. Other markings were similar to those of the 141st TFS.

The red fin stripe on this NJ ANG F-105B marks it out as a 119th TFS aircraft. An Air Force OUA ribbon is also visible on the tail, immediately above the ANG emblem.

121st TFS/113th TFG, District of Columbia ANG

Based at Andrews AFB, Maryland, the 121st Tactical Fighter Squadron, District of Columbia ANG, converted from F-100C Super Sabres to F-105D/Fs in July 1971. In July 1981, the 121st TFS began conversion to the F-4D Phantom, completing the transition in March 1982.

Markings on 121st TFS F-105s included the legend 'District of Columbia' on a thin blue stripe with white borders on the vertical tail, a yellow radar reflector, the ANG badge on the middle of the vertical fin, and the 121st TFS squadron emblem on the fuselage below the cockpit.

Other than the ANG emblem and a 'District of Colombia' legend, this F-105F is comparatively free of markings. Like the other two ANG squadrons equipped with F-105Ds and Fs, the 121st TFS was equipped with 'Thuds' for much of the 1970s.

As the Thunderchief was being retired from the front-line USAF, the Kansas ANG was given the important role of F-105 RTU, tasked with training new crews on the type. A 127th TFTS F-105F is pictured.

149th TFS/192nd TFG, Virginia ANG

The 149th Tactical Fighter Squadron, Virginia ANG, converted from Republic F-84F Thunderstreaks to F-105D/Fs at Byrd Field ANGB in February 1971. The 149th TFS was the first F-105D unit in Air Guard service, flying the type until 1982 when the squadron converted to LTV A-7Ds.

Markings on 149th TFS Thunderchiefs included a yellow tail stripe with white trim and a black lightning bolt superimposed. The ANG badge was displayed in the middle of the fin, and the legend 'Virginia' was painted above the serial number. Some aircraft had the canopy rail painted yellow. Many of the 149th TFS Thunderchiefs had 'nose art' painted under the intakes, very similar to that which adorned the combat Thunderchiefs in SEA.

128th TFS/116th TFG, Georgia ANG

The 128th Tactical Fighter Squadron, Georgia ANG converted from F-100D Super Sabres to F-105Gs transferred from the 35th TFW, beginning in early 1979, the transition being completed in the summer. Based at Dobbins ANGB, Georgia, the 128th TFS was the only ANG unit to operate the F-105G Wild Weasel. In August 1983, the 128th TFS completed conversion into the F-4D.

Some aircraft had yellow canopy rails and a yellow band with white borders on the tail fin. Many aircraft also had a 'sharkmouth' nose marking and wrap-around camouflage.

This 128th TFS F-105G, while devoid of squadron colours, sports a 'sharkmouth' and wrap-around camouflage. An AGM-45 acquisition round is also carried.

127th TFTS/184th TFTG, Kansas ANG

The McConnell AFB-based 127th Tactical Fighter Training Squadron, Kansas Air National Guard, converted from F-100C Super Sabres into F-105D/Fs in March 1971. The mission of the 127th TFTS was as a Replacement Training Unit to train F-105 air crews for service with the US Air Force, Air Force Reserve and Air National Guard. In November 1979, the 127th TFTS was redesignated as a TFS and began conversion to the F-4D.

Markings on 127th TFTS F-105D/Fs included a red tail stripe with white trim, red canopy sills and radar reflector. The ANG badge was displayed in the middle of the vertical fin. Some aircraft had the 127th TFS squadron emblem painted on the fuselage under the cockpit.

Air Force Reserve

Three AFRES units operated Thunderchiefs between 1972 and 1984. All operated a mixture of F-105Ds and Fs; the 466th TFS had earlier employed F-105Bs between 1973 and 1981.

305th Tactical Fighter Squadron (465th Tactical Fighter Squadron)

On 20 May 1972, Air Force Reserve converted the 305th Military Airlift Squadron to a Tactical Fighter Squadron equipped with F-105D/Fs, the first squadron in Air Force Reserve to operate with the F-105. The first F-105D arrived at Tinker AFB, Oklahoma, on 14 April 1972. On 20 May 1973, the 305th TFS was renumbered as the 465th TFS. On 10 January 1980 the 465th TFS completed transition from F-105Ds to F-4D Phantoms.

Aircraft markings applied to 305th TFS F-105s included a blue tail band with the legend 'Sooners' in white, and the tail code of 'UC'. When the 305th TFS was renumbered as the 465th, the code was changed to 'SH', other markings remaining unaltered.

The first AFRES unit to receive Thunderchiefs, the 305th TFS 'Sooners' was equipped with F-105Ds some eight months before the 466th TFS converted to F-105Bs passed down from the New Jersey ANG.

457th Tactical Fighter Squadron

The 457th Tactical Fighter Squadron, Air Force Reserve, was activated at Carswell AFB, Texas, in July 1972. The 457th TFS was assigned 38 F-105D/F aircraft, including 29 of the advanced 'Thunderstick II' F-105Ds, the only unit to be equipped with the 'T-Stick II' systems. In October 1981, the 457th TFS began a transition to the McDonnell F-4D Phantom.

Markings included red canopy rails and a red band on the tail with white trim. Below the latter was a map of Texas in white. Most aircraft had 'AFRES' painted on the aft fuselage in either white or black, with codes letter 'TH' painted on the tail. Some aircraft had standard SEA camouflage with light grey undersides, while later aircraft had a wrap-around camouflage scheme.

466th Tactical Fighter Squadron

The 466th TFS was organised at Hill AFB, Utah, on 1 January 1973, with 18 F-105Bs transferred from the 119th TFS, New Jersey ANG. In April 1980, the 466th TFS began the transition from F-105Bs to F-105D/Fs; the last flight by an F-105B took place on 5 January 1981. Initially equipped with 28 F-105Ds and four F-105Fs, the re-equipped 466th was to be the last unit to fly the F-105D/F, the final flight taking place on 25 February 1984; the 466th then flew F-16A/Bs.

Unit markings included a yellow canopy rail, nose stripe and tail stripe, the latter with white trim. Initial camouflage was to T.O.1-1-4 standard, though some aircraft later wore a wrap-around camouflage. At least one aircraft was finished in a tan/brown scheme; another carried 'European One' green/green/grey. A rattlesnake emblem was painted in the tail of some aircraft; tail codes were 'HI'.

Pictured at Hill AFB in January 1974, 466th TFS F-105B 57-5814 was a veteran of the 'Thunderbirds' display team.

Boeing B-47 Stratojet

Boeing's first jet bomber and America's first swept-wing jet bomber, the Stratojet was the most important aircraft in the US Air Force's Strategic Air Command (SAC) fleet during the unsettled days of the mid-1950s. Though its service life was unspectacular and comparatively short, the B-47 was a highly significant machine, for this successful marriage of two new technologies – the jet engine and the swept wing – set Boeing up as a pre-eminent force in military and commercial aviation.

In the months after VE-Day the Allies became aware of the incredibly advanced work that German aeronautical engineers had done during the war's final years. Naturally, there was an eagerness to see and evaluate the data from their work. Among the many Allied technical fishing expeditions into the battered Reich was the one led by Dr. Theodore von Karman, the great Hungarian-born aerodynamicist who was, by then, a leading figure at the California Institute of Technology in Pasadena.

Von Karman's entourage during the spring of 1945 included George Schairer, the chief aerodynamicist for the Boeing Airplane Company in Seattle. During World War II, Boeing had emerged as a world leader in long-range heavy bombers, and Schairer had been a key player in designing the heaviest and longest-ranging – the B-29 Superfortress.

Schairer's now famous evaluation of German wind tunnel data confirmed what many in the United States had suspected from their own studies – that swept wings would greatly increase the speed of an aircraft. What Schairer had discovered was that the Germans had done a great deal more work on the subject, and that they were way ahead of the rest of the world.

Indeed, Messerschmitt had already demonstrated the principle with the Me 262 jet fighter, where the 18.5° sweep improved performance to a significant extent. Of course, that was another thing that had been discovered – swept wings could only work their magic on jets. Beyond a certain speed, propellers created excess aerodynamic drag of their own, preventing the aircraft from going faster. The notion that jet propulsion was the way of the future

B-47E-65-BW 51-5257 is engaged in maximum gross weight take-off trials at Edwards AFB using the new 33-bottle, externally-mounted 'horse collar' RATO pack in this view. Judging by the amount of black smoke emitted by the aircraft's J47 engines, the pilot appears to be using its water/alcohol injection system to further boost take-off power. By the time the B-47E-IV, with its 230,000-lb (104326-kg) MTOW, entered service any performance improvements offered by the injected -25A engines had been cancelled out by weight gains.

Above: Test pilots Bob Robbins (centre) and Scott Osler chat with Boeing flight test chief Norman D. Showalter around the time of the XB-47's first flight on 17 December 1947. Lasting 27 minutes, this first 'hop' from Seattle saw Robbins take 46-065 over the Cascade Mountains to Boeing's Moses Lake test facility.

Left: Such was the demand for Stratojets that a production pool, comprising three manufacturers, was established – a unique occurrence in peacetime. This aircraft was the product of that arrangement; B-47E-50-LM 52-3363 was built at Lockheed's Marietta, Georgia, plant.

Inset left: In one of the enduring images from the 1950s, a B-47 stands under guard and on alert at a SAC base. From 1 October 1957 one-third of the SAC bomber and tanker force was kept on ground alert at all times; by then there were no fewer than 28 wings (nominally) equipped with 45 aircraft each.

Steps toward the Stratojet

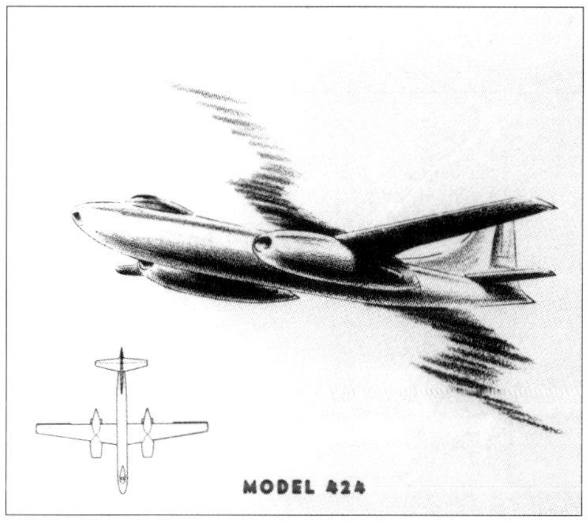

Submitted to the USAAF in January 1944, the Model 424 was a scaled-down B-29 powered by four GE TG-180 axial-flow engines housed in 'high-speed', underwing nacelles. However, the latter impaired wing efficiency, causing unacceptable drag and lift distribution irregularities; the design was dropped.

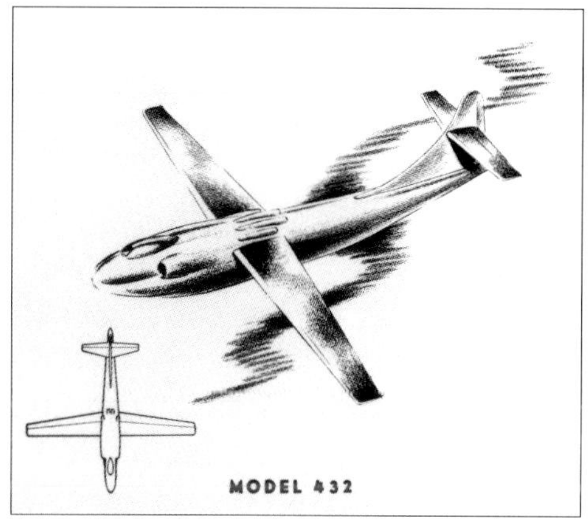

In April 1944 the USAAF issued refined requirements for a jet-powered medium bomber, including a 500-mph (805-km/h) top speed and a 1,000-mile (1609-km) combat radius. Boeing's submission was another straight-winged design, with four turbojets mounted in the upper fuselage.

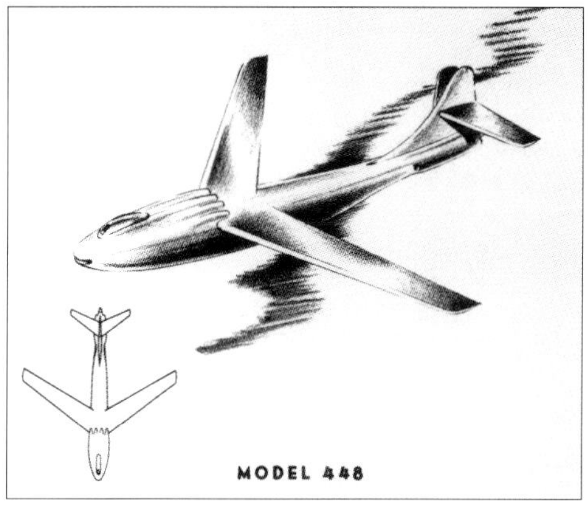

In March 1945 the USAAF elected to award mock-up contracts to four manufacturers, stipulating that Boeing (and Martin) produce designs powered by six engines. Benefiting from wartime German research, Boeing reworked the Model 432 to incorporate wings with 35° of sweep. Four engines were buried in the upper forward fuselage and two were mounted below the tailplane – a layout the USAAF judged unsatisfactory; it did not favour buried engines on safety grounds and the jet pipes above the wing caused air flow problems.

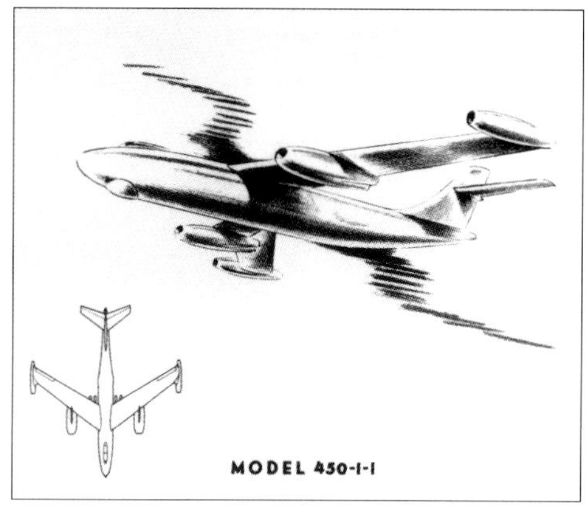

After prolonged investigations into alternative engine locations, Boeing decided to group a pair of engines in pods under each wing with a single engine on each wing tip. The Model 450-1-1 was presented to the Air Force in October 1945; a mock-up was inspected in April 1946 (the outer engines having been moved inboard and beneath the wing in the meantime) and a contract for the construction of two prototypes awarded shortly thereafter. The first refined Model 450-3-3 (XB-47) was rolled out on 12 September 1947 and flew on 17 December.

was already understood, but what George Schairer was discovering was that swept wings were an equally significant part of the equation.

As Schairer was making his discovery, the American aircraft industry was already well into the jet age. Lockheed P-80 jet fighters were already equipping US Army Air Forces (USAAF) squadrons, and more jets were under development at every major airframe manufacturer. The first USAAF jet bomber would be the Douglas B-43 Jetmaster, although it would not make its first flight until 1946. Under a contract issued on the last day of March in 1944, the Jetmaster would be a straight-winged medium bomber, and only a single prototype was ordered for evaluation purposes.

All-jet medium bomber proposals

In the meantime, though, the USAAF had an eye toward the future and operational jet bombers. In November 1943, the USAAF had requested proposals for all-jet medium bombers with a range of 3,500 miles (5633 km), a service ceiling of 45,000 ft (13716 m) and a top speed of 550 mph (885 km/h). The development of such aircraft moved slowly at first, as jet engines with sufficient range evolved.

In April 1944, however, the Allison (later General Electric) J35 turbojet demonstrated that

Few could have foreseen the impact that the B-47's design would have on the design of large jet aircraft. Most of today's airliners owe their general layout to the pioneering work of Boeing's designers. This view of the first XB-47 shows its swept wings and tail and podded engines to good advantage.

such an engine was possible, and the USAAF sent out a request for proposals for that first operational jet bomber. By December, four companies had responded, all of them experienced builders of wartime medium or heavy bombers – Consolidated, Martin, North American Aviation and Boeing. Though Douglas would make a first flight with the small XB-43 in 1946, the company would not create a large jet bomber design for this round of competition.

First designs

At Boeing, the design process began with a series of studies that involved hanging four J35 turbojets on the long straight wing of the B-29, Boeing's Model 345. These studies ranged from the Model 413, a scaled down medium bomber, to the larger Model 424. Finally, the company settled on the Model 432, which clustered the four engines within the fuselage, as was then being done with jet fighter designs.

By the end of 1944, the USAAF had issued Phase I design study contracts to all four companies. George Schairer had not yet made his discovery and the Boeing Model 432 had straight wings. During 1945, however, the Model 432 gradually evolved into the swept wing Model 448. Schairer had become such a convert to swept wings that he proposed 35° of sweep when German data had been tested at a maximum angle of just 29°. The Boeing team picked up where the Germans had left off. By now, the group of converts included Bill Cook and Bob Withington at the company's wind tunnel, as well as George Martin, who had been with Schairer in Germany.

Schairer sold his boss, corporate vice president Ed Wells, on the idea of the swept wings, and Wells sold the concept to the initially sceptical military. The result was the Model 448, which emerged in September 1945. The USAAF had finally accepted the swept wings, but

Engineers prepare the second XB-47 for its first flight, which took place seven months after that of the first aircraft. Comparison with the 1,000th Stratojet (pictured opposite) reveals some of the detail changes made to the initial design during the course of production.

Boeing went back to the drawing board over the issue of the engines buried within the fuselage. The result this time would be the Model 450-1-1, which embodied the eventual production configuration of the B-47.

In December 1945, Boeing received the green light to proceed with a full scale mock-up. This was inspected and approved in April 1946, and the USAAF officially placed an order for two Model 450s.

By summer, all four of the competing companies had cut metal for the first generation of jet bombers. The Boeing and Martin aircraft would each be powered by six J35 turbojets, while the others used four each of the same engine type. One feature that the huge bombers would all

Stratojet bombers in service

The 306th Bombardment Wing at MacDill AFB in Florida was earmarked by SAC as the first operational Stratojet unit, and the wing that would train cadres of aircrew for other units. The 306th operated a few B-47As as early as May 1951, and became operational later in the year.

On 23 October 1951, the US Air Force took delivery of its first operational B-47B (50-008), as Colonel Michael N.W. McCoy, commander of the 306th Bombardment Wing, flew the aircraft from Wichita to MacDill AFB. By the end of the year, the first dozen had been delivered, and two SAC wings were converting to Stratojets. Within two years, more than 80 percent of the 399 B-47Bs had been delivered, and 13 wings were fully or partially equipped. These wings would each be tasked with maintaining at least one squadron on 24-hour alert status.

Regular SAC bombing competitions kept the Stratojet crew's skills honed. 321st Bomb Wing B-47E* City of Winter Park *joined another 321st aircraft (*City of Orlando II*) in a one-two finish in the 1957 event.

Typically, the B-47 wings were organised with three squadrons, but in 1958 a fourth squadron was added because of the four-phase alert cycle. This cycle called for one day each of flight planning, flying, ground alert status and time off.

In 1953 SAC began operational deployments of entire wings overseas. The 2,000-mile (3219-km) unrefuelled combat radius of the Stratojet meant that it could not hit targets in the eastern Soviet Union from the United States. Boeing KC-97 tankers were available, and KC-135s were on the way, but planners felt that the use of fixed bases on a temporary rotational basis would be the best option. Fixed bases near the targets would shorten response time and also eliminate the complexity of having to have a fleet of fast bombers slow down to refuel from sluggish KC-97s under combat conditions.

The basing scenario called for 90-day rotations of entire wings rather than stationing them overseas on a permanent basis. In this way, in the event of World War III, more wings would have gone through the complicated experience of deploying overseas, and more wings could be moved overseas faster.

Because the war plans accepted as a given that Western Europe would be overrun by Soviet ground forces in the event of war, the B-47s would not be operated routinely from fields on the Continent, except for brief deployments to Spain. It was decided that at least one wing would be based in the United Kingdom at all times, and this would be routine from 1953 through 1958. Meanwhile, the French bases at Sidi Slimane and Ben Guerir, in what was then French Morocco, would also be used. These would continue to be active B-47 bases for about a year after Morocco became independent in 1956.

The wings would initially make the move with each aircraft staging through Harmon AFB in Newfoundland or Limestone AFB in Maine. The B-47s would be accompanied by the wing's assigned KC-97 aircraft, which would carry maintenance support personnel, as well as their equipment. The first overseas deployment was of the 306th Bombardment Wing, which made the trip to the United Kingdom between 3 and 6 June 1953. Once there, the aircraft were dispersed to RAF Fairford and RAF Mildenhall.

It was also in 1953 that the Stratojets participated in the annual SAC Fairchild Trophy Bombing Competition for the first time. The results were disappointing, and blamed on various problems, including poor navigation. A Convair B-36 wing won the Fairchild Trophy, with Boeing B-50 wings rounding out the rest of the top five places. The best that a B-47 unit did was ninth place. The following year, a B-47 wing would place fourth, and in 1955, the Stratojets took the trophy.

In 1954, SAC began routine, very-long-range B-47 training missions using aerial

Boeing-Wichita delivered the 1,000th Stratojet (B-47E-105-BW 52-0609) to the USAF in December 1954. Of 2,042 B-47s constructed, 1,392 were eventually built by Boeing, all but the two prototypes and the 10 B-47As at its Wichita factory.

have in common would be narrow, fighter-style cockpits and canopies, which were adopted in the interest of streamlining. Typical side-by-side seating, as was done in previous bombers, would create more aerodynamic drag and eat into the speed budget.

Except for the Boeing aircraft, another feature that all had in common was the familiar straight wings that had always been standard on the piston-engined aircraft of the past. The North American Aviation XB-45 made its first flight in March, followed by the Convair XB-46 in April, and the Martin XB-48 in June. The Boeing Model 450-3-3, the first of two XB-47s, rolled out at Boeing's Plant 2 at Boeing Field in Seattle on 12 September, six days before the USAAF officially became the independent US Air Force.

The XB-47 was larger than the B-29, the superbomber of the war that had just ended. Yet it looked like a fighter, with its sleek, 'built-for-speed' fuselage, razor-thin wings and the fighter-style cockpit with pilot and co-pilot seated in tandem. The futuristic lines of the aircraft that Boeing would christen 'Stratojet' set it apart from the straight-winged aircraft that had been rolling through the big doors at Plant 2 for the past three decades.

The first flight was scheduled for 17 December 1947, 44th anniversary of the Wright Brothers' first powered flight. With Boeing test pilots Bob Robbins and Scott Osler at the controls, the XB-47 made its debut at 2:02 pm. An earlier take-off had been scrubbed by a malfunctioning fire warning light. Having lifted off from Boeing Field, the first flight took them out to Moses Lake in the high desert country of Washington state. Here, Boeing had taken over a sprawling former USAAF training base for use as a test flight facility.

Flight testing would reveal many things about the behaviour of jet engines in big aircraft. Though turbojets made aircraft go fast, they made them go fast slowly. In other words, the pace of acceleration was slow. This meant a very long take-off roll, and it meant that the process of throttling up to go around after an aborted landing was potentially disastrous.

The take-off problem on the Stratojet would be resolved through the addition of 18 rocket-assisted take-off (RATO) bottles built into the sides of the fuselage. The latter would be resolved by the use of a ribbon drogue parachute that would slow the aircraft prior to landing without having to throttle the engines back so far. If a landing had to be aborted, the pilot would be closer to full power.

Also during early 1948, the more powerful General Electric J47 turbojet became available,

refuelling. In June, three B-47s of the 22nd Bombardment Wing made a non-stop, 6,700-mile (10783-km) flight from March AFB in California to Yokota AB in Japan. The idea was to train for strikes against the Soviet Union that could be launched from the United States, and recovered at safe overseas bases. The latter would include the United Kingdom and Morocco, as well as other locations. Such a scenario was first tested during Operation Leap Frog in August 1954, when 308th Bombardment Wing Stratojets took off from Hunter AFB in Georgia, flew a simulated mission and recovered in Morocco.

The Stratojet's point-to-point distance record of 11,450 miles (18427 km) was set on 14 August 1957 with a non-stop flight (supported by four aerial refuellings) between Andersen AFB on Guam and Morocco. The greatest duration Stratojet flight came two years later in November 1959, when one aircraft stayed aloft for three days, eight hours and 36 minutes. In this flight, the aircraft logged more than 39,000 miles (62764 km).

Peak strength

By 1956, SAC had taken delivery of 1,306 B-47Bs and B-47Es – along with 254 RB-47s – and was ready to conduct the largest American bomber exercises since World War II. Reminiscent of the 'thousand-plane raids' of 1944 and 1945, Operation Power House and Operation Road Block brought together more than 1,000 B-47s and KC-97s in a massive display of force across upper North America and the Arctic.

This year also marked the apogee of B-47 operational strength, with 1,560 Stratojets of all types in service. However, the number of bombers would continue to rise to a peak of 1,367 in 1958 as the number of RB-47s declined to 176. In 1959, the numbers remained relatively unchanged at 1,366 and 174 respectively.

The proportion of Stratojets in the SAC fleet was remarkable. In 1957, for example, more than 55 percent of all aircraft and nearly 78 percent of the bomber force were Stratojets. In 1959, as SAC reached its own all-time peak strength, the numbers would be 48 percent and 76 per cent, respectively. By 1958, the 90-day overseas rotations of B-47s would be terminated in favour of shorter duration Reflex operations that used US Air Force bases such as Andersen AFB on Guam and Elmendorf AFB in Alaska, as well as foreign-owned sites. Activities at the bases in Morocco would end, but United Kingdom bases at Brize Norton and Upper Heyford would continue to be used, and both Torrejon AB and Moron AB in Spain would also be part of the Reflex cycle.

In October 1962, during the Cuban Missile Crisis, SAC B-47s were dispersed at Air Force bases – as well as at civilian airports – throughout the southern United States. Based on flying time, this was probably the closest that B-47 bombers ever came to actually going to war.

The Stratojet was in the last months of its SAC bomber career in May 1965 when B-47E 52-0326 of the 100th Bombardment Wing was photographed at McGuire AFB. Operational with B-47s from 1 January 1956, this 8th Air Force wing operated B-47s until 11 February 1966, when its last Stratojet (B-47E 53-2286), with another from the 9th SAW, made its final flight to Davis-Monthan AFB, bringing the B-47's career as a SAC bomber to an end. Based at Portsmouth AFB, New Hampshire (renamed Pease AFB in 1957), the wing deployed overseas to RAF Brize Norton between December 1957 and April 1958. Constituent squadrons during this period were the 349th, 350th and 351st Bombardment Squadrons; the 418th BS joined the wing in 1959.

This view of a 6 May 1957 demonstration of the LABS technique shows a Stratojet near the 'top' of the manoeuvre, close to the point of weapon release. LABS placed enormous strains on the B-47's airframe and, after a spate of crashes, wing-strengthening modifications were hastily made under Project Milk Bottle.

and this engine was used in the second XB-47, which made its first flight on 21 July 1948. The Air Force was still sceptical of the B-47 because of its radical design, so early in the summer of 1948, Boeing President William McPherson 'Bill' Allen invited General K. B. Wolfe, head of procurement for the Air Materiel Command, to come out to Moses Lake and see for himself. During World War II, Wolfe had been the Materiel Command's point man on adapting the B-29 for combat duty, and he had seen that great aircraft evolve from a trouble-plagued hangar queen into a war-winning weapons system.

Wolfe, who had never before flown in a jet, was offered a ride. Once airborne, the Air Force test pilot suggested that Wolfe might want to take the controls. As the story goes, Boeing was given the nod for a production contract the following day. Of the four new prototype jet bombers, only the North American XB-45 – the smallest of the four – and the Boeing XB-47 – the largest of the four – earned production contracts.

Production in Kansas

Boeing's contract, actually issued on 28 October 1948, called for 10 service test aircraft, which were ordered as B-47A, rather than as YB-47. According to Boeing company legend, one caveat that Wolfe had specifically stipulated was that the production series should be not be built in Seattle. The summer of 1948 was the summer of the Berlin Blockade. The Cold War had begun, and the fear of an enemy attack – which seemed to have vanished in 1945 – was back. Wolfe perceived that Seattle was potentially vulnerable to air attack, so he insisted that Boeing move Stratojet production south and far inland to the big facility in Wichita that Boeing had expanded for B-29 production.

None of the production B-47s would be built in Seattle, but when it later came to the B-52, most of the early production aircraft were built in Seattle. The B-52 was, however, the last major Seattle production programme. On 14 November 1949, as work on setting up the tooling proceeded at Wichita, the US Air Force issued a supplementary order for 17 examples of the first true production series Stratojet under the designation B-47B. The B-47B production aircraft would be engined with the more powerful General Electric J47-GE-23 with 5,800 lb (25.8 kN) of thrust. These aircraft would have their gross weight increased from the 121,080 lb (54921 kg) of the prototypes to 184,908 lb (83872 kg). They would also be equipped with defensive armament. During World War II, bombers had bristled with gun turrets, but by the end of the war, it had been discovered that if a bomber was fast enough, it was vulnerable to fighters only when attacked from the rear. Thus, the only defensive weaponry in the B-47 would be a remotely controlled tail turret. It would be equipped with a pair of .50-calibre machine-guns operated by the Emerson A-2 fire control system.

Toss bombing trials were conducted by Boeing test pilots Jack Funk and Dick Taylor in B-47E 51-7069; the red/white striping is presumed to have aided visual tracking.

Nuclear weapons delivery

The typical offensive scenario for the Stratojet fleet in the event of war initially called for delivery of nuclear weapons from high altitude. The choice of weapons included Mk 28, Mk 36, Mk 42, Mk 53 and B43 free-fall nukes. Offensive missiles were tested but never deployed.

The undersides of the Stratojets, like those of early operational B-52s, were painted gloss white to reflect light, heat and radiation from the nuclear blast. However, the altitude of the attack and speed of the aircraft were also a solid safety factor.

The altitude capability of the Stratojet would also make it relatively safe from Soviet anti-aircraft artillery and interceptor attacks. In the mid-1950s, however, the US military underwent a major tactical shift with regard to the delivery of nuclear weapons. Improvements in Soviet SAM capability now dictated that low-level operations should be considered and adopted. Both the US Air Force and US Navy began training for the Low-Altitude Bombing System (LABS).

LABS, or 'toss bombing', was actually more of a technique than a 'system'. At low altitude, simply dropping the bomb and accelerating away at the fastest speed possible would not be enough to ensure that the aircraft would not be damaged or destroyed by a powerful nuclear blast. Using LABS, the B-47 pilot approached the target at low level before pulling up into a half loop, or 'Immelman' turn, during which the weapon was released. Having, literally, 'thrown' its bomb at the target, the B-47 half-rolled right side up and departed on a reciprocal heading, having gained several thousand feet of altitude. A rapid descent to the safety of low level would follow.

Stratojets began running LABS test flights in June 1955, and SAC operational training began in December of the same year under Project Hair Clipper. Eventually, more than 100 aircraft would be dedicated to the LABS mission. Of course, the B-47 airframes had never been designed for repeated and routine 2-*g* and 3-*g* manoeuvring. As stress cracks began creeping into those airframes, routine LABS training for the Stratojet fleet was curtailed early in 1958.

The development of the arsenal of larger and heavier hydrogen bombs occurred after the B-47's configuration had been defined, so the use of the Stratojet as a platform for such weapons was an afterthought. Because of the enormous size of the blast, it was originally believed that only unmanned systems – such as ICBMs – could deliver hydrogen bombs. However, long-range autonomous guidance systems were in their infancy in the early 1950s, the best solution seemed to be a piloted aircraft that would be crewed most of the way to the target and then abandoned for the final bomb run. A similar scenario had been used sparingly against German hardened targets in northern France during World War II under the code name Aphrodite. Under a project designated as Brass Ring (MX-1457), two B-47Bs were intended to be converted as robot hydrogen bomb carriers with the designation MB-47. After completing all necessary aerial refuellings, the crew would bail out over friendly territory and the aircraft would fly on. It would then be guided by remote control from an accompanying DB-47 drone director, or by itself, using an Autonetics guidance system. At the time the North American Aviation Autonetics Division was the world leader in such guidance systems. Indeed, it would build the guidance control for the Minuteman programme. Though studied and planned, the MB-47 aircraft were never actually converted.

Had the Bell GAM-63 Rascal strategic air-to-surface missile performed to specification and not been overtaken by more advanced weapons, the B-47 may well have served in large numbers as a missile launch platform. One of the two YDB-47Es is pictured, complete with Rascal test round mounted on its fuselage pylon.

The arrangement of the Stratojet's landing gear – like that of the later B-52 – was of the bicycle type. Because the wing was so thin, the main gear was housed within the fuselage, with small outrigger gear at the wing tips, more for balance than to support weight. Initially, pilots found this hard to get used to and many mishaps resulted. However, as crews got used to the Stratojet, it earned one of the best safety records in the Air Force.

By the spring of 1950, the B-47As were finally taking shape on the assembly line at Wichita. On 25 June 1950, the first one taxied out to make its first flight. At that same moment Communist armies were pouring across the 38th parallel into South Korea. The Cold War had turned hot.

Preparing for war

A backdrop to the US Air Force planning for acquisition of the B-47 was the larger issue of planning by the United States Joint Chiefs of Staff for fighting a possible – then considered probable – World War III against the Soviet Union.

The Korean War had begun less than a year after the Soviet Union had exploded its first nuclear weapon. The intelligence community had predicted neither, and both events seriously quickened the heart rates of American military planners. For these planners – especially the ones at SAC – the Cold War was a real war. World War II had begun with the United States totally unprepared. The American military establishment felt duty-bound to see that this situation not recur a decade later. For men such as General Curtis LeMay, who would command SAC from 1948 to 1957, the United States had to be prepared to fight World War III at a moment's notice. To accomplish this, he ran his command as though it was already at war.

Meanwhile, the events in Europe that culminated in the Berlin Blockade convinced American war planners that a massive Soviet invasion of Western Europe was probable, and that it was unstoppable by conventional means. Because the United States had dismantled the enormous conventional ground force that had defeated Germany in World War II, the only option in reacting to such an invasion would be a massive airborne nuclear counterattack.

In 1948 Pentagon planners came up with Operation Bushwacker, the general plan for fighting World War III. Bushwacker called for a massive nuclear air offensive to hold the Soviet forces down while the United States began the slow process of mobilising for conventional warfare. The Air Force planning for Bushwacker continued to evolve into 1949 via War Plan Fleetwood and War Plan Trojan, which set a goal of stockpiling 400 nuclear weapons by January 1953. Much of the evolution of SAC which would take place in the 1950s – including the huge build-up of B-47s – can be understood more clearly in the context of these plans, which were not made known until after the end of the Cold War half a century later.

In 1950, the need for a fast jet bomber was suddenly a matter of great urgency. The Stratojet was the right plane in the right place on its production curve at the right time. The initial 'buy' of 17 production series B-47Bs – ordered with fiscal year 1949 funds – was expanded to 88 with 1950 funding, and eventually to more than 600 aircraft through the 1951 budget. The operational target date of 1953 dovetailed with War Plans Fleetwood and Trojan.

Production of the definitive B-47E

Late in 1950, Boeing and the US Air Force began to create the specifications for what would be the definitive Stratojet. The B-47E would be powered by six J47-GE-25A engines, delivering 7,200 lb (32.03 kN) of thrust, and it would be capable of carrying a nuclear payload, as or many as 18 1,000-lb conventional bombs. The first 861 aircraft would be delivered with the K-4A bombing system, while later aircraft carried the newer MA-7A equipment. Rather than the machine-guns of the B-47B, the B-47E's defensive armament would consist of a pair of M24A 20-mm cannon.

To meet the needs of its war plans, the US Air Force needed the B-47E, a lot of them, and fast. As had been the case with many high-priority World War II aircraft, especially Boeing's B-17 and B-29, the need to get a large number of B-47s into service quickly led to the US Air Force adopting a multi-source acquisition for the B-47. It was to be the only post-war production combat aircraft that would be manufactured for the Air Force by multiple planemakers.

In addition to the Boeing plant at Wichita, two other sites would build Stratojets. First, Lockheed would build them at the factory in Marietta, Georgia, where Bell Aircraft had built Boeing B-29s during World War II. Next, the Douglas-operated, Air Force-owned facility in Tulsa, Oklahoma, would also be brought into the programme. To facilitate the incorporation of the Lockheed and Douglas assembly lines, Boeing shipped eight unassembled B-47Bs to Marietta and 10 to Tulsa. Working on these two projects helped bring the Lockheed and Douglas people up to speed on the coming project. The first B-47E made its debut flight at

From early in its career the Stratojet fulfilled a number of second-line functions, including service as an aerial target for surface-to-air missiles under development. Most were eventually destroyed, though two were spared and later employed as targets during trials of the AIM-47 air-to-air missile intended for the Lockheed YF-12. JQB-47E 53-4256 (built as an RB-47E and later converted to QB-47E standard) was one of these aircraft, its duties indicated by Bomarc and YF-12 silhouette markings on its nose.

Wichita on 30 January 1953, and over the coming three years, Boeing would build 691 of the aircraft, while Lockheed and Douglas would turn out 386 and 264, respectively. The last B-47E, and the last B-47 bomber to be produced, was a Wichita-built aircraft (63-6244) that was delivered to the Air Force on 24 October 1956.

Reconnaissance types

As the B-47Es were being manufactured, the Wichita facility would also build 240 examples of a reconnaissance variant. The RB-47E's nose and belly bulged with 11 optical camera systems and other equipment. An electronics countermeasures version, designated as EB-47E, was also created, but through conversions from bomber airframes rather than as factory-fresh aircraft.

As Soviet radar evolved, so too did efforts to defeat it. Starting in June 1955, the reconnaissance Stratojets rolling out at Wichita would change substantially. The RB-47E would evolve into the RB-47H, a special variant designed to identify enemy ground-based radar sites and operate as a jammer and pathfinder for formations of Stratojet bombers on strike missions. To accommodate additional crewmembers tasked with reconnaissance duties, a crewed pod was designed to fit into the bomb bays of both the EB-47E and the RB-47H. It was similar in concept to the manned reconnaissance pod designed for the RB-52B that was being produced by Boeing at the same time. A total of 32 RB-47Hs would be built new by Boeing at Wichita, and three of these were converted as ERB-47H aircraft for special operations missions. Another branch of the RB-47E family tree led to the RB-47K. These were essentially a continuation of the RB-47E production series, in which the last 15 aircraft off the line were fitted with both photo reconnaissance and weather reconnaissance equipment and designated as RB-47K. Meanwhile, the Air Weather Service also operated more than 30 B-47Es – and at least one B-47B – that had been converted as WB-47 weather reconnaissance aircraft.

The last Stratojet variant would be the EB-47L, a series of 35 B-47Es that were converted in 1963 for use as command and control relay aircraft.

Bill Yenne

Most B-47s ended their careers with a one-way trip to the MASDC at Davis-Monthan AFB. Of over 1,000 B-47s and RB-47s placed in storage, most were scrapped during 1968. A handful survived a little longer; the hulks of B-47E 51-5251 and RB-47H 53-4304 were photographed at 'DM' during 1976.

The Stratojet in combat

Under his 1955 Open Skies proposal President Dwight D. Eisenhower suggested that the world could ensure peace through a practice of open aerial reconnaissance conducted by all of the world's major powers. Despite opposition from the Soviet Union, the US continued to conduct reconnaissance missions that included overflights of the Soviet Union, though these were officially ended after the shoot-down of Francis Gary Powers in a Lockheed U-2 on 1 May 1960. However, missions around the periphery of the Soviet Union, and similar Soviet missions against the United States, would continue for another three decades.

RB-47s and EB-47s were involved in many such flights, including some that involved penetrations of Soviet air space. Several aircraft were attacked and two Stratojets were shot down by Soviet forces.

■ 8 May 1954
An RB-47 based at RAF Fairford penetrated Soviet air space near the naval base at Murmansk and photographed a number of Soviet facilities. As it was turning home, the RB-47 encountered Soviet MiG-15s, as well as MiG-17s, which were not known to be operational at the time. The MiG-17s attacked, and the RB-47 returned fire. Moments later, the Stratojet passed into Finnish air space and the Soviets broke off their assault. The RB-47 suffered several hits, but no aircraft on either side was lost.

■ 17 April 1955
An RB-47 of the 26th Strategic Reconnaissance Wing operating out of Eielson AFB in Alaska was shot down by MiG-15s near Cape Lopatka, at the southern end of the Kamchatka Peninsula. According to Soviet documents circulated after the end of the Cold War, the aircraft was acquired by radar near Cape Vasiliev and did not actually violate Soviet air space. It was intercepted east of Cape Kronotski and attacked for two minutes. The RB-47 disappeared from radar screens and the crew was reported missing.

■ Autumn 1958
Three non-fatal attacks against RB-47s are reported to have occurred during the autumn of 1958 in widely separated locations. These were on 31 October over the Black Sea, on 7 November near Gotland Island in the Baltic Sea, and on 17 November over the Sea of Japan. In each case, the Stratojet involved eluded the Soviet interceptors and returned safely to its base.

■ 1 July 1960
RB-47H 53-4281 was shot down over the Barents Sea. The United States government insisted that the RB-47 had never been closer than 30 miles (48 km) to Soviet territory, and the matter was discussed and debated in the UN Security Council in the weeks following the incident.

Assigned to the 38th SRS of the 55th Strategic Reconnaissance Wing, the RB-47H was operating out of RAF Brize Norton. It was piloted by Major Willard G. 'Bill' Palm, with Captain Freeman 'Bruce' Olmstead as co-pilot and Captain John R. McKone as navigator. In the reconnaissance pod were Major Eugene Posa, Captain Oscar Goforth and Captain Dean Phillips. The aircraft was reportedly shot down by a MiG-15 piloted by Vasili Poliakov. Both Olmstead and McKone ejected and survived. Palm's body was recovered and was returned to the United States on 25 July. Olmstead and McKone were tried and convicted of espionage in a Soviet court and jailed. However, they were eventually released on 25 January 1961. Reportedly, Posa's body was also recovered, but it was never returned to the US.

■ 27 April 1965
In what was probably the last shooting incident involving a Stratojet, an ERB-47H was intercepted by a pair of North Korean MiG-17s over the Sea of Japan. Extensive damage resulted in the loss of two engines, although the Stratojet escaped to make an emergency landing at Yokota AB in Japan. Though relatively few Stratojets were sent into harm's way, they were used operationally for a decade, and the fact that only two were shot down is a credit to their inherent robustness as well as to the skill of their crews.

Boeing B-47E-125-BW Stratojet

380th Bomb Wing (Medium) Plattsburgh AFB, New York, 1964

B-47E 0-32399 is depicted in markings worn during a Reflex Action deployment to RAF Brize Norton in 1964. The aircraft has been fitted with Phase II ECM equipment (which included the circular antennas fitted to the rear fuselage, below the tailplane) as part of the initial B-47E-I modification programme.

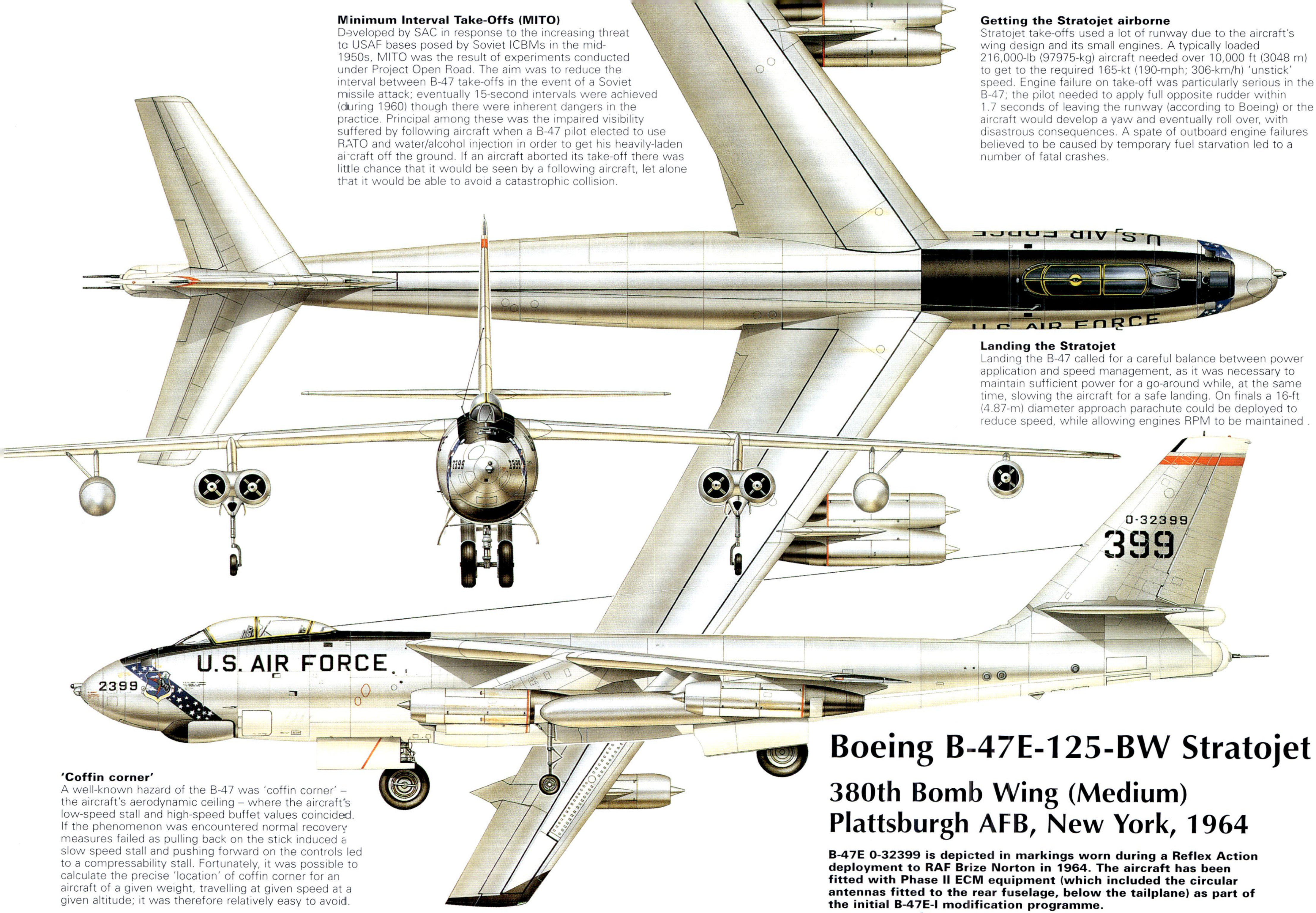

Minimum Interval Take-Offs (MITO)
Developed by SAC in response to the increasing threat to USAF bases posed by Soviet ICBMs in the mid-1950s, MITO was the result of experiments conducted under Project Open Road. The aim was to reduce the interval between B-47 take-offs in the event of a Soviet missile attack; eventually 15-second intervals were achieved (during 1960) though there were inherent dangers in the practice. Principal among these was the impaired visibility suffered by following aircraft when a B-47 pilot elected to use RATO and water/alcohol injection in order to get his heavily-laden aircraft off the ground. If an aircraft aborted its take-off there was little chance that it would be seen by a following aircraft, let alone that it would be able to avoid a catastrophic collision.

Getting the Stratojet airborne
Stratojet take-offs used a lot of runway due to the aircraft's wing design and its small engines. A typically loaded 216,000-lb (97975-kg) aircraft needed over 10,000 ft (3048 m) to get to the required 165-kt (190-mph; 306-km/h) 'unstick' speed. Engine failure on take-off was particularly serious in the B-47; the pilot needed to apply full opposite rudder within 1.7 seconds of leaving the runway (according to Boeing) or the aircraft would develop a yaw and eventually roll over, with disastrous consequences. A spate of outboard engine failures believed to be caused by temporary fuel starvation led to a number of fatal crashes.

Landing the Stratojet
Landing the B-47 called for a careful balance between power application and speed management, as it was necessary to maintain sufficient power for a go-around while, at the same time, slowing the aircraft for a safe landing. On finals a 16-ft (4.87-m) diameter approach parachute could be deployed to reduce speed, while allowing engines RPM to be maintained.

'Coffin corner'
A well-known hazard of the B-47 was 'coffin corner' – the aircraft's aerodynamic ceiling – where the aircraft's low-speed stall and high-speed buffet values coincided. If the phenomenon was encountered normal recovery measures failed as pulling back on the stick induced a slow speed stall and pushing forward on the controls led to a compressability stall. Fortunately, it was possible to calculate the precise 'location' of coffin corner for an aircraft of a given weight, travelling at given speed at a given altitude; it was therefore relatively easy to avoid.

XB-47

There were two prototypes (Boeing Model 450-3-3) ordered in 1945 and constructed at Boeing's Plant 2 at Boeing Field in Seattle. The first (46-065) was completed on 12 September 1947, and flown on 17 December. The second (46-066) made its debut on 21 July 1948. They were the only two Stratojets that were built by Boeing in Seattle. The first XB-47 was first flown with six General Electric J35-GE-7 turbojets, each delivering 3,750 lb (16.68 kN) of thrust, but the latter used six J47-GE-3s with 5,200 lb (23.13 kN). In October 1949, the first XB-47 would be retrofitted with J47s.

Because of high wing loading and the slow acceleration characteristics common to early jet engines, 18 jet-assisted take-off (JATO) units would be built into the sides of the fuselage. The XB-47s were 107 ft 6 in (32.8 m) in length, with a tail height of 27 ft 8 in (8.5 m). As such, they were slightly longer than the wartime Boeing B-29 Superfortress. The XB-47 design established the wing span of 116 ft (35.36 m) that would be common to all subsequent Stratojets. The wing area was 1,427 sq ft (132.6 m²).

The XB-47s weighed 74,623 lb (33849 kg) empty and had a maximum take-off weight of 162,500 lb (73710 kg). Their top speed was 580 mph (930 km/h) at 115,000 lb (52164 kg) at 15,000 ft (4,572 m) – or 568 mph (914 km/h) at sea level. The cruising speed was 466 mph (750 km/h), and the service ceiling calculated at 41,000 ft (12500 m). The range was 2,650 miles (4465 km) with a 10,000-lb (4535-kg) payload.

XB-47-BO: 46-065/066 (2)

Rolled out during September 1947 with little fanfare and in the presence of fewer than 100 people, the first XB-47 (pictured) was dismissed by many as a research vehicle that would never see full-scale production. The radical nature of the Stratojet's layout is clear in this view of the first prototype, seen with a number of C-54/DC-4 airliners during the early 1950s.

XB-47

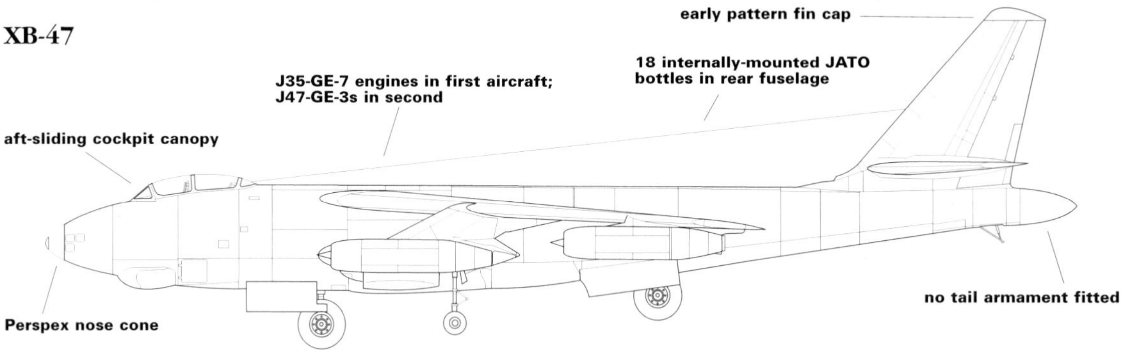

Below: Outrigger wheels retracting into the inner engine nacelles supported the B-47's wings on the ground. 46-065 is pictured at the time of its roll-out.

Above: USAF Majors Russell Schleeh and Joseph Howell land the first XB-47 at Andrews AFB on 8 February 1949 having completed a record-breaking 3-hour, 46-minute flight from Moses Lake AFB, a distance of 2,289 miles (3684 km). The 'white-walled' tyres on the aircraft's main gear wheels are an interesting feature!

Below: Pictured at Chanute AFB during May 1959, the second XB-47 had, by this stage, been fitted with a modified nose cone of the type fitted to the B-47A.

Left: Three of the XB-47's 18 RATO bottles are test fired. Each produced 1,000 lb (4.45 kN) of thrust. Later installations were externally mounted and could be jettisoned after use.

Below: The Stratojet's thin wing was unsuitable for the carriage of fuel; tanks above the weapons bay and in the rear fuselage held 17,000 US gal (64350 litres).

B-47A

Ten service test aircraft, Boeing Model 450-10-9, were ordered in November 1948 (with Fiscal Year 1949 funding) under the designation B-47A. All 10 were built at the Boeing Wichita facility. They were essentially identical to the XB-47 in most respects, although their empty weight was 73,240 lb (33222 kg), and their maximum take-off weight was 151,324 lb (68641 kg). The powerplants were General Electric J47-GE-11 turbojets with 5,200 lb (23.13 kN) of thrust. The B-47As were used, among other things, to test the tail armament and the A-2 and A-5 fire control systems. The first B-47A (49-1900) made its first flight on 25 June 1950 and deliveries continued through the ensuing year.

B-47A-BW	49-1900/1909 (10)

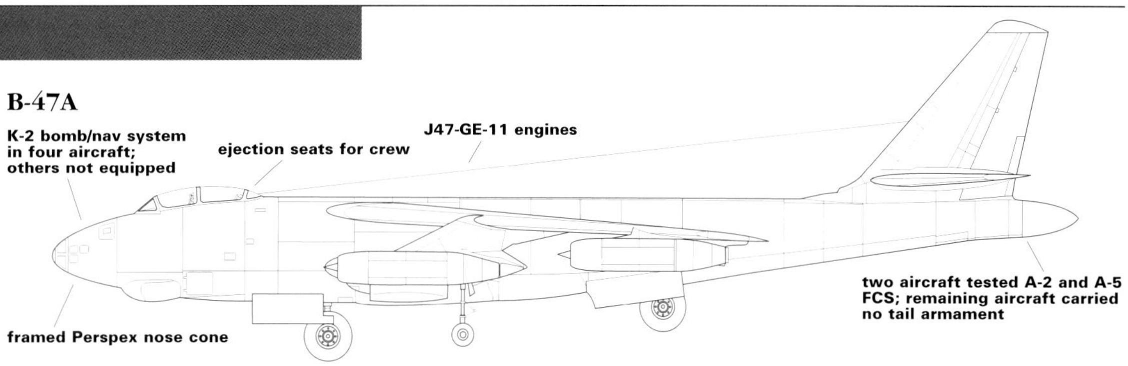

Below: B-47A 49-1909 formates with an F-86 chase aircraft during testing. These were, respectively, the fastest fighter and bomber in the world at the time.

Above: B-47A 49-1908 was one of the two examples actually fitted with tail armament. Also evident in this view are the aircraft's large Fowler flaps, tandem main landing gear, tail-mounted twin 50-calibre machine-guns and internally-mounted RATO gear.

DB-47A

Designation of projected director aircraft for MB-47B unmanned hydrogen bomber. One B-47A was earmarked for conversion, though it is believed that the work was never carried out.

B-47B

The first true production series Stratojet was generally similar to its predecessors except for an increase in weight to 76,130 lb (34533 kg), and the more robust J47-GE-23 with 5,800 lb (25.8 kN) of thrust. It was just 3 in (7.62 cm) taller in the tail than the previous Stratojets, and 8 in (20.32 cm) shorter in overall length.

A total of 399 B-47Bs was built, all at Wichita. More than 1,700 were planned – although not actually ordered – during 1951. However, this strategy was pared back as the overall acquisition programme refocussed on the B-47E. The first 87 were built as Boeing Model 450-11-10, with the next 202 as Boeing Model 450-67-27, as refinements and modifications were introduced. The last B-47Bs were built as Model 450-157-27.

One important issue in the Model 450-11-10 and later types was the crew escape system. The B-47As had been equipped with ejection seats that were shown to be inadequate and even dangerous, so they were replaced by an escape hatch system in the early B-47Bs. This was also deemed unacceptable, but it would be late in the B-47B's evolution before satisfactory ejection seats were finally incorporated.

The B-47Bs were built fully armed with a pair of .50-calibre tail guns, and the Emerson A-2 gun control system. When the latter was found to be unsuitable, a decision was made to upgrade to the General Electric A-5 gun control system. Because the A-5 was not immediately available, many early B-47Bs had their A-2s replaced by optical sights.

The B-47B was equipped with AN/APS-54 radar, AN/APT-5A electronic countermeasures gear, and the K-2 bombing navigation system. When the latter proved to be hard to maintain, it was superseded by the K-4A.

Because both Douglas and Lockheed had been chosen for future Stratojet production work, Boeing shipped a small number of partially-completed airframes to those two companies to try their hand at final assembly. Nine were sent to Lockheed at Marietta, Georgia (51-2145, 51-2197, 51-2204, 51-2210, 51-2217, 51-2224, 51-2231, 51-2237, 51-2243) and 10 went to the Douglas plant at Tulsa, Oklahoma (51-2141, 51-2150, 51-2155, 51-2160, 51-2165, 51-2170, 51-2175, 51-2180, 51-2185 and 51-2190.

B-47B-1-BW	49-2642/2645 (4)
B-47B-5-BW	49-2646, 50-001/012 (13)
B-47B-10-BW	50-013/025 (13)
B-47B-15-BW	50-026/050 (25)
B-47B-20-BW	50-051/082 (32)
B-47B-25-BW	51-2045/2081 (37)
B-47B-30-BW	51-2082/2136 (55)
B-47B-35-BW	51-2137/2191 (55)
B-47B-40-BW	51-2192/2246 (55)
B-47B-45-BW	51-2247/2301 (55)
B-47B-50-BW	51-2302/2356 (55)

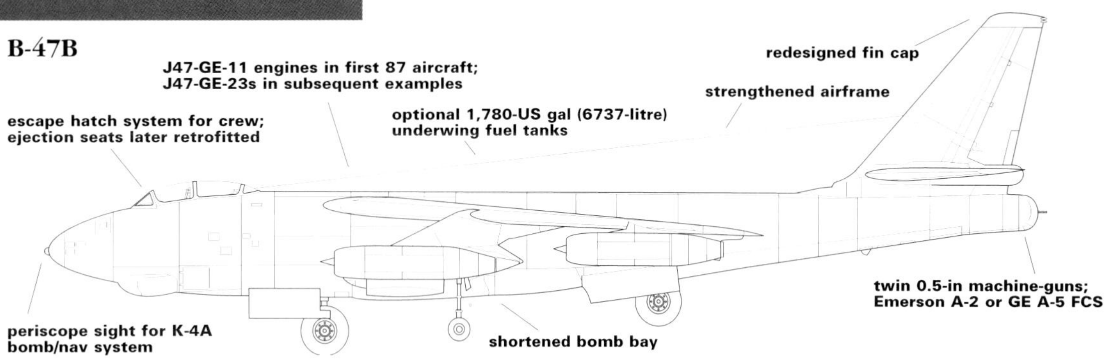

Below: This B-47B was employed as a flying test bed for the P&W J57 turbojet destined to power the Stratojet's successor, the B-52. B-47B-1-BW 59-2643 carried a J57 in each outboard position.

Left: The first B-47B-25-BW, 51-2045, rolls to a halt with assistance from its 32-ft (9.75-m) diameter braking parachute. Landing was the most difficult aspect of flying the Stratojet, which had a high approach speed (due to its clean airframe), unresponsive engines (which could take as long as 20 seconds to spool up from idle) and unorthodox undercarriage. 'Porpoising' on landing caused the loss of a number of aircraft and their crews.

YDB-47B/DB-47B

The DB-47B was conceived as the launch aircraft for the Bell GAM-63 (previously B-63) Rascal stand-off, surface-to-air missile. Had it not been cancelled in November 1958, the missile would have equipped the 445th Bomb Sqn, 321st Bomb Wing, at Pinecastle AFB.

The YDB-47B designation is believed to apply to 51-2186, a B-47B modified during 1953 with a XGAM-63 on a mounting under the starboard wing root. Subsequently, 73 B-47Bs (51-2160, 51-2162/2185, 51-2187/2234) were earmarked for conversion to 'production' DB-47B standard, though it is unclear exactly how many were actually converted.

The first successful Rascal launch from a DB-47 was made in July 1955 (from a DB-47E – see entry), but SAC's concerns regarding the missile's reliability, performance degradation suffered by DB-47 carrier aircraft and the costs of conversion and crew training sealed its fate and the programme was cancelled.

51-2186, seen carrying a GAM-63 Rascal test round, is the aircraft believed to have been designated YDB-47B and appears to have been converted to 'B-47B-II' standard. Rascal was an acronym for RAdar SCAnning Link, the missile's guidance system installed in the DB-47B's fuselage.

B-47B crew stations

Left: The tandem layout of the cockpit was unlike that of any previous bomber type and offered excellent visibility as well as the impression that the crew was in command of a fighter, rather than a bomber. The front cockpit is depicted here; the co-pilot, who had a similar set of controls, was able to swivel his seat through 180° to operate the tail armament via a separate panel.

Right: In this view of the navigator/bombardier's position in the nose of a B-47B the two scopes are (left) for the K-2 or K-4A bomb/nav system's main radar set and (right) the periscopic bomb sight. The panel above includes navigation instrumentation (including an ASI, altimeter and compass). The angled uppermost panel is the radar power monitoring panel.

MB-47B

In a programme known as Brass Ring, which began in April 1951, the US Air Force considered converting two B-47Bs as unmanned hydrogen bomb carriers. A single B-47A was to be employed as a DB-47A director aircraft (though autonomous guidance was considered initially). In the event the programme came to nought; the B-47Bs were never converted and Brass Ring was cancelled in April 1953.

YRB-47B/RB-47B

In need of an aircraft with better speed and range performance than existing RB-29 and RB-50 aircraft, the USAF initiated the design of the RB-47B in 1951. The RB-47B was to incorporate many of the improvements intended for the B-47E, including the A-5 defensive FCS and uprated J47-GE-25 engines, but would not be ready for service until 1954. By this time the B-47E bomber was due to enter service and it was decided to designate the final design RB-47E instead, given that it had more in common with the B-47E.

Subsequently, the USAF ordered the modification of 24 B-47Bs (including YRB-47B 51-2292, below) to an interim RB-47B standard, equipped with an eight-camera, climate-controlled reconnaissance pod in the bomb bay. These were used by several units for training during 1953/55.

TB-47B

The TB-47B was introduced to satisfy the need for a means to train the large number of crews that would be needed for the Stratojet fleet. It was created by adding a fourth crew position to the cockpit of 66 B-47Bs (some sources say 108, possibly as part of the Ebb Tide programme). The first 48 of the TB-47s were converted by Douglas at Tulsa, but the US Air Force did the remainder of the work itself at its materiel depot at Tinker AFB in nearby Oklahoma City. Modifications included deletion of all armament provisions and ECM equipment.

TB-47B 50-057 carries the Air Training Command badge on its nose; it served with the 3520th FTW at McConnell AFB. Engine nacelle close-off doors prevented engines from 'windmilling' (and increasing drag) when shut down in flight.

ETB-47B

This designation was applied to an unknown number of TB-47Bs, including 50-037, 50-054 and 51-2198, during service as flight test aircraft with Air Research and Development Command at the Wright Air Development Center.

These examples of the ETB-47B each carry a Wright Air Development Center emblem and appropriate 'Flight Test' legend (right) on the forward fuselage.

WB-47B

A harbinger of later weather reconnaissance Stratojets was the WB-47B, of which one (51-2115, below) was converted in 1954, and used for hurricane-hunting and other weather-related missions before the WB-47Es became available. Between 1957 and 1965, when it was replaced, it was assigned to the 55th WRS, 9th Weather Reconnaissance Group .

'B-47B-II'

Beginning in 1954, after the B-47E replaced the B-47B as the standard production model, 165 B-47Bs were brought up to B-47E standard under a programme called High Noon. As well as standard IRAN (inspect and repair as necessary) work, modifications/additions included ejection seats for all three crew, increased fuel capacity and structural strengthening, increased bomb bay capacity for conventional bombs, new water-injected J47-GE-25 engines with 6,000 lb (26.69 kN) of thrust, the A-5 fire control system (MD-4 in the last 14 aircraft), new ECM and radio equipment and an approach 'chute.

A second phase upgrade programme called Ebb Tide was conducted during 1956 and 1957, this bringing a further 66 aircraft up to High Noon standard. One source suggests that an additional 30 aircraft were brought up to a similar standard as DB-47Bs.

The High Noon/Ebb Tide Stratojet was widely referred to within the factory as the B-47B-II, although this designation was not officially endorsed by the US Air Force.

B-47Bs brought up to B-47E standard were virtually indistinguishable from the production B-47E. B-47B-II 51-2355 was the second last B-47B-50-BW. Notable features are the lack of internal RATO gear (replaced by jettisonable RATO gear), a reworked nose and a partial application or anti-flash white paint.

YB-47C (XB-56)

The idea of the XB-56 was for a 'Super' Stratojet powered by four, rather than six, Allison YJ71-A-5 engines, each delivering 10,090 lb (44.88 kN) of thrust. One B-47B was earmarked for conversion as the XB-56, but the designation was changed to YB-47C. The project lagged when the engines were not available. The Pratt & Whitney J57 powerplant, which was planned for the B-52, was suggested as a substitute, but the programme was terminated in December 1952 without a YB-47C having been built.

XB-47D

The most radical conversion actually carried out on a Stratojet airframe was probably the pair of B-47Bs that were retrofitted with huge Wright YT49-W-1 turboprop engines in place of the four inboard turbojets. The two outboard jet engines would be retained. This programme was undertaken by Boeing as Model 450-162, and designated by the US Air Force as XB-47D. The project was initiated in April 1951 when turboprop engines were considered to be a viable alternative to the early turbojets in terms of range. Essentially, the US Air Force wanted to hedge their bets on the Stratojet by undertaking the parallel development of a variant whose primary power source was not a jet engine. Ironically, development problems with the turboprop engines would delay the effort for more than four years.

The first XB-47D (Model 450-162-48), 51-2046, made its debut flight on 26 August 1955. Its sister ship (51-2103), designated as Model 450-162-49, first flew on 15 February 1956. The empty weight of the XB-47Ds was 79,800 lb (36197 kg) and they had a gross weight of 184,428 lb (83657 kg). In tests, the XB-47D achieved a top speed of 597 mph (961 km/h) at 13,500 feet (4115 m). This was the fastest yet by a propeller-driven aircraft, and on a par with the official top speeds of any all-jet B-47 variant. Tests were limited to the two prototype XB-47Ds, and no production series was ordered.

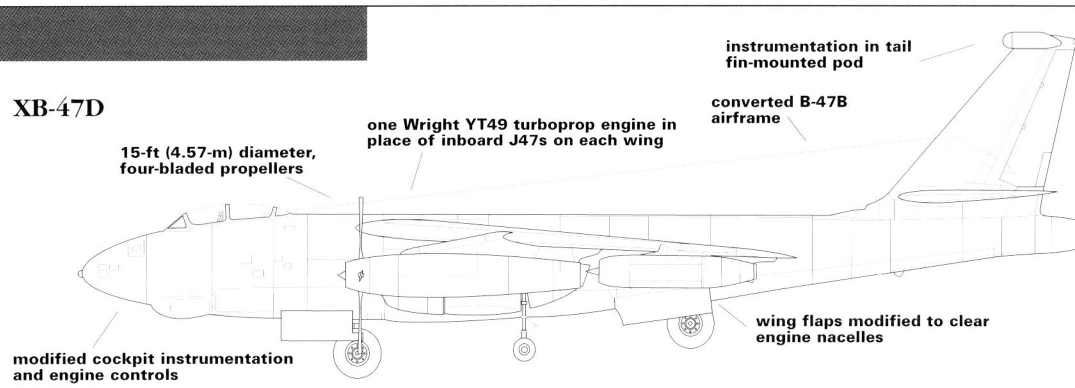

The XB-47D's propellers were some 15 ft (4.57 m) in diameter, each paddle blade measuring 24 in (61 cm) in width. Surprisingly, the XB-47D's performance was similar to that of the conventionally-powered Stratojet.

Landing performance was improved with the addition of the XB-47D's fully-reversible propellers. As the Stratojet was not equipped with air brakes, Stratojet pilots often lowered the aircraft's undercarriage to reduce speed, as seen in this view.

B-47E

The definitive Stratojet was the B-47E (Boeing Model 450-157-35). Three out four of all Stratojets built would be of this type. Like all other Stratojets, it had a wing span of 116 ft (35.36 m), and a wing area of 1,428 sq ft (132.7 m²). It was 107 ft 1 in long (32.6 m) and had a tail height of 28 ft (8.53 m). It weighed 80,756 lb (36631 kg) empty, and had a gross weight of 198,180 lb (89895 kg). The engines were six General Electric J47-GE-25s delivering up to 6,000 lb (26.69 kN) of thrust. As many as 33 JATO units were also available, but unlike the JATO units on previous Stratojets, these were jettisonable.

The B-47E had an unrefuelled ferry range of 4,035 miles (6495 km) and a combat radius of 2,013 miles (3240 km) with 10,845 lb (4919 kg) of payload. The service ceiling was 33,100 ft (10090 m), with a combat ceiling of 40,500 feet (12350 m). The cruising speed was 498 mph (802 km/h), with a top speed of 607 mph (977 km/h). Actually, under ideal conditions, the aircraft cruised much faster. In January 1957, a transcontinental average speed record of 710 mph (1143 km/h) was set flying from March AFB in California – with strong tail winds – to Hanscom AFB in Massachusetts in 3 hours 47 minutes.

The B-47E incorporated all of the structural and avionics upgrades that had been cobbled piecemeal into the B-47B at various stages of its production lifetime, including the accommodation for an aerial refuelling receptacle. Other detail changes included the A-5 gun control as standard equipment, and 20-mm cannons, rather than .50-calibre machine-guns Like the later B-47B, however, the later B-47Es contained refinements not present in the first of their breed.

Progressive modification resulted in four sub-variants of the B-47E. The **B-47E-I** introduced water/alcohol-injected -25A engines, delivering 7,200 lb (32.03 kN) of thrust, Phase II ECM equipment (denoted by the small circular antenna on the rear fuselage of aircraft so-equipped), 33-bottle RATO gear and a 200,000-lb (90718-kg) maximum take-off weight. **B-47E-II** modifications included the addition of Phase III ECM gear. This comprised an AN/ALT-6 jammer housed in a bulged fairing beneath the rear fuselage and antennas housed in wingtip extensions. A three-phase AC electrical power supply was the main modification introduced in the **B-47E-III**, while the **B-47E-IV** added an avionics upgrade (including AN/APS-54 and AN/APG-39 search and targeting radar, as well as the MA-7A bombing and navigation system) and an increased gross weight of 230,000 lb (104326 kg). The latter allowed a greater fuel load to be carried (made possible by the use of a shorter bomb bay) and a maximum warload of 25,000 lb (11340 kg).

The B-47E was the only Stratojet variant that was mass-produced using the Boeing-Douglas-Lockheed production pool. The three companies produced 691, 264 and 386, respectively. The last B-47E was delivered to the SAC on 24 October 1956.

B-47 production was undertaken by Boeing (Wichita), Douglas (Tulsa) and Lockheed (Marietta), as follows:

B-47E-55-BW	51-2357/2411 (55)
B-47E-60-BW	51-2412/2445 (34)
	51-5214/5234 (21)
B-47E-65-BW	51-5235/5257(23)
	51-7019/7050 (32)
B-47E-70-BW	51-7051/7064 (14)
B-47E-75-BW	51-7065/7083 (19)
	51-17368/17386 (19)
B-47E-80-BW	52-394/431 (38)
B-47E-85-BW	52-432/469 (38)
B-47E-90-BW	52-470/507 (38)
B-47E-95-BW	52-508/545 (38)
B-47E-100-BW	52-546/583 (38)
B-47E-105-BW	52-584/620 (37)
B-47E-110-BW	53-2261/2296 (36)
B-47E-115-BW	53-2297/2331 (35)
B-47E-120-BW	53-2332/2367 (36)
B-47E-125-BW	53-2368/2402 (35)
B-47E-130-BW	53-2403/2417 (15)
	53-4207/4244 (38)
B-47E-135-BW	53-6193/6244 (52)
B-47E-1-DT	52-019/028 (10)
B-47E-5-DT	52-029/041 (13)
B-47E-10-DT	52-042/054 (13)
	52-055/058 (4)
B-47E-15-DT	52-059/081 (23)
B-47E-20-DT	52-082/111 (30)
B-47E-25-DT	52-112/120 (9)
	52-146/176 (31)
B-47E-30-DT	52-177/201 (25)
B-47E-35-DT	52-1406/1417 (12)
	52-2028/2040 (13)
B-47E-40-DT	53-2090/2103 (14)
B-47E-45-DT	53-2104/2117 (14)
B-47E-50-DT	53-2118/2131 (14)
B-47E-55-DT	53-2132/2144 (13)
B-47E-60-DT	53-2145/2157 (13)
B-47E-65-DT	53-2158/2170 (13)
B-47E-70-DT	53-2041/2089 (49 cancelled)
B-47E-5-LM	51-15804/15810 (7)
B-47E-10-LM	51-15811/15812 (2)
	52-202/207 (6)
B-47E-15-LM	52-208/220 (13)
B-47E-20-LM	52-221/235 (15)

B-47E (late production)

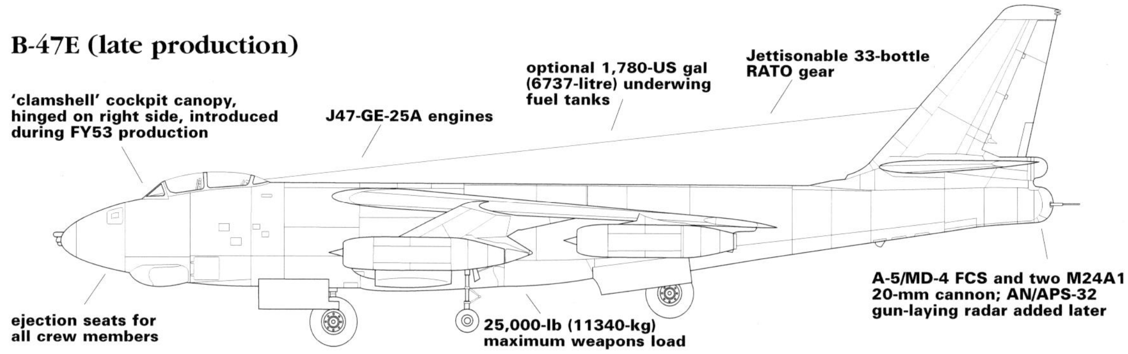

B-47E-25-LM	52-236/260 (25)
B-47E-30-LM	52-261/292 (32)
B-47E-35-LM	52-293/330 (38)
B-47E-40-LM	52-331/362 (32)
B-47E-45-LM	52-363/393 (31)
B-47E-50-LM	52-3343/3373 (31)
B-47E-55-LM	53-1819/1849 (31)
B-47E-60-LM	53-1850/1880 (31)
B-47E-65-LM	53-1881/1911 (31)
B-47E-70-LM	53-1912/1942 (31)
B-47E-75-LM	53-1943/1972 (30)
B-47E-80-LM	53-1973/2027 (55 cancelled)

Below: In-flight refuelling capability, as introduced in the B-47E, was an important addition. Seen from the boom operator's position aboard a KC-97, a B-47E tops up its tanks in this view. The speed differential between the two types made refuelling problematic.

Differences between the first B-47Es and later examples are clear in these views. The B-47E-55-BW (above) bore a strong resemblance to the B-47B and lacked many 'E' features; these were added under Project Turnaround from 1953 (resulting in the B-47E-I) and included ejection seats, external RATO gear, a new bomb/nav system, reinforced landing gear, in-flight refuelling equipment and an approach parachute. The B-47E-65-BWs (below) were equipped to the later standard and were outwardly representative of the bulk of B-47E production.

ECM equipment of one type or other was added to most B-47Es. 53-1884 (above) has Phase III ECM gear, with extended wingtips and a bulged jammer fairing under the aft fuselage. During 1955 around 100 B-47E-Is were equipped with pairs of fuselage-mounted Tee Town ECM pods, each carrying four AN/ALT-6B jammers, for use by the 303rd and 509th Bomb Wings (below).

Government-owned plants produced B-47E-DTs (upper left) and B-47E-LMs. Lockheed-built 51-15808 (left) was the fourth example to emerge from the Georgia factory; note the internal RATO gear, later deleted.

YDB-47E/DB-47E

Four B-47Es were experimentally converted as launch platforms for the Bell GAM-63 (previously B-63) Rascal air-to-surface missile. The first two (51-5219 and 51-5220, right) were converted as Boeing's Model 450-167-50 and delivered in January 1954 under the designation YDB-47E. The next two (53-2345 and 53-2346) were retrofitted as Model 450-172-50 aircraft, and designated as DB-47E. (These later served as drone directors with the 3205th Drone Group at Eglin AFB.) Plans to produce a large number of DB-47Es were abandoned in 1958.

EB-47E

As it became clear that more powerful ECM equipment would be needed to ensure that an attacking force of B-47s reached its target, a requirement for a dedicated ECM variant emerged. A number of different EB-47E configurations existed, the first equipped with the Phase IV ECM package, comprising 16 AN/ALT-6B jammers on a bomb bay-mounted cradle. These so-called Blue Cradle aircraft were operated by the 376th BW.

Blue Cradle was followed by the Phase V EB-47E, operated by the 301st BW, which carried a pressurised capsule in its bomb bay for two EWOs assigned the task of operating 13 jammers of various types. These aircraft offered the ability to focus their jammers on to specific 'threats', rather than providing blanket jamming coverage.

It is believed that a total of 40 aircraft was converted to Phase IV or V standard.

Two other aircraft assigned the EB-47E designation were ex-376th BW B-47Es 52-410 and 52-412, loaned to the US Navy in 1965. Flown by Douglas flight crews these aircraft were employed in support of the Fleet Electronic Warfare Support Group (FEWSG), chiefly from NAS Point Mugu and Oceana, testing ECM equipment and acting as aggressors in fleet exercises. The aircraft were assigned the BuNos 24100 and 24102, respectively, and were the last operational Stratojets, finally retiring in December 1977.

Below: BuNo. 24100 was one of the two EB-47Es operated by Douglas in support of the US Navy's FEWSG. Equipped with various items of ECM equipment, these aircraft often carried external pods on the underwing auxiliary fuel tank pylons. Other antennas and chaff dispensers were also carried.

Above: In their role as telint-gathering aircraft, two of the somewhat enigmatic EB-47E(TT)s stood on alert at Incirlik (one primary crew and a back-up crew), ready to scramble at a moment's notice in the event of a Soviet rocket or IRBM launch. Missions were flown over the Black Sea (for Kapustin Yar) or northeastern Iran (for Tyuratam) at as high an altutude as possible.

EB-47E(TT)

Operated by Detachment 4, 55th SRW, and based at Incirlik AB, Turkey, the Tell-Two EB-47E(TT)s were telemetry intelligence (telint) gatherers, employed to collect data from the IRBM facility at Kapustin Yar and the Soviet space centre at Tyuratam during vehicle launches.

Three B-47Es were converted to Tell-Two standard (including 53-2316 and 53-2320) with a capsule in the bomb bay for a pair of EWOs and an external 'armrest' fairing fitted either side of the nose of the aircraft. The latter covered much slimmer, so-called 'chopstick' antennas and were later removed as they did little to improve antenna performance and produced discernible drag.

The EB-47E(TT)s served between about 1958 and 1967.

JB-47E

This designation was applied to a number of B-47Es modified for various test programmes, the 'J' prefix indicating temporary test status. Among the aircraft employed was JB-47E 53-2280, the first USAF aircraft to fly using a 'fly-by-wire' primary flight control system. Successfully tested by the Wright Air Development Center between 1967 and 1969, the aircraft has since been preserved at the USAF Museum.

NB-47E

The sole NB-47E was another aircraft bailed to the US Navy for a specific task, in this case employment as a flying engine testbed. B-47E 53-2104 (right) carried a General Electric TF34-GE-2 (on the fuel tank pylon under the port wing) destined for service in the Lockheed S-3 Viking. Operated between 1969 and 1975, this aircraft was among the last to be despatched to the MASDC at Davis-Monthan AFB and was later preserved.

QB-47E

Fourteen RB-47Es were converted as remotely-controlled drones between 1959 and 1960. Operated by the 3205th Drone Director Group at Eglin AFB, the QB-47s were employed as targets for surface-to-air missile development programmes, including the Boeing IM-99 (originally F-99) Bomarc. Controlled by DB-47E or other director aircraft, these aircraft were not initially intended to be destroyed, though at least one was unintentionally downed by a Bomarc. Most of these drone aircraft were later destroyed during missile testing; two survived to be redesignated JQB-47E.

In a costly conversion programme the 14 QB-47Es were equipped with control equipment, six 16-mm movie cameras and other data collection equipment, ECM gear, chaff dispensers and explosives (for use if control of the aircraft was lost).

JQB-47E

Two QB-47Es (including 53-4256) were involved in test work surrounding the development of the Hughes AIM-47 long-range air-to-air missile (and the associated AN/ASG-18 radar set) for use by the Lockheed YF-12.

The pod fitted to the fuel tank pylon on the wing of 53-4256 contained 16-mm movie cameras used to record 'near misses' by missiles under test. This aircraft survived two AIM-47 test shots.

RB-47E

Built by Boeing at Wichita, the RB-47E was the factory-built reconnaissance version of the B-47E. Begun as RB-47Bs, the RB-47Es incorporated many of the upgrades introduced in the E-model and were soon redesignated as such.

The type's main distinguishing feature was its nose. This was extended by 34 in (86 cm) to accommodate an air conditioned compartment containing four types of camera. Additional cameras and 10 photoflash bombs could be accommodated in the bomb bay. Cameras were operated by the navigator/bombardier, now known as the navigator-photographer.

The first RB-47E made its debut flight in August 1953. A total of 240 RB-47Es was built, with 52 manufactured as Model 450-126-29, and 188 as Model 450-158-36. the type was withdrawn from 1957.

RB-47E-1-BW	51-5258/5264 (7)
RB-47E-5-BW	51-5265/5270 (6)
RB-47E-10-BW	51-5271/5276 (6)
	51-15821/15827 (7)
RB-47E-15-BW	51-15828/15853 (26)
RB-47E-20-BW	52-685/719 (35)
RB-47E-25-BW	52-720/754 (35)
RB-47E-30-BW	52-755/789 (35)
RB-47E-35-BW	52-790/825 (36)
RB-47E-40-BW	52-3374/3400 (27)
RB-47E-45-BW	53-4245/4264 (20)

Below: The second RB-47E-1-BW (nearest the camera) formates with a B-47E-65-BW. The lengthened nose of the RB-47 is clearly evident.

RB-47E

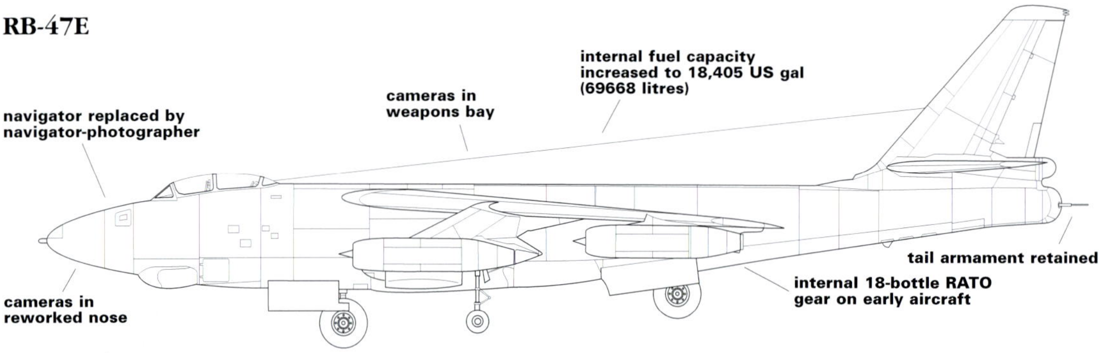

WB-47E

A total of 34 B-47Es was converted as weather reconnaissance aircraft, becoming operational with the Military Air Transport Service's Air Weather Service in March 1963.

Modifications, carried out by Lockheed's Marietta plant, included the fitting of nose-mounted cameras (for cloud formation recording) and a sensor pod in the bomb bay. The latter was equipped with air sampling and data recording equipment. The aircraft's defensive fire control system and tail gun were removed.

Gradually replaced, from 1965, by such types as the Lockheed WC-130 and Boeing WC-135, the WB-47E was finally retired on 31 October 1969; the WB-47Es were the last operational Stratojets in the USAF.

Outwardly there was little to distinguish a WB-47E from its bomber brethren, apart from the 'WEATHER' tail band. WB-47Es were operated by the 53rd and 55th Weather Reconnaissance Sqns; a 55th WRS aircraft is pictured.

JRB-47E

One heavily modified RB-47E (53-4262) was assigned to the Rome Air Development Center (RADC) at Griffiss AFB, New York, during the late 1950s. Fitted with a deeper undernose radome, a number of external antennas on the lower and upper fuselage and a nose probe, the aircraft may (it has been suggested) have been employed to test equipment for the EB-47E(TT).

ETB-47E

This designation was applied to an unknown number of B-47Es that were used for training duties during the late 1950s and early 1960s. Equipped with a fourth crew seat for an instructor, these aircraft are belived to have replaced older TB-47B training aircraft.

JTB-47E

Two aircraft with this designation (but also sometimes referred to as JRB-47Es) were employed by the Rome Air Research and Development Center (RADC) in support of classified ECM programmes; each aircraft was equipped with an array of electronic countermeasures devices. Some sources suggest that these aircraft were used to test the effectiveness of CONUS early warning radar operated by NORAD.

YB-47F/KB-47G

In order to evaluate the British 'hose-and-drogue' aerial refuelling configuration (as an alternative to the Boeing 'flying boom' employed by KB-29s, KB-50s and KC-97s), two B-47s were converted during 1953.

The YB-47F was a modified B-47B (50-009) configured with a nose probe as the receiver, while the KB-47G was also a reworked B-47B (50-040), converted to tanker configuration with a fuel tank and hose/drogue gear in its bomb bay.

As well as refuelling the YB-47F during trials of the system at the Wright Air Development Center, the KB-47G was also used to experimentally refuel Republic F-84G and F-101 fighters.

The programme was abandoned in 1954; the B-47 was an impractical choice as a tanker as it simply lacked sufficient fuel capacity. A two-drogue KB-47 was proposed during 1956 for use by TAC-operated, probe-equipped fighters that were experiencing problems refuelling from KB-50s, though this too came to nought and was cancelled in July 1957.

The tanker trials aircraft, seen here during a refuelling trial, were known as 'Ma' (KB-47G) and 'Paw' (YB-47F);

RB-47H

Similar, at least in appearance, to the RB-47E, the RB-47H was a spcialised Elint-gathering platform designed for such missions as locating and identifying enemy ground-based radar sites and their emissions. It carried an optional three-man pressurised compartment in the bomb bay. The RB-47H weighed 83,462 lb (37859 kg) empty and had a maximum take-off weight of 220,000 lb (99790 kg). The first RB-47H made its debut in June 1955, and a total of 44 were ordered, although a dozen of these were cancelled.

Most RB-47Hs were upgraded from 1961 to 'Mod 44' or 'Silverking' standard, with more capable sensors.

RB-47H 53-4296 of the 55th SRW made its last flight, to Davis-Monthan AFB, on 29 December 1967, marking the end of SAC Stratojet operations.

RB-47H-1-BW	53-4280/4309 (30)
RB-47H-1-BW	53-4310/4321 (12 cancelled)
RB-47H-1-BW	53-6247/6248 (2)

RB-47H (as built)

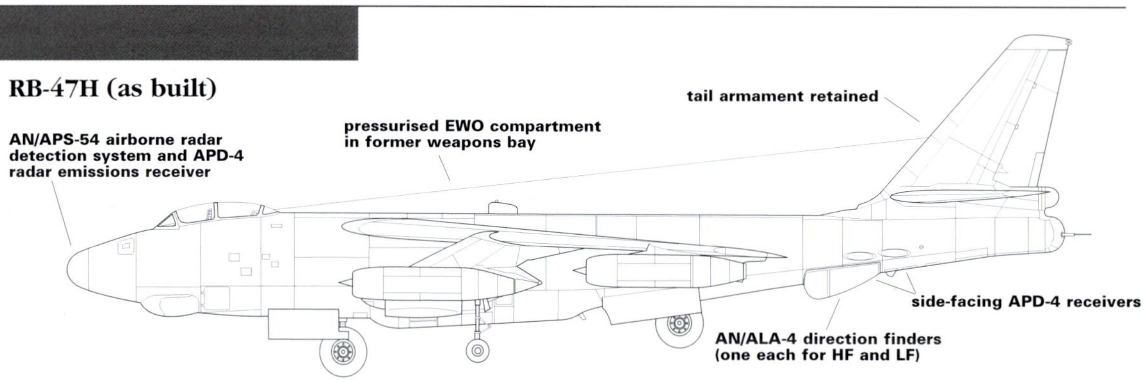

Below: This RB-47H, pictured at RAF Upper Heyford during 1967, carries an ALD-4 receiver pod, on a pylon attached to the starboard side of its rear fuselage. ALD-4 was fitted as part of the 'Silverking' upgrade.

Above: The last RB-47H to be retired was soon resurrected from the 'Boneyard' and used as a testbed for avionics destined for the General Dynamics F/FB-111. An F-111 nose cone was fitted to the aircraft, as well as underwing pods. The unique aircraft is pictured at Eglin AFB during 1976; it has since become an exhibit at the base's museum, its 'needle nose' having been replaced by a conventional nose cone.

ERB-47H

Three B-47Es – 53-6245 (right), 53-6246, and 53-6249 – were converted for specialised operations under the designation ERB-47H. As the ERB-47H carried more mission equipment in its bomb bay pod, the EWO complement was reduced to two. Only select crews were selected to fly these aircraft; the type was retired concurrently with the RB-47H.

YB-47J

One B-47E (identity unknown) was retrofitted with the MA-2 radar bombing/navigation system, developed for the B-52, under the designation YB-47J. Some sources suggest that there may have been more than one aircraft equipped with MA-2 during testing, though one aircraft is known to have been redesignated.

RB-47K

The last 15 RB-47Es ordered by the USAF were equipped with an additional high-resolution SLAR and air sampling equipment, while retaining the E's photo-reconnaissance equipment. Designated RB-47K (right), these aircraft were assigned to the 338th SRS and had meteorological data collection and weather reconnaissance as their primary missions.

EB-47L

In 1963, a total of 35 B-47E aircraft was modified with AN/ARC-89 communications relay transceivers for service as command and control relay aircraft. Designated EB-47L, these relayed communications between USAF 'Looking Glass' EC-135 command posts and US Navy EC-130Qs and ground stations. New technologies saw the EB-47Ls phased out during 1965/66.

Canadair CL-52

Only one Stratojet ever flew in foreign markings – former TB-47B 51-2059 (right) – loaned to the Royal Canadian Air Force in 1956 as a flying test-bed for the 26,000-lb (115.7-kN) thrust Orenda Iroquois turbojet, developed in parallel with the ill-fated Avro-Canada CF-105 Arrow, which it was to have powered.

On 8 February 1956 Canadair was awarded a contact by Orenda to convert TB-47B 51-2059 for the task; the aircraft was assigned the company designation CL-52.

Eight days later 51-2059 arrived at Canadair's facility at Cartierville Airport, just outside Montreal, and was officially taken on strength by the RCAF on 5 March 1956, assigned the non-standard serial number, X-059, a truncation of its USAF serial.

The starboard side of the aft fuselage was selected as the location for the engine and its nacelle, this position allowing the engine's exhaust to clear the aircraft structure and offered the least risk to aircraft performance, handling and structure. That said, a number of modifications were required. The dry thrust of the Iroquois exceeded that of four J47s combined and the torsional loads generated were compounded further by the engine's offset location, which tended to twist the aircraft about its vertical axis. This necessitated reinforcement of the rear fuselage, which consisted of doubling the aft fuselage skins, reinforcing the existing bulkheads and adding new internal bulkheads and external longerons running the full length of the fuselage.

Mounting an offset engine presented Canadair engineers with a number of problems, the first of which was balancing the detrimental effect of asymmetric thrust. It was first thought that yawing moments caused by the Iroquois could be balanced solely by using differential thrust from the J47s. But it was also feared that the cumulative thrust of all J47 engines could, under certain flight conditions, accelerate the aircraft to near or beyond its buffet boundary. Alternative means of achieving yaw balance were thus investigated and this included applying rudder, and differential use of spoilers, extending the landing gear and wind-milling drag from shut-down engines. Conversely, instantaneous loss of Iroquois thrust presented problems of its own.

A wind tunnel programme was therefore launched to investigate the CL-52's performance and handling, flight envelope, aerodynamic loads, and optimisation of the nacelle/pylon shape.

By the end of 1956, work on the CL-52 was essentially complete. The flight crew assigned to the CL-52 were all Orenda personnel and consisted of Mike Cooper-Slipper (Orenda chief test pilot), Len Hobbs (co-pilot) and Johnny McLaughlin (flight test engineer). In 1956 all were sent on a 10-week B-47 course at McConnell AFB. On 15 April 1957 the CL-52 made its first flight, from Canadair to Orenda in Malton near Toronto.

The Iroquois accumulated 31 hours of in-flight running time on the CL-52 before the programme was terminated. Most flights were routine, though on its only full-power flight the Iroquois suffered a fan blade failure. The CL-52 sustained damage to elevator and rudder controls, as blades entered the fuselage, but landed safely.

On 20 February 1959 the Arrow/Iroquois project was cancelled by the Canadian government. On 22 May 1959 the CL-52 was struck-off charge by RCAF and during June and July Orenda, Tinker AMC and Boeing personnel worked to restore the aircraft to near TB-47 configuration for its return to the US; in August 1959, Cooper-Slipper flew the aircraft to Davis-Monthan AFB where it was eventually scrapped.

Terry Panopalis

In order to maintain the aircraft's centre of gravity, the aft-mounted Iroquois (4,680 lb/2123 kg) and nacelle (1,380 lb/626 kg) were counterbalanced by 8,160 lb (3701 kg) of lead ballast in the nose section, radar compartment and in the cavities left vacant by the removal of two fuel cells. A 2,000-lb (907-kg) instrumentation package was installed in the bomb bay.

The CL-52's pilot was tasked with flying the aircraft to and from the test range, but control during Iroquois tests rested with the co-pilot and his station was modified accordingly. An Iroquois throttle quadrant was installed and had four stops: cut-off, start and idle, 100 per cent RPM and maximum afterburner. Monitoring of the Iroquois was the responsibility of the flight test engineer seated in the bomb/navigator's station and appropriately equipped with test instrumentation and electrical control panels.

The last Starfighters

Lockheed F-104s of the Aeronautica Militare Italiana

Above: Finished in recent years in an air superiority grey colour scheme, the AMI's surviving F-104S/ASA-Ms will soldier on until the Eurofighter Typhoon becomes available in about 2004. These are 5° Stormo F-104S/ASA-Ms based at Cervia.

On 3 March 1962, a Lockheed F-104G Starfighter (serial MM6501) took off from Palmdale airport, California, on an exhibition flight, over the base. The aircraft wore an Italian roundel and was flown by Captain Bonazzi of the Aeronautica Militare Italiana (AMI, Italian air force). A few minutes before the fight, the aircraft had been officially handed over to the AMI; after the flight, it was disassembled for shipment to Italy, where it served as a pattern aircraft for local production. The career of the Starfighter in Italy began the following year and today, almost 40 years on, the Starfighter is still the country's primary air defence aircraft.

In fact, the Starfighter has covered not only air defence, but also the tactical bombing and nuclear strike roles for three of the last four decades; it was only with the introduction into service of the Panavia Tornado in the mid-1990s that the AMI's F-104s relinquished their bomber tasking.

The choice of the Starfighter dates back to the mid-1950s, when NATO countries (with the exception of the UK and France, which had their own aircraft in development) were looking to purchase a first-generation, supersonic multi-role fighter capable of delivering a tactical nuclear weapon. The decision regarding a new fighter for the AMI was not an easy one, as the other aircraft on offer included the best available in Europe (the French Mirage III, British Lightning and Swedish Draken) and the US (the F-4 Phantom, F-8 Crusader, F-104 Starfighter, F-106 Delta Dart and F-101 Voodoo). The Italian air force urgently needed to replace two different types – fighters (F-86E) and fighter-bombers (F-84F and RF-84F) – though finances dictated that only one type could be selected.

Most contenders were ruled out because of their cost or because they were not truly multi-mission aircraft. Another matter for consideration was that the NATO member air forces of Germany, Belgium and Holland had selected the F-104 as their main fighters, and established an international European consortium for its eventual production; the AMI command thought that including Italy in this undertaking would further extend the concept of NATO standardisation, reduce production costs and bring expertise to Italian industry.

Contract signed

In December 1960 signatures were put to a contract; the Starfighter was selected and Italy would enter the newly-formed European industry consortium. This was organised into four production groups; Italy, being associated with Belgian firms Avions Fairey and SABCA, become a member of Arbeitsgemeinschaft (ARGE, working group) West. Co-operation between consortium members included the exchange of experience, components and materials including complete airframes. Some 88 of the first 199 aircraft built in Italy were delivered to West Germany and 25 to the Netherlands, and several components produced in Italy flew in Belgian Starfighters.

The distinctive shape of the Starfighter seems likely to finally disappear from the front-line ranks of the world's air forces in the near future, though not before the type completes 40 years service with the AMI.

Numerous Italian companies were involved in the Starfighter project and, in November 1960, formed a separate Italian working group. FIAT led the group and was in charge of the final assembly line, fuselage production and, in conjunction with Alfa Romeo, engine production. Aerfer was charged with producing the wings and the tail, while Aermacchi, Piaggio, Officine Aeronavali, SACA and SIAI-Marchetti produced minor components, and avionics and electric systems. The first F-104G built in Italy (from US components) was flown by test pilot Vittorio Sanseverino from Torino-Caselle on 9 June 1962. The first completely Italian Starfighter flew on 5 October the same year.

The first AMI pilots were qualified in Germany at the Jever-based Waffenschule, flying the two-seat F-104F. They then moved to Grosseto, home of 4° Stormo, the first unit equipped with the new fighter. At Grosseto the new pilots flew 10 solo sorties, chased by an F-104G flown by a qualified pilot. This training syllabus lasted until February 1965, when the first 12 Lockheed-built TF-104Gs were delivered and 20° Gruppo Addestramento Operativo, similar to an operational conversion unit, was established at Grosseto.

To mark the 70th anniversary of 4° Stormo in 2001, one of the unit's surviving F-104S/ASA aircraft was specially marked with a 4° Stormo tail fin badge. Much of the aircraft's airframe has been stripped of its grey paint as a reminder of the Starfighter's appearance when the type first entered AMI service in 1963.

First deliveries - F-104G/TF-104G

Of the 200 single-seat F-104Gs assembled in Italy by Fiat (later Aeritalia), 125 were delivered to the AMI (the balance going to West Germany and the Netherlands – 50 and 25, respectively). The AMI aircraft were serialled as follows: MM6501/6599, 6601, 6603, 6608/6611, 6631/6638, 6643/6651 and 6658/6660 (MM6658 and 6660 were later rebuilt as F-104S/ASA prototypes). Of these aircraft 20 were later converted to RF-104G standard; known serials inlcude MM6514, 6525, 6529, 6540, 6545, 6550 and 6576. For conversion training 24 two-seat TF-104Gs were also acquired. Assembled by Fiat, these aircraft were serialled MM54226/54237 and 54250/54261. Six ex-Luftwaffe examples were added in the mid-1980s (MM54553/54558).

Italy's F-104s were operated only briefly in this natural metal finish (seen on aircraft from the first AMI F-104 unit – 9° Gruppo, 4° Stormo – above, and 51° Stormo, below left); NATO standard dark green/dark sea grey/light aircraft grey tactical camouflage was applied from July 1964, though high-visibility unit codes in black or white (as on the 51° Stormo machines, below right) were retained until the 1980s.

AMI HQ planned to equip eight flying groups with the Starfighter, replacing the four Canadair CL-13 Sabre fighter groups and four Republic F-84F Thunderstreak fighter-bomber groups. The first group of ground technicians was formed at Torino-Caselle, where the 4ª Aerobrigata Command created the 'Nucleo 4° A/B'; this was to address the need to train, with the aid of FIAT Aviazione engineers, the first group of technicians who would work on the Starfighter at Grosseto. The 9° Grupo of the 4ª Aerobrigata Intercettori, also at Grosseto, was the first flying unit to be equipped with the F-104G, receiving its first aircraft on 13 March 1963. The unit was declared combat ready in 1964, the same year that the second group of the 4ª Aerobrigata – the 10° Gruppo – received the F-104G. During 1965, the 4ª Aerobrigata was fully operational on Starfighters and amassed 5,000 flying hours. Delivery of Starfighters to other units was also underway.

In 1963 21° Gruppo, 51ª Aerobrigata began to convert from F-86Ks to Starfighters, and for this reason was deployed temporarily to Grosseto from its usual base at Cameri. This group's F-104Gs were the only operational Italian Starfighters to wear the group number painted in white under the cockpit, in place of the wing number. (This practice was also adopted by the 20° Gruppo OCU for some years when it operated as an autonomous training group). The two other flying groups of the 51ª Aerobrigata – the 22° and 23° Gruppi – flew the 'Kappone', as the F-86K was known in Italy, from Istrana and Rimini air bases, respectively, for a number of years.

In 1964 the 154° Gruppo, 6ª Aerobrigata Caccia Bombardieri, based at Ghedi, began converting to the F-104G and used the Starfighter in the fighter-bomber mission.

In May of that year, the 101° and 102° Gruppi of the 5ª Aerobrigata began to exchange their F-84F Thunderstreaks for F-104Gs. The 101° Gruppo flew the Starfighter for only three years as, in September 1967, the groups left the 5ª Aerobrigata and the base of Rimini to go to Cervia air base to form the new 8° Stormo, flying again with the old F-84F.

At the time of their delivery to 4° Stormo the AMI's first F-104Gs carried a natural metal finish with 'anti-radiation' white upper fuselage and wing surfaces, the latter in recognition of the Starfighter's nuclear role. By early 1964, however, only upper wings were painted and, in any case, by the end of the year tactical camouflage was being applied across the fleet.

Starfighters replace Thunderflashes

Not until February 1970 did the 3ª Aerobrigata Ricognitori, the reconnaissance unit of the AMI, equipped with the RF-84F Thunderflash, begin to convert to the RF-104G Starfighter. This version of the Lockheed fighter was equipped with a trimetrogon camera system in the forward fuselage, comprising one vertical and two lateral cameras. The first group to be equipped with the RF-104G was the 28° Gruppo; the two other flying groups of the 3ª Aerobrigata, the 132° and the 18° Gruppi, received the Starfighter only in 1972 and 1974, respectively, and flew the RF-84F Thunderflash until those dates. Most Starfighters delivered to these two groups were normal F-104Gs, received from former fighter-bomber groups that converted to the F-104S, with minor modifications to allow the use of the external Dutch-built Orpheus infra-red line scan camera pod.

The complexity of the new aircraft was responsible for low serviceability rates in the first years of its use, as the Starfighter needed 40 maintenance hours and 12 ground specialist hours for every flight hour, though efficiency improved eventually. The main problem was the age of other fighters and bombers in the AMI fleet and the fact that all the F-84Fs, F-86Ks and RF-84Fs needed replacement in the space of a few years. Another consideration was the NATO requirement of 18 aircraft for each group: the AMI ordered a total of 145 Starfighters for nine groups, leaving a shortfall of around 20 aircraft. Italy ordered 125 F-104G single-seaters, including 20 RF-104Gs, and 24 TF-104G two-seaters; in 1984 the AMI purchased six TF-104Gs formerly operated by the Luftwaffe.

4 June 1994 saw the retirement of the last of the F-104Gs operated by the AMI. On this date the Starfighter conversion unit – 20° Gruppo, 4° Stormo – flew the last 'G' sortie in '4-49', pictured here.

Right: The AMI's original 24 two-seat Starfighters were all locally-built TF-104Gs and have served exclusively with 20° Gruppo at Grosseto. Six ex-Luftwaffe aircraft were added in the mid-1980s; in all 15 survived to see conversion to TF-104G-M standard in the late 1990s.

BVR fighter requirement

The search for a new fighter capable of BVR (beyond visual range) operations began as soon as the Starfighter was introduced into service. The AMI sought a new-technology war machine that could enter the inventory around 1970.

The Starfighter's career in Europe had not been a particularly happy one, especially in Germany (where, in 1965, 28 aircraft were lost in crashes attributed to a number of causes), and many air forces that operated the F-104G decided to replace it with other types – in the case of the Luftwaffe, the F-4F Phantom. For the Italian government, questions of local manufacture and the cost of the aircraft were of primary concern. The Italian aeronautical industry received considerable technical benefits from the Starfighter programme, reaching new operational and technical levels of expertise, and the government did not want to lose this newly acquired capacity. Also of importance was the enormous logistic effort required in supporting the F-104G; this needed to be set against the complexity of new procedures, structures and training systems that would need to be established if the choice was made to 'start again' with a new aircraft. The opportunity to retain some of the benefits but to obtain a more powerful aircraft with only few technical differences from the Starfighter was an important factor in the final choice.

Italy evaluated the McDonnell F-4C Phantom, Dassault Mirage IIIC-2, Northrop F-5A and Lockheed CL-901, plus a new Starfighter version with the new J79-GE-19 engines capable of 79.61 kN (17,901 lb) thrust, some 9.8 kN (2,205 lb) more powerful than the earlier J79s of the F-104G.

The new Starfighter promised enhanced accel-

Above: Twenty of the F-104Gs were converted to RF-104G standard, with internally-mounted cameras. In this view of a 102° Gruppo/5° Stormo aircraft its camera ports are visible in a fairing on the underside of the forward fuselage.

Right: Dutch-built Orpheus reconnaissance pods were introduced in 1977 and carried by F-104Gs of 132° Gruppo, 3° Stormo (pictured) and 18° Gruppo.

Sparrow-equipped F-104S

F-104S production (excluding the two F-104Gs converted to 'S' standard in the US), ran to 246 examples, 205 of which were delivered to the AMI (MM6946 replacing 6766, which crashed prior to delivery) and 40 to the Turkish air force. AMI serial numbers were as follows: MM6701/6850, 6869/6881, 6886, 6887, 6890 and 6907/6946.

***Left:** Two F-104Gs were shipped to Lockheed's Palmdale factory for conversion to F-104S standard. MM6658 '5-41' was the first to be reworked (as an F-104S/CB fighter-bomber) and is seen here during an early post-conversion test flight, carrying six 750-lb bombs. Two of the M117s are carried on fuselage-mounted pylons, added under the engine intakes, during conversion to 'S' standard.*

***Below:** In otherwise clean configuration and bereft of tip tanks, prototype '5-41' shows off the most important aspect of the F-104S's new capability – its AIM-7E Sparrow AAMs.*

***Below:** This production F-104S/CI interceptor carries Sidewinders on its fuselage pylons. An extra pair of outboard underwing stations was also added to the F-104S for Sparrow carriage.*

eration and manoeuvrability, improved avionics and, particularly, more effective weapons, as it could accommodate the medium-range AIM-7 Sparrow. It was evident that the F-4 Phantom was the best fighter in terms of its load capacity, performance, range and avionics, but the Phantom was not as fast (Mach 2.16 compared to 2.4 of the Starfighter), even with two engines. The F-4 also required two crew members and was very expensive, though the strongest factor against it was that the Phantom programme offered little benefit in terms of production and technology transfer to Italian industry.

The Mirage III was bypassed because it cost 57 per cent more than the F-104G, the proposed IIIC-2 version was not in production at the time of the selection, and the performance of its new medium-range MATRA R.530 missiles was, in any case, similar to that of the Sparrow. The F-5A was not selected because it was in the same class as the Fiat G91Y and was incapable of all-weather operations. Although it cost 15 per cent less than the F-104G, it would bring insufficient work to Italian industry and offered no new technology.

Costs win the day – F-104S selected

In light of all these factors, the Lockheed proposal was thought to be the best, considering the overall benefits it would bring and that the total cost of the programme was 225 million lire for the planned 165 aircraft. The aircraft's per-hour cost was estimated to be 9,880 lire, of which 4,750 lire would be required for foreign material. One hundred per cent of the airframe, 50 per cent of the accessories and the electronic equipment, and 43 per cent of the engine components were to be produced in Italy. The only negative aspect seemed to be that the F-104G was not held in high regard in Europe – but that was considered to be a price worth paying for such a new and advanced machine.

The AMI requested that Lockheed make certain changes to the CL-901 to better suit its requirements, Fiat working on the project directly with Lockheed. Two F-104Gs, the last aircraft produced by Fiat (serial numbers 6660 and 6658), were sent to the US to be modified as the first examples of the F-104S – 'S' indicating the capability to launch a Sparrow III missile.

Although the F-104S and G have similar dimensions, the S has two small fins in addition to the single underfuselage fin to the rear of the aircraft, and can carry an increased load of 3400 kg (7,496 lb) of stores, thanks to its nine hardpoints (including the wingtip stations). Another difference is the air intakes: in order to increase airflow these were moved aft 23 mm (0.9 in) and their diameter increased slightly. Their interiors were coated with steel, allowing higher temperatures that enabled the aircraft to reach a maximum speed of Mach 2.2 at

F-104S avionics and armament fit differed according to the role assigned to the aircraft. F-104S/CI interceptors, equipped to fire Sparrow AAMs, had their M61A1 cannon and a fuel tank removed to make room for the missile's guidance electronics.

11000 m (36,090 ft), compared to the F-104G's Mach 1.92 at the same altitude. The aircraft would be powered by a J79-GE-19 engine, designated J79-J1Q in NATO service.

Two configurations

The F-104S version was introduced in two different configurations: one dedicated to the air defence role and one to the fighter-bomber role.

In the air defence version, known as F-104S/CI, the radar was replaced with a new version of the FIAR/NASARR F15G, interfaced with a continuous wave illuminator to guide the Raytheon AIM-7E Sparrow. The missile's guidance systems were accommodated by removing the M61A1 gun and a 462-litre (102-Imp gal) fuel tank. The fighter-bomber F-104S/CB was fitted with the FIAR/NASARR R21G-H radar, underwent avionics changes including the installation of a radar altimeter, had internal armament consisting of the M61A1 Vulcan 20-mm cannon with a maximum 725 rounds, and carried an extra 462-litre (102-Imp gal) fuel tank in the fuselage.

To compare the payloads of the F-104G and F-104S, the former could carry three M117 340-kg (750-lb) bombs over a maximum range of 689 km (428 miles) in Lo-Hi-Lo profile or 474 km (295 miles) in Lo-Lo-Lo profile; the F-104S can carry the same weapons in the same mission profile over 990 and 682 km (615 and 424 miles). The normal weapons load of the F-104S/CB included BLU-1/B 340-kg bombs, Mk 81 (113 kg/250 lb), Mk 82 (226 kg/500 lb), Mk 83 (454 kg/1,000 lb), and M117 340-kg bombs; the F-104S can also carry a single Mk 84 weighing 907 kg (2,000 lb), an SUU-21A training pod and the 'special bomb load' (i.e. nuclear weapon) on the fuselage hardpoint. All F-104Ss are equipped with a Martin-Baker IQ-7A 'zero-zero' ejection seat.

Apart from the front-line units, 311° Gruppo, Reparto Sperimentale di Volo (RSV, experimental flight unit) at Pratica di Mare has operated Starfighters since the early days of their AMI service. 'RS-04' is an F-104S, complete with nose-mounted 'barber's pole' instrumentation probe and the RSV badge on its tail.

On 30 December 1966, F-104S MM6658 took off from Palmdale airport; the first Italian-built F-104S (6701) took off from Torino-Caselle on 30 December 1968. The AMI initially bought 165 F-104Ss, in two batches totalling 82 and 83 aircraft. In 1974 the AMI purchased another production batch of 40 machines, bringing the overall total to 205 F-104Ss (not including one additional airframe built to replace an earlier machine which crashed before delivery).

Low-rate 'S' production

The production rate of the new version was very low, due to financial constraints and the intention to keep the assembly line open for as long as possible in the hope that new orders may be forthcoming, in addition to a Turkish air force order for 40 F-104Ss placed in 1974. Thus, the 100th F-104S was delivered to the AMI on 16 January 1973 and the 200th on 18 May 1976; production continued until March 1979.

The first delivery of the new fighter to operational AMI groups was in June 1969 to the 22° Gruppo CIO (Caccia Intercettori Ognitempo – All Weather Fighter Interceptor) of the 51° Stormo. Normally based at Istrana, the unit was deployed to Cameri until September of that year. The group was not declared combat ready until January 1972. The second unit to receive the F-104S was the 9° Gruppo CIO, 4° Stormo, in 1970; during the same year, the 36° Stormo based at Gioia del Colle, near Bari, began to convert its 12° and 156° Gruppi to the Starfighter. 21° Gruppo of 53° Stormo, based at Cameri, was declared operational at the end of 1972, and during 1973 23° Gruppo, 5° Stormo, based at Rimini, converted from the last F-86Ks to the F-104S. The 10° Gruppo, 9° Stormo, based at Grazzanise, was declared combat ready in 1974.

Upgraded F-104S/ASA

Finished in an attractive shade of blue with yellow Aeritalia logos and an 'ASA: Aggiornamento Sistemi d'Armam' legend on either side of the tail, the F-104S/ASA prototype (right) was completed in 1984; the 147th and last example was redelivered to the AMI in 1993.

Below: This 9° Gruppo/4° Stormo F-104S/ASA shows off the armament improvements introduced in this upgraded variant, which was equipped with the locally-produced Selenia Aspide 1A and all-aspect AIM-9L AAMs. Not visible is the reinstated M61A1 cannon, mounted on the port side of the aircraft. Plans existed for the integration of the German AS34 Kormoran anti-ship missile with the ASA, but these come to nought, presumably on cost grounds.

Below: Publicity photographs of the 100th F-104S/ASA included the prototype; note the 'bolt on' refuelling probe, test flown but not adopted on production machines due to the drag penalty it imposed.

The F-104S was originally conceived as an interceptor, but the AMI decided to use it also in the fighter-bomber role, replacing the last of the F-84Fs. Upon completion of delivery to fighter groups, the F-104S began to equip a number of fighter-bomber groups. In 1974 the 155° Gruppo moved to Istrana, having previously been assigned to the 50° Stormo based at Piacenza. The unit had equipped with the F-104S only two years before and was the last AMI unit to fly the F-84F.

In 1973 the 102° Gruppo began to receive F-104Ss, making the 5° Stormo the third wing to have two groups equipped with F-104s assigned differing missions. The other two were the 51° Stormo (the 22° and 155° Gruppi) and the 36° Stormo (the 12° and 156° Gruppi).

Of course, the 311° Gruppo of the Reparto Sperimentale Volo, the AMI's test wing, also operated F-104G and F-104S aircraft. These had been permanently assigned for trials and test work since the beginning of the programme.

Two groups – the 154° Gruppo, 6° Stormo and the 102° Gruppo, 5° Stormo – were assigned the nuclear strike mission – their primary tasking until 1971.

F-104S/ASA armament upgrade

In the early 1980s, the AMI lost the opportunity to purchase a new fighter to fill the gap until the arrival of the EF2000. After an evaluation of the fighters available, the AMI selected an 'upgrade' of the F-104S, largely for political reasons, that increased its operational capacity and improved flight safety. Initially called F-104S Upgraded Avionics, but officially renamed F-104S/ASA (Aggiornamento Sistemi d'Armam, weapons system upgrade), the programme eventually included 147 aircraft (of 153 originally envisaged). The first F-104S/ASA was completed in 1984. Principal changes included a new look-down/shoot-down version of the FIAR R21G/M Setter radar, compatible with the use of the Italian-built, improved version of the Sparrow (the Selenia Aspide 1A) and the improved IR-guided AIM-9L missile.

Unfortunately, the performance of the new Aspide was limited by the restricted range of the F-104S/ASA's radar (about 100 km/62 miles), which allowed target tracking to only 40 km

On F-104A/ASA aircraft Aspide 1A (the larger of the two missiles in the above view) replaced the AIM-7E Sparrow. With a burn-out speed of Mach 4 thanks to a more powerful motor and all-round performance, and guidance claimed to be better than that of the AIM-7F, Aspide is equipped with a 35-kg (77.2-lb) warhead.

Left: 22° Gruppo, 51° Stormo operated ASA aircraft until October 1998. '51-03' is pictured at the unit's Istrana base, 'armed to the teeth' with AIM-9Ls and Aspide 1As.

F-104S/ASA-M and TF-104G-M

As had been the case with the F-104S/ASA, aircraft were selected for conversion to final ASA-M standard on the basis of airframe life and condition; in all, 64 aircraft were reworked at the former Fiat/Aeritalia factory at Torino-Caselle – now operated by Alenia – comprising 49 single-seaters and 15 two-seaters. Pictured are a pair of F-104S/ASA-Ms aircraft (above) of 5° Stormo and a TF-104G-M of 4° Stormo, during May 2001.

(25 miles) and the guidance of only one missile.

Other modifications were made to the electrical and electronic systems, and new advanced communications equipment, new ECM systems, a chaff and flare dispenser, and new RWR and IFF systems were installed. During flight tests, it was decided not to fit an in-flight refuelling probe, as this would have limited speed to Mach 1.7.

Delivery of the new version began in 1986 and was completed in 1993. Although some modified aircraft were S/CB fighter-bombers, all the F-104S/ASAs equipped AMI air defence groups, as the fighter-bomber fleet comprised the Tornado IDS and the AMX. Units receiving the F-104S/ASA were 22° Gruppo CI, 51° Stormo based at Istrana, 12° Gruppo, 36° Stormo based at Gioia del Colle, 9° Gruppo, 4° Stormo based at Grosseto, 23° Gruppo, 5° Stormo based at Rimini, 10° Gruppo, 9° Stormo based at Grazzanise, and the 18° Gruppo, 37° Stormo based at Trapani.

The 18° Gruppo had been formed on 1 September 1983 as an autonomous group called the Nucleo Operativo di Difesa Aerea, equipped initially with F-104S/CBs; it was the only AMI Starfighter group simultaneously assigned both air defence and fighter-bomber roles. For some years, it had been equipped with F-104Ss armed with the internal gun and unable to fire Sparrows. Only after it received the F-104S/ASA was the 18° Gruppo tasked solely with air defence, over southern Italy.

Some F-104S/ASAs were also issued to the operational conversion unit, the 20° Gruppo of the 4° Stormo at Grosseto, which operated mainly unmodified TF-104Gs until they were updated to G-M standard from 1997.

F-104S/ASA-M

The F-104S/ASA maintained a minimum level of operations in peacetime air defence, as it

A tight formation comprising a pair of F-104S/ASA-Ms, a TF-104G-M and 4° Stormo's '70th anniversary' F-104S/ASA executes a loop. With the exception of the two-seater, these aircraft are powered by the uprated P&W J79-GE-J1Q introduced in the F-104S to offset the extra weight and to allow an increased warload.

Above: Widely expected to be the first AMI Typhoon unit, 9° Gruppo, 4° Stormo was not originally slated to receive the F-104S/ASA-M. However delays plaguing Eurofighter forced the AMI to reconsider and the unit was issued with 'Ms' in January 2002. This 9° Gruppo F-104S/ASA is pictured getting airborne from Grosseto during 2001.

Left: The S/ASA-M modification programme was confined to communications and navigation equipment improvements; budgetary constraints precluded any further improvements to armament fit. This pair carries the well-known 5° Stormo badge of 'Diana Cacciatrice' (Diana the Hunter) on their tails.

was obsolete and unsuitable for integration with multinational operations, its technology being far from comparable with that of NATO's F-16s, F/A-18s and Tornados. However, in the late 1990s the AMI, hamstrung by delays in the EF2000 programme, decided that it needed to keep the Starfighter in service beyond 2000. Aware of the F-104S/ASA's limitations, the air force also elected to lease 24 ex-RAF Tornado ADVs for the air defence mission. The decision to retain the F-104 also had a political dimension, as Italian industry had declared that the purchase of an American fighter (such as the F-15 or F-16, which were among the AMI's first choices to replace the Starfighter) could lead to the loss of thousands of Italian jobs. In 1994, the AMI officially announced the lease of the Tornado ADV and gave the go-ahead for the F-104S/ASA-M ('M' for modified) programme.

This life-extension programme was intended primarily to keep the remaining Starfighters airworthy and operating with acceptable efficiency. It involved only the 64 best airframes, comprising 49 single-seaters and 15 two-seaters. The cost of this new update programme was around $US650 million and offered more than 130,000 further flying hours over a 10-year period.

All modifications to S/ASA-M standard were carried out at Alenia's factory at Torino-Caselle. The main intention was to improve flight safety, to which end changes included replacing the radio and navigation equipment with a new TACAN and HSI, installing an LN-39A2 INS (the same as that of the AMX but adapted to the F-104) and a GPS and radar altimeter. Unfortunately, the TACAN and the GPS are isolated from the inertial platform, limiting the full effectiveness of the navigation system.

For financial reasons, no modifications were made to the radar or the Aspide guided missile systems, and in order to avoid a redesign of the avionics compartment, the AN/ALQ-70 self-defence system was removed. Some minor modifications were made to the instrument arrangements in the cockpit, such as removing the gun sight. To enhance flight safety, the electrical cabling, control wiring and hydraulic connections were replaced.

The TF-104G-M modifications allowed the two-seaters to be used more effectively, though only as training aircraft, with no operational capability. All armament was removed, and the radar can be used only for ground mapping; the aircraft lack the MIRC (Missile-In-Range Computer), IR sight, and lock-on capability.

The first S/ASA-M prototype, MM6945, flew at Torino-Caselle for the first time on 31 July 1995. Aircraft deliveries began on 18 December

Pictured during 2002, one of 23° Gruppo's F-104S/ASA-Ms rests at Cervia between sorties. NATO's Operation Allied Force saw aircraft detached to Amendola on QRA and to provide a CAP 'back-up' for the main NATO CAP force aircraft.

On active duty – Ace Guard and Allied Force

Operation Ace Guard marked the Italian Starfighter's combat debut, when F-104Gs of the 3° Stormo Ricognitori were deployed to Erhac, Turkey.

This operation began on 5 January 1991 when six F-104Gs took off from Gioia del Colle for Turkey. They were to supply reconnaissance support to the coalition and to prevent an attack by Iraqi forces on NATO-member Turkey. The Starfighters were employed mainly on recce tasks, using the external centreline Orpheus pod. The six aircraft incorporated modifications to allow the use of the ALE-40 chaff and flare dispenser. The F-104Gs flew 12 sorties daily, seven days per week for more than two months, until the end of the air campaign over Iraq.

The aircraft's other combat deployment was in early 1999, in support of NATO's Operation Allied Force. The Starfighters were deployed to Amendola air base, in Puglia, and cells of aircraft from the interceptor groups were also deployed from their bases. Only the 23° Gruppo/5° Stormo operated from its own base of Cervia, until mid-June 1999 when it deployed to Amendola to relieve the other groups' F-104S/ASA-Ms. The main mission for the Starfighters was QRA, for which they retained their peacetime configuration of wingtip tanks, one Aspide 1A and one AIM-9L. The Starfighters also flew many missions to cover holes in CAP coverage, such as when the original CAP fighters were sent to intercept a target under AWACS control. The F-104S/ASA-Ms were highly dependent on advanced controllers such as AWACS. During these CAP missions, the Starfighters were occasionally used to identify and intercept non-hostile targets. In the course of Operation Allied Force, the Starfighters made two to six operational sorties daily, for a total of more than 300 flight hours.

Now very much in their twilight years, the AMI's Starfighters soldier on in the face of severe operational limitations imposed by their short endurance and the age of their radar and avionics. The arrival of Typhoon cannot come too soon for the AMI's interceptor force.

1997 when the first F-104S/ASA-M (6935, coded '5-30') was delivered to 23° Gruppo, 5° Stormo based at Cervia. After eight months, the group had received all its ASA-Ms and the 18° Gruppo, 37° Stormo began to take delivery of its modified Starfighters. In January 1999 the 10° Gruppo, 9° Stormo transitioned to the new version and at the end of that year the 20° Gruppo OCU began to receive the first of four single-seat F-104S/ASA-Ms. Delivery of two-seat TF-104G-Ms began in November 1997 with the arrival of MM54254 (coded '4-36'); five two-seaters were also converted to the new variant by the 4° Reparto Manutenzione Velivoli. The last TF-104G-M was delivered to the 20° Gruppo in August 2000.

Initially 9° Gruppo, 4° Stormo did not receive the new variant, as it was intended to be the first group to re-equip with the EF2000. However, delays to the EF2000 programme coupled with logistic problems in keeping the F-104S/ASA in service made the 9° Gruppo's transition to the new Starfighter variant inevitable, and in January 2002 the first ASA-M was delivered to the group.

AMI Starfighters today

In 2002 the F-104S/ASA-M continues to perform quick reaction alert (QRA) duties; normally two aircraft are on QRA in standard interceptor configuration, i.e. carrying one all-aspect AIM-9L and one Aspide 1A. If a hostile target is detected by a ground station, aircraft are scrambled and airborne within minutes, accelerating to supersonic speed within 10 minutes with the aim of intercepting the target some 250 km (400 miles) from base, adopting a head-on attack profile.

However, while the Italian Starfighters are still able to provide a credible QRA capability, their usefulness in the combat air patrol (CAP) role is limited by their short range and outmoded radar; the latter means that ground or airborne control is a necessity. Other aircraft with more powerful long-range radar, such as the Tornado ADV, F-15 or F-16, can also detect targets and 'hand off' to a Starfighter for interception, using the F-104 as an advanced missile platform, though this approach has its limitations.

Thus, it may be seen that the Starfighter continues to retain some credible operational capability, as it can be airborne in five minutes after the scramble alert, but its radar and avionics limitations are a definite handicap. Nonetheless, the F-104 is keeping impatient Italian pilots current as they wait for their state-of-the-art fighter, and the F-104's replacement, the Eurofighter Typhoon.

Luigino Caliaro

AMI Starfighter units, 1963-2002

Unit: **9° Gruppo CI, 4° Stormo**
(under 4ª Aerobrigata until Sept 1967)
Base: Grosseto
Variants (dates): F-104G (March 1963-1970)
F-104S (1970-1986)
F-104S/ASA (1986-January 2002)
F-104S/ASA-M (January 2002-date)

Unit: **10° Gruppo CI, 9° Stormo**
(under 4ª Aerobrigata until Sept 1967)
Base: Grazzanise
Variants (dates): F-104G (1964-1972)
F-104S (1972-1990)
F-104S/ASA (1990-1999)
F-104S/ASA-M (1999-date)

Unit: **12° Gruppo CB, 36° Stormo**
Base: Gioia del Colle
Variants (dates): F-104S (1971-1988)
F-104S/ASA (1988-1995)

Unit: **18° Gruppo CI/CB, 37° Stormo**
(under 3ª Aerobrigata until 1977, when it disbanded; see below)
Base: Trapani
Variants (dates): F/RF-104G (1974-1977)
F-104S (1983-1987)
F-104S/ASA (1987-1998)
F-104S/ASA-M (1998-date)

Before it reformed as 18° Gruppo, 37° Stormo, the group operated briefly as a NODA (Nucleo Operativo Difesa Aerea) assigned to 36° Stormo, with a detachment at Trapani (between September 1983 to June 1984) and then Gioia del Colle. On 1 October 1984, NODA became 18° Gruppo under 37° Stormo.

Unit: **20° Gruppo AO, 4° Stormo**
(20° Gruppo Addestramento Operativo Autonomo until 1985)
Base: Grosseto
Variants (dates): F-104G (late 1963-June 1994)
TF-104G (late 1965-1999/2000)
F-104S (1989-1994)
F-104S/ASA (1994-2001)
F-104S/ASA-M (2001-date)
TF-104G-M (late 1999-date)

Unit: **21° Gruppo CI, 53° Stormo**
(under 51ª Aerobrigata until September 1965, see below)
Base: Cameri
Variants (dates): F-104G (April 1964-1971)
F-104S (1971-1989)
F-104S/ASA (1989-1996)

Operated as autonomous 21° Gruppo Autonomo between September 1965 and March 1967, then under newly formed 53° Stormo.

Unit: **22° Gruppo CI, 51° Stormo**
Base: Istrana
Variants (dates): F-104S (1969-1989)
F-104S/ASA (1989-October 1999)

Unit: **23° Gruppo CI, 5° Stormo**
Base: Cervia (Rimini-Miramare until March 1995)
Variants (dates): F-104S (March 1973-1987)
F-104S/ASA (1987-1997)
F-104S/ASA-M (1997-date)

Unit: **28° Gruppo CRO, 3° Stormo**
(3ª Aerobrigata until September 1967)
Base: Villafranca
Variants (dates): F/RF-104G (February 1970-end 1992)

Unit: **101° Gruppo CRO, 5ª Aerobrigata**
Base: Rimini-Miramare
Variants (dates): F/RF-104G (1964-June 1966)

Unit: **102° Gruppo CB, 5° Stormo**
(under 5ª Aerobrigata until Sept 1967)
Base: Rimini-Miramare
Variants (dates): F-104G (May 1964-1973)
F-104S (1973-1987)
F-104S/ASA (1987-July 1993)

Unit: **132° Gruppo CRO, 3° Stormo**
(3ª Aerobrigata until September 1977)
Base: Villafranca
Variants (dates): F/RF-104G (1972-1990)

Unit: **154° Gruppo CB, 6° Stormo**
(6ª Aerobrigata until 1 September 1967)
Base: Ghedi
Variants (dates): F-104G (1963-1982)

Unit: **155° Gruppo CB, 51° Stormo**
(50° Stormo until 1973)
Base: Istrana (Piacenza until 1973)
Variants (dates): F-104S (1972-1 January 1985)

Unit: **156° Gruppo CB, 36° Stormo**
Base: Gioia del Colle
Variants (dates): F-104S (1970-late 1983)

Abbreviations:

AO	Addestramento Operativo (operational conversion)
CI	Caccia Intercettori (fighter-interceptor)
CB	Caccia Bombardieri (fighter-bomber)
CRO	Caccia Ricognitori Ognitempo (fighter/all-weather reconnaissance)

INDEX

Page numbers in **bold** refer to illustrations

Index

Picture acknowledgments

Front cover: Sergey Skrynnikov, 82 Wing Photographic via Nigel Pittaway, Zale Collection via Warren Thompson. **4:** Air Force Flight Test Center (three), Emiel Sloot. **5:** US Navy, Pilatus via Peter R. March, Boeing. **6:** Boeing, Peter R. Foster, Marnix Sap. **7:** Peter R. Foster, Air Force Flight Test Center, MBDA. **8:** Indian Navy, Nigel Pittaway, Ivan Siminic (two). **9:** David Donald (two), Sikorsky via Tom Kaminski, Eric Katerburg. **10:** Boeing, Boeing via Tom Kaminski, Tieme Festner, Peter R. Foster. **11:** Dick Lohuis, Marnix Sap, Yaso Niwa. **12:** Marnix Sap, Paul Kievit and Carlo Kuit (three). **13:** Chris Ryan, USAF via Tom Kaminski. **14:** Eric Katerburg, Marnix Sap, Yaso Niwa (two). **15:** Yaso Niwa, USMC, Ministry of Defence. **16:** Tom Kaminski (three). **17:** Tom Kaminski (two), Shlomo Aloni (two). **18:** 82 Wing Photographic via Nigel Pittaway (NP), David Donald, via Terry Panopalis. **19:** RAAF Museum via NP, Defence PR via NP. **20:** via Terry Panopalis, Nigel Pittaway. **21:** RAAF Museum via NP (two), via Terry Panopalis. **22:** RAAF PR via NP, RAAF Museum via NP. **23:** RAAF Museum via NP (two), 82 Wing Photographic via NP. **24:** RAAF Museum via NP, Defence PR via NP, Flt Lt D. Riddel via NP (two), Nigel Pittaway (four). **25:** Nigel Pittaway (four), RAAF Museum via NP, David Donald. **26:** David Donald (two), Flt Lt D. Riddel via NP (two). **27:** 82 Wing Photographic via NP, Nigel Pittaway, RAAF Museum via NP. **28:** RAAF PR via NP. **29:** David Donald, Nigel Pittaway, AC S. Hobbs/82 Wing Photographic via NP. **30:** Nigel Pittaway, Sgt Brent Tero/South Queensland Defence Public Affairs. **31:** 82 Wing Photographic via NP (two). **32-39:** Marnix Sap/MIAS. **41:** Konstantinos Dimitropoulos (five), Timm Ziegenthaler. **43-47:** Konstantinos Dimitropoulos. **49:** Konstantinos Dimitropoulos (eight), Anthony Mylonakis. **51-55:** Konstantinos Dimitropoulos. **56-57:** US Navy, Sergey Skrynnikov. **58:** Lt Col Anatoliy Artemyev, Jeremy Flack/API. **59:** Sergey Skrynnikov, via Piotr Butowski. **60:** Aerospace, US Navy. **61:** Aerospace, US Navy. **62:** Aerospace, via Sergey Skrynnikov, Cees-Jan van der Ende, Hugo Mambour. **63:** Victor Drushlyakov, via Eric Hehs/Code One. **64:** Gennadiy Petrov, Piotr Butwoski. **65:** US Navy, Jeremy Flack/API. **66:** Lt Col Anatoliy Artemyev (two), Gennadiy Petrov, Piotr Butowski (two), Andrey Salnikov. **67:** Victor Drushlyakov, Jeremy Flack/API (two). **68:** Victor Drushlyakov, Lt Col Anatoliy Artemyev (two). **71:** Victor Drushlyakov, Lt Col Anatoliy Artemyev (nine), Piotr Butowski. **72:** USAF, Aleksey Mikheyev. **73:** Sergey Skrynnikov, MoD. **74:** Indian Navy (two), via Terry Panopalis. **75:** Indian Navy, Simon Watson. **76:** US Navy, Victor Drushlyakov. **77:** Piotr Butowski, Aleksey Mikheyev. **78:** Piotr Butowski (two), Sergey Sergeyev via Piotr Butowski. **79:** Victor Drushlyakov, Piotr Butowski. **80:** via Piotr Butowski, US Navy (two). **81:** Victor Drushlyakov, Aleksey Mikheyev. **82:** via Piotr Butowski (four). **83:** via Terry Panopalis, Sergey Skrynnikov. **84:** Mikhail Kuznetsov, Piotr Butowski, USAF. **85:** Hugo Mambour, David Donald, Sergey Skrynnikov. **86:** Gary Bihary, USAF, David Donald, Cees-Jan van der Ende. **87:** Sergey Skrynnikov (two), Eugène Gadet (two). **88:** USAF, via Piotr Butowski (two). **89:** Piotr Butowski (six). **90:** Piotr Butowski (two), Mikhail Kuznetsov. **91:** Sergey Skrynnikov, Hugo Mambour, Eugène Gadet. **92-102:** Cees-Jan van der Ende and Marco Muntz. **103:** Cees-Jan van der Ende and Marco Muntz (three), Brazilian navy via Santiago Rivas (two). **104:** Oakley Allen via Warren Thompson (WT). **105:** Paul Wilkins via WT, via WT, Mario Prevosti via WT. **106:** Orrin Fox via WT, Jerry Dore via WT, Sam Forbert via WT. **107:** James Hamilton via WT, via WT (two), B.C. Reed via WT, Zale Collection via WT. **108:** James Harris via WT, Homer Hansen via WT, Herbert Mondy via WT, Bob Dewald via WT. **109:** Wayne Jacskon via WT, Milt Tarr via WT, Glen Wold via WT. **110:** Glen Wold via WT, Jim Peck via WT, Earl Plesia via WT, John Lent via WT. **111:** via WT (four), Norman Thomas via WT, T.C. Rogers via WT, Al Wagner via WT. **112:** Glen Wold via WT, Ed Mason via WT. **113:** Wayne Jackson via WT, James Harries via WT, Bill Crampton via WT, Charles Schreffler via WT, Tom Shockley via WT. **114:** Charles Morehouse via WT, William O'Donnell via WT, Howard Price via WT. **115:** Joe Dishongh via WT, Gene Derrickson via WT. **116:** Charles E. Brown/RAF Museum, via Andrew Thomas. **117:** MoD (two), Aerospace. **118:** via Andrew Thomas (two), Aerospace. **119:** Aerospace (two). **120-121:** Richard Kierbow via Larry Davis (LD). **122:** Republic via LD, Republic, AFFTC History Office via Terry Panopalis (TP). **123:** AFFTC History Office via TP, Aerospace, Republic. **124:** Aerospace (two), Mike Machat Collection via TP, San Diego Aerospace Museum via TP. **125:** Tom Brewer via LD, Aerospace, Republic. **126:** Republic via LD, Republic, via LD (three), Merle Olmsted via LD. **127:** USAF via LD (two), Cradle of Aviation Museum Collection via Ken Neubeck via TP. **128:** via LD, Warren Bodie via LD, Tom Brewer via LD. **129:** Cradle of Aviation Museum Collection via Ken Neubeck via TP, via Robert F. Dorr, San Diego Aerospace Museum via TP (two). **130:** San Diego Aerospace Museum via TP, Mike Machat Collection via TP, Aerospace. **131:** via LD (two), Aerospace, Tom Brewer via LD. **132:** Dr J.G. Handleman via LD, H. Tate via LD, Republic via LD. **133:** Dave Graben via LD, Mick Roth Collection via TP, Dave Menard via LD, USAF, USAF via LD. **134:** USAF via David Donald, USAF, Aerospace. **135:** Joe Bruchs via LD, Republic via TP (two). **136:** Robert Ames via LD, USAF (three). **137:** Aerospace, USAF via LD, Joe Bruchs via LD, via TP. **143:** via LD (two), David Donald. **144:** Jim Sullivan via LD. **145:** Paul Freeburger via LD, via LD, Cradle of Aviation Museum Collection via Ken Neubeck via TP, Fairchild-Hiller. **146:** Tom Brewer via LD (two), Duane Kasulka via LD. **147:** via LD (two), Dave Ostrowski via LD. **148:** Steven Miller via LD (two), Don Linn via LD. **149:** Ron Thurlow via LD (two), Robert Pickett via LD. **150:** Frank MacSorley via LD, Tom Brewer via LD, USAF, via LD. **151:** via LD, Tom Brewer via LD (two), Steven Miller via LD, Roger Warren via LD. **152:** USAF via LD, Tom Brewer via LD, Thomas Waller via LD. **153:** D. Jay via TP, Art Krieger via LD, Tom Brewer via LD, Fairchild-Hiller. **154:** Dave Hansen via LD, via LD, via TP, Nick Williams via LD. **155:** Frank MacSorley via LD, Lars Soldens via LD, Terry Love via LD, Tom Brewer via LD, B. Knowles via Robert F. Dorr. **156:** via Bill Yenne (two), Aerospace. **157:** via Bill Yenne (six). **158:** via Bill Yenne, via Larry Davis. **159:** via Bill Yenne, via Larry Davis. **160:** via Bill Yenne (two). **161:** AFFTC via Terry Panopalis. **162:** via Larry Davis (two). **164:** Museum of Flight (Seattle)/Geroy Williams Collection via Terry Panopalis, Boeing, Aerospace, via Bill Yenne (two), via Robert F. Dorr. **165:** Aerospace, Museum of Flight (Seattle)/Geroy Williams Collection via Terry Panopalis, Boeing (two). **166:** via Bill Yenne (two), D.W. Lucabaugh via Terry Panopalis, via Robert F. Dorr (three), Museum of Flight (Seattle)/Geroy Williams Collection via Terry Panopalis, McDonnell via Robert F. Dorr. **167:** Aerospace, via Bill Yenne (two). **168:** Boeing via Robert F. Dorr, Boeing, Aerospace (two), USAF, Lockheed via Robert F. Dorr, USAF Museum, via Bill Yenne. **169:** via Terry Panopalis, via Larry Davis (three), AFFTC via Terry Panopalis, Boeing. **170:** via Robert F. Dorr, via Bill Yenne, H. Holmes via Terry Panopalis, R.D. Boyd via Terry Panopalis. **171:** via Bill Yenne, Boeing via Robert F. Dorr, Canadair via Terry Panopalis (two), Orenda via Robert F. Dorr. **172:** Luigino Caliaro. **173:** Aerospace, Luigino Caliaro. **174-175:** Luigino Caliaro. **176:** Aerospace, Lockheed, Aeritalia, Luigino Caliaro. **177:** Luigino Caliaro (two). **178:** Aerospace, Luigino Caliaro (three), Aeritalia. **179-181:** Luigino Caliaro.